7th Young Researcher Meeting – Torino 2016

Journal of Physics: Conference Series Volume 841

Torino, Italy
24 – 26 October 2016

ISBN: 978-1-5108-4239-7
ISSN: 1742-6588

Printed from e-media with permission by:

Curran Associates, Inc.
57 Morehouse Lane
Red Hook, NY 12571

proceedings.com

Some format issues inherent in the e-media version may also appear in this print version.

Printed by Curran Associates, Inc. (2017)

For permission requests, please contact the Institute of Physics at the address below.

Institute of Physics
Dirac House, Temple Back
Bristol BS1 6BE UK

Phone: 44 1 17 929 7481
Fax: 44 1 17 920 0979

techtracking@iop.org

Additional copies of this publication are available from:

Curran Associates, Inc.
57 Morehouse Lane
Red Hook, NY 12571 USA
Phone: 845-758-0400
Fax: 845-758-2633
Email: curran@proceedings.com
Web: www.proceedings.com

Table of contents

Volume 841

7th Young Researcher Meeting
24–26 October 2016, Torino, Italy

Accepted papers received: 4 May 2017
Published online: 7 June 2017

Preface

011001
OPEN ACCESS
7th Young Researcher Meeting

011002
OPEN ACCESS
Peer review statement

Astrophysics and Cosmology

012001
OPEN ACCESS
Optimization of the half wave plate configuration for the LSPE-SWIPE experiment

A Buzzelli, G de Gasperis, P de Bernardis, S Masi and N Vittorio.....1

012002
OPEN ACCESS
Sterile neutrinos with non-standard secret interactions imprints on Cosmic Microwave Background anisotropies

F Forastieri.....7

012003
OPEN ACCESS
Hyperspectral camera based on a Fabry-Pérot with varying beam incidence

L Giovannelli, F Berrilli and D Del Moro.....13

012004
OPEN ACCESS
Derivation of the Hubble parameter using galaxy clusters

A Kozmanyan, H Bourdin, G de Gasperis, P Mazzotta and N Vittorio.....19

012005
OPEN ACCESS
Magnetic fields in galaxy clusters in the SKA era

F Loi, F Govoni, M Murgia, V Vacca, H Li, L Feretti and G Giovannini.....25

012006
OPEN ACCESS
Anisotropies in the flux of cosmic ray electrons and positrons

S Manconi, M Di Mauro and F Donato.....31

012007
OPEN ACCESS
Constraints on a k-dependent bias from galaxy clustering

E Menegoni.....37

012008
OPEN ACCESS
Cosmic Microwave Background large-scale directional anomalies as seen by Planck and WMAP

L Polastri.....43

Biomedical Physics

012009
OPEN ACCESS
Development and implementation of a non Gaussian model for the lateral dose prediction in a proton therapy treatment planning system

V E Bellinzona.....49

012010
OPEN ACCESS
Characterization of physicochemical and colloidal properties of hydrogel chitosan-coated iron-oxide nanoparticles for cancer therapy

E Catalano and A Di Benedetto.....55

012011
OPEN ACCESS
The INSIDE project: on-line monitoring and simulation validation with the in-beam PET scanner

V Ferrero on behalf of the INSIDE collaboration.....61

012012
OPEN ACCESS
Optimal configuration of a low-dose breast-specific gamma camera based on semiconductor CdZnTe pixelated detectors

B Genocchi, O Pickford Scienti and DG Darambara.....67

012013
OPEN ACCESS
Computer-aided detection systems to improve lung cancer early diagnosis: state-of-the-art and challenges

A Traverso, E Lopez Torres, M E Fantacci and P Cerello.....73

Condensed Matter Physics

012014
OPEN ACCESS
Thermal fatigue life evaluation of SnAgCu solder joints in a multi-chip power module

C Barbagallo, G L Malgioglio, G Petrone and G Cammarata.....79

012015
OPEN ACCESS
Optical lattice clocks towards the redefinition of the second

F Bregolin, G Milani, M Pizzocaro, B Rauf, P Thoumany, F Levi and D Calonico.....85

012016
OPEN ACCESS
Low energy quantum regimes of 1D dipolar Hubbard model with correlated hopping

S Fazzini, L Barbiero and A Montorsi.....89

012017
OPEN ACCESS
Adaptive CFD schemes for aerospace propulsion

A Ferrero and F Larocca.....95

012018
OPEN ACCESS
Effects of pre- and post-electrospinning plasma treatments on electrospun PCL nanofibers to improve cell interactions

M Asadian, S Grande, R Morent, A Nikiforov, H Declercq and N De Geyter.....101

012019
OPEN ACCESS
Thermodynamics of a qubit undergoing dephasing

S Marcantoni.....107

012020
OPEN ACCESS
Investigation of ferrofluid nanostructure by laser light scattering: medical applications

E K Nepomnyashchaya, E N Velichko, I V Pleshakov, E T Aksenov and E A Savchenko.....113

012021
OPEN ACCESS
Theory of photon-driven correlated electrons in one dimension

M Puviani and F Manghi.....119

012022
OPEN ACCESS
ICRF Traveling Wave launcher for fusion devices

R Ragona.....125

Detectors for particle physics

012023
OPEN ACCESS
Investigating the low-energy K^- interactions in nuclear matter with AMADEUS

R Del Grande, M Bazzi, G Belloti, C Berucci, A M Bragadireanu, D Bosnar, A D Butt, M Cargnelli, C Curceanu, L Fabbietti, C Fiorini, F Ghio, C Guaraldo, R S Hayano, M Iliescu, M Iwasaki, P Levi Sandri, J Marton, M Miliucci, S Okada, D Pietreanu, K Piscicchia, A Romero Vidal, A Scordo, H Shi, D L Sirghi, F Sirghi, I Tucakovic, O Vazquez Doce, E Widmann, S Wycech, J Zmeskal, G Mandaglio, M Martini and P Moskal.....131

012024
OPEN ACCESS
Dimuon production in Pb-Pb collisions at 20-160 AGeV at the CERN SPS: Mapping the QCD phase diagram in the transition region with a new experiment

E Incani and G Usai.....137

012025
OPEN ACCESS
^{arch}PbMoO4 scintillating bolometer as detector to searches for the neutrinoless double beta decay of ^{100}Mo

S Nagorny, L Pattavina, M B Kosmyna, B P Nazarenko, S Nisi, L Pagnanini, S Pirro, K Schäffner and A N Shekhovtsov.....143

012026
OPEN ACCESS
The first pion-Ar cross-section measurement with the LArIAT experiment

I Nutini on behalf of the LArIAT Collaboration.....149

Invited speaker

012027
OPEN ACCESS
Assess the neutrino mass with micro and macro calorimeter approach

A Giachero.....155

Geophysics

012028
OPEN ACCESS
Heterogeneous nucleation of ice in the atmosphere

A Nicosia, M Piazza, G Santachiara and F Belosi.....161

012029
OPEN ACCESS
Detection of terrestrial gamma-ray flashes with the AGILE satellite

A Ursi, M Marisaldi, M Tavani, P Sanò, D Casella and S Dietrich.....167

Metrology

012030
OPEN ACCESS
Review of techniques for 2D camera calibration suitable for industrial vision systems

G D'Emilia and D Di Gasbarro.....173

012031
OPEN ACCESS
Cryocooled Josephson standards for AC voltage metrology

P Durandetto, A Sosso, E Monticone, B Trinchera, M Fretto and V Lacquaniti.....179

012032
OPEN ACCESS
Metrological traceability of carbon dioxide measurements in atmosphere and seawater

F Rolle, E Pessana and M Sega.....185

012033
OPEN ACCESS
Optimal estimation of parameters of an entangled quantum state

S Virzì, A Avella, F Piacentini, M Gramegna, G Brida, I P Degiovanni and M Genovese.....191

Theoretical Physics

012034
OPEN ACCESS
Constraining auto-interaction terms in α-attractor supergravity models of inflation

A Di Marco, P Cabella and N Vittorio.....197

012035
OPEN ACCESS
Testing strong gravity with gravitational waves and Love numbers

E Franzin, V Cardoso, P Pani and G Raposo.....203

012036
OPEN ACCESS
Available evidence of a missing theory? An empirical, thermodynamical approach to quantum gravity through an alternative to dark matter

D Mamone.....209

012037
OPEN ACCESS
Tree-level scattering amplitudes from the amplituhedron

L Ferro, T Lukowski, A Orta and M Parisi.....214

012038
OPEN ACCESS
The emergent gravity's entropy Ansatz from an augmented variational principle

M Tuveri, L Fatibene and M Ferraris.....220

IOP Conf. Series: Journal of Physics: Conf. Series **841** (2017) 011001 doi:10.1088/1742-6596/841/1/011001

Young Researcher Meeting, Torino 2016

F Agostini, C Antolini, A Avella, G Cattani, R Cuccaro, A Cultrera, M Di Stefano, G Fragione, L Lolli, M Migliaccio, L Pagnanini, E Pessana, F Piacentini, D Pietrobon, E Pusceddu, R Romeo, M Serra, E Simonetto, F Stellato

YRM
YOUNG RESEARCHER MEETING

IPN
INTERNATIONAL PHYSICISTS NETWORK

PREFACE

The Young Researcher Meeting (www.yrmr.it) has been established as a forum for students, postdoctoral fellows and young researchers determined to play a proactive role in the scientific progress. Since 2009 we run itinerant yearly meetings to discuss the most recent developments and achievements in Physics, as we are firmly convinced that sharing expertise and experience is the foundation of the research activity. One of the main purposes of the conference is actually to create an international network of young researchers, both experimentalists and theoreticians, and fruitful collaborations across the different branches of Physics.

The format we chose is an informal meeting primarily aimed at students and researchers at the beginning of their scientific career, who are encouraged to present their work in brief presentations able to provide genuine engagement of the audience and cross-pollination of ideas.

The 7th edition of the Young Researcher Meeting was held at the Istituto Nazionale di Ricerca Metrologica (INRiM) in Turin, the Italian National Metrology Institute responsible for the national standards of the International System of Units.

The conference took place from Monday 24th to Wednesday 26th October 2016. This edition gathered 120 participants belonging to universities and research centres from all over the world. The plenary talk sessions covered several areas of pure and applied Physics, and they were complemented by an extremely rich and interactive poster session, which was also extended to the coffee breaks.

IOP Conf. Series: Journal of Physics: Conf. Series **841** (2017) 011001 doi:10.1088/1742-6596/841/1/011001

The programme included a "poster & wine" session on the first evening, a guided tour at the Turin Astrophysical Observatory and a very enjoyable and friendly conference dinner in a typical Piedmont-cuisine restaurant. The participants had furthermore the unique opportunity to visit the INRiM laboratories, which host a wide spectrum of Physics experiments. This year's edition of the meeting featured a *lectio magistralis* by Walter Bich[1], entitled "Metrology, why not?", which provided an all-around picture of the connections between metrology and several other physics-related research fields. Three invited speakers (Dr. E. Palazzi[2], Prof. F. Sciarrino[3], and Dr. A. Giachero[4]), with different scientific backgrounds, kindly accepted to give broad-audience talks pertaining their research fields (Dr. A. Giachero also contributed to this publication).

This edition of the Young Researcher Meeting gave the organisers the opportunity to present to the attendees, and officially launch, two projects of interest for the wide scientific community:

- ✓ the recently established International Physicists Network (IPN)[5], *an Italian non-profit association having the main purpose to create and promote an international network of young researchers in Physics, both experimental and theoretical, in order to initiate collaborations between different branches of Physics*. IPN will be responsible for running the organisation of the YRM in the coming years, as it is the natural continuation of an almost decade-long collaboration between the funders and organisers of the YRM;

- ✓ the partnership code-named *Dandelion* with the Consortium for Technology Transfer (C2T) and specifically with its top-tier service designed for professional researchers, Find Your Doctor (FyD). Dr. Eva Ratti[6] gave the talk "What meaning for a PhD nowadays? Knowledge workers as a drive for change" to introduce FyD goals and methods to help researchers to connect with companies looking for their expertise. The project *Dandelion* has been presented during the "poster & wine" session. It aims to *"let competences, ideas, resources and company requirements meet together in an online/offline shared area and exploit the best of both sides"* adding a bottom-up approach where *"young researchers can now submit their own ideas and research/application projects directly to possible interested companies"*. It is mainly based on the *Elevator Pitch* paradigm, complementary to the tradition-

[1] W. Bich was the responsible for the national copy of the international prototype of the kilogram until 2015, date of his professional retirement.

[2] Institute of Atmospheric Sciences and Climate-National Research Council - ISAC-CNR, Torino, IT.

[3] Università di Roma La Sapienza, Roma, IT.

[4] Istituto Nazionale di Fisica Nucleare - INFN, Università di Milano Bicocca, Milano, IT.

[5] http://www.iphysnet.com/

[6] Find your Doctor, http://www.findyourdoc.org/

al top-down approach where companies advertise their openings/research challenges and the interested candidates submit their proposals.

This volume consists in a selection of the contributions presented at the conference as either talks or posters. These cover topics in astrophysics and cosmology, particle and theoretical physics, soft and condensed matter, medical physics, quantum information and quantum technologies, and of course metrology.

Organising and Editorial Committee

Fabio Agostini
(fabioagostini31@gmail.com)
NAIS

Claudia Antolini
(cla.antolini@gmail.com)

Alessio Avella
(a.avella@inrim.it)
INRiM

Giordano Cattani
(giordano.cattani@gmail.com)

Rugiada Cuccaro
(r.cuccaro@inrim.it)
INRiM

Alessandro Cultrera
(a.cultrera@inrim.it)
INRiM

Marco Di Stefano
(marco.distefano@cnag.crg.eu)
Centre Nacional d'Anàlisi Genòmica
- Centre de Regulació Genòmica (CNAG-CRG)

Giacomo Fragione
(giacomo.fragione90@gmail.com)
Hebrew University of Jerusalem

Lapo Lolli
(l.lolli@inrim.it)
INRiM

Marina Migliaccio
(migliaccio@asdc.asi.it)
ASDC – ASI Science Data Center

Lorenzo Pagnanini
(lorenzo.pagnanini@gssi.infn.it)
Gran Sasso Science Institute

Enrica Pessana
(e.pessana@inrim.it)
INRiM

Fabrizio Piacentini
(f.piacentini@inrim.it)
INRiM

Davide Pietrobon
(daddeptr@gmail.com)

Emanuela Pusceddu
(emanuela.pusceddu@gmail.com)
National Research Council-CNR

Raffaella Romeo
(r.romeo@inrim.it)
INRiM

Matteo Serra
(matteoserra83@gmail.com)

Enrico Simonetto
(e.simonetto@inrim.it)
INRiM

Francesco Stellato
(francesco.stellato@roma2.infn.it)
INFN Roma Tor Vergata

Silvia Cavallero
(Local Organising Committee)
(s.cavallero@inrim.it)
INRiM

ACKNOWLEDGEMENTS

The organisers of the 7th Young Researcher Meeting, held in Torino, would like to thank all the authors for their scientific contributions and all the people who made this conference smooth and pleasant.

We thank the Istituto Nazionale di Ricerca Metrologica (INRiM) for hosting the conference. INRiM and the commercial sponsors (Anton Paar, CalRef, Crisel Instruments, ID Quantique, LOT-Quantum Design, Pfeiffer Vacuum, Sistemi HS, Teledyne-Lecroy) for covering the organisational costs, the expenses for the publication of the proceedings, and providing travel grants to selected speakers.

We are grateful to Dr. W. Bich for filling his lecture with his contagious enthusiasm, Dott. Alberto Cora (and his colleagues) at the Istituto Nazionale di Astrofisica enrolled at the Astrophysical Observatory of Pino Torinese for the interesting guided tours. Many thanks also to the colleagues at INRiM who guided the extensive tour of the laboratories.

The event was broadcast live on the conference website, and the recordings are available on YRM Youtube channel at
http://www.youtube.com/c/YoungResearcherMeeting.

IOP Conf. Series: Journal of Physics: Conf. Series **841** (2017) 011001 doi:10.1088/1742-6596/841/1/011001

TELEDYNE LECROY

CalRef

IDQ FROM VISION TO TECHNOLOGY

PFEIFFER VACUUM

IOP Conf. Series: Journal of Physics: Conf. Series **841** (2017) 011002 doi:10.1088/1742-6596/841/1/011002

Peer review statement

All papers published in this volume of *Journal of Physics: Conference Series* have been peer reviewed through processes administered by the proceedings Editors. Reviews were conducted by expert referees to the professional and scientific standards expected of a proceedings journal published by IOP Publishing.

IOP Conf. Series: Journal of Physics: Conf. Series **841** (2017) 012001 doi:10.1088/1742-6596/841/1/012001

Optimization of the half wave plate configuration for the LSPE-SWIPE experiment

A Buzzelli[1,4], G de Gasperis[3,4], P de Bernardis[1,2], S Masi[1,2] and N Vittorio[3,4]

[1] Dipartimento di Fisica, "Sapienza" Università di Roma, piazzale Aldo Moro 5, I-00185, Roma, Italy
[2] Sezione INFN Roma 1, piazzale Aldo Moro 5, I-00185, Roma, Italy
[3] Dipartimento di Fisica, Università di Roma "Tor Vergata", via della Ricerca Scientifica 1, I-00133, Roma, Italy
[4] Sezione INFN Roma 2, via della Ricerca Scientifica 1, I-00133, Roma, Italy

E-mail: alessandro.buzzelli@roma2.infn.it

Abstract. The search for the B-mode polarization of Cosmic Microwave Background (CMB) is the new frontier of observational Cosmology. A B-mode detection would give an ultimate confirmation to the existence of a primordial Gravitational Wave (GW) background as predicted in the inflationary scenario. Several experiments have been designed or planned to observe B-modes. In this work we focus on the forthcoming Large Scale Polarization Explorer (LSPE) experiment, that will be devoted to the accurate measurement of CMB polarization at large angular scales. LSPE consists of a balloon-borne bolometric instrument, the Short Wavelength Instrument for the Polarization Explorer (SWIPE), and a ground-based coherent polarimeter array, the STRatospheric Italian Polarimeter (STRIP). SWIPE will employ a rotating Half Wave Plate (HWP) polarization modulator to mitigate the systematic effects due to instrumental non-idealities. We present here preliminary forecasts aimed at optimizing the HWP configuration.

1. Introduction

The Cosmic Microwave Background (CMB) is the oldest light in the universe, freely streaming since the cosmic recombination epoch, characterized by a nearly isotropic black body spectrum with temperature anisotropies of only few parts out of 10^5. The CMB has represented the most important tool to constrain the initial stage, evolution and content of the universe. In fact, the CMB anisotropy pattern has provided tight constraints on the cosmological parameters, such as the curvature and the expansion of the Universe and the relative abundances of baryons, Dark Matter and Dark Energy [1]. However, to break the degeneracies among some parameters, an accurate analysis of the CMB (linear) polarization pattern is required. CMB polarization is generated at last scattering surface by Thomson scattering of quadrupolar anisotropies. A complete description of the polarization field can be achieved with the Stokes parameters Q and U, as I describes only temperature and $V = 0$ as Thomson scattering induces no circular polarization. Since the polarization field can be encoded in a 2-rank symmetric trace-free tensor, we can decompose it into a gradient (E-modes) and a curl (B-modes) component. E-modes have been widely observed, while B-modes are still buried into instrumental noise and foreground contamination. Primordial scalar (i.e. density) perturbations produce only E-modes, while tensor perturbations produce both E and B-modes. Therefore, a detection of primordial B-modes would provide a definitive confirmation to the existence of a stochastic Gravitational Wave (GW) background, as predicted in the inflationary paradigm. Several experiments have been designed or planned with the ultimate goal of detecting B-mode polarization, either from ground, balloons or space. In this work, we focus on the forthcoming LSPE mission that will be devoted to the accurate measurement of CMB polarization at large angular scales [2]. LSPE will consist of a high frequency balloon-borne bolometric polarimeter, SWIPE [3], and of a low frequency ground-based coherent polarimeter, STRIP [4]. The peculiarity of SWIPE will be the presence of a rotating half wave plate (HWP) polarization modulator to mitigate the instrumental systematics. The work is organized as follows: in Section 1 we briefly overview the CMB B-mode polarization; in Section 2 we introduce the scientific case of the LSPE experiment; in Section 3 we present preliminary simulations aimed at optimizing the SWIPE HWP configuration; finally, we draw our conclusions in Section 4.

2. CMB B-mode polarization

The amplitude of B-modes is usually parametrized by the tensor-to-scalar ratio, r, that expresses the relative amplitude of tensor and scalar primordial perturbations. The satellite mission Planck, together with the ground based experiments BICEP2 and Keck Array, set the most stringent upper limit to date of $r < 0.07$ at 95% C.L. [5], but we do not have precise constraints on its lower limit. However, theoretical arguments suggest $r > 10^{-3}$ [6]. The LSPE experiment is dedicated to observe CMB polarization at large angular scales, where most of the primordial B-mode information is encoded, with the primary aim to constrain r down to around 0.02. This tight contraint will allow to rule out several inflationary scenarios, thanks to the strict relation between r and the inflation energy scale: $E = 3.3 \times 10^{16} r^{1/4}\ GeV$. Inflation is a stage of exponential expansion in the very early Universe that lasted for a fraction of second. This theory was introduced to solve the classical puzzles of the Hot Big Bang scenario, such as the horizon, the flatness and the monopole problems. From the Friedmann equations, a period of accelerated expansion implies an exponential growth for the scalar factor a, i.e. $a(t) \propto e^{Ht}$, where H is the Hubble parameter, and an unusual relation for the equation of state, $p = w\rho$ with $w < 1/3$, where p and ρ indicate the Universe mean pressure and density, respectively. Moreover, inflation provides a mechanism for the production of perturbations in the early Universe, both scalar and tensor, from primordial quantum fluctuations. According to the current understanding, cosmic structures have been originated by gravitational instability of density (scalar) primordial perturbations, which are "adiabatic" and form a random Gaussian field with a nearly scale-invariant spectrum. Imprints of scalar perturbations are in both temperature and E-mode polarization patterns. Tensor perturbations are associated to a stochastic GW background, which impressed specific traces in the temperature, E-mode and, unequivocally, B-mode patterns. For this reason, a B-mode detection would represent a priceless breakthrough in Cosmology and fundamental physics. Moreover, B-mode polarization can shed light on some critical aspects of cosmic reionization, eventually associated to the first cosmic structure formation, which is expected to move part of E-modes into B-modes at very large angular scales. Different inflationary scenarios do exist. The canonical single-field slow-roll model is described by only two parameters, ϵ and η [6]. The spectral index of the power spectrum of scalar perturbations is defined as $n_s = 2\eta - 6\epsilon + 1$: inflation predicts a value of n_s slightly less than 1. The most recent constraint by Planck is $n_s = 0.968 \pm 0.006$ [1]. The tensor spectral index is given by $n_t = -2\epsilon$, and it is expected to be negative. CMB temperature (T) and polarization (E, B) fields can be described in terms of their auto (TT, EE, BB) and cross (TE, TB, EB) angular power spectra C_l, representing the power in fluctuations at each angular scale, which is expressed in terms of the multipole moment ℓ in harmonic space. According to the standard cosmological model, $C_l^{TB} = C_l^{EB} = 0$ due to the parity properties of the CMB physics. Recent

IOP Conf. Series: Journal of Physics: Conf. Series **841** (2017) 012001 doi:10.1088/1742-6596/841/1/012001

estimates of C_l^{TT}, C_l^{TE} and C_l^{EE} can be found in [1]. Notice that the prediction of Gaussian primordial fluctuations implies that also CMB fluctuations are expected to be Gaussian: this means that the power spectra fully describe the statistical anisotropy pattern. The spectrum C_l^{BB} is characterized by a peak at $\ell \leq 10$ (reionization bump), due to the conversion from E-modes to B-modes occurred at redshift $z \sim 8$, and by a peak at $\ell \sim 80$ (recombination bump), that is the imprint of the inflationary GW background. We only mention here three important effects that may introduce a spurious (i.e. non primordial) B-mode component: systematic effects, Gravitational Lensing (GL) and Galactic foreground contamination. Systematic effects are due to instrumental non-idealities or data analysis artifacts. GL is caused by large scale mass inhomogeneities between us and the last scattering surface and its effect is dominant at small angular scales. Polarized foreground contamination is mainly due to synchrotron radiation, below 40 GHz, and thermal dust emission, above 90 GHz. These effects must be carefully controlled and possibily suppressed as they may introduce a B-mode component in general larger than the primordial B-modes, degrading thus the performance of any experiment.

3. The LSPE experiment

LSPE is a next-generation CMB experiment [2], devoted to the observation of large scale polarization, with the primary aim of constraining the B-mode signal due to reionization (reionization bump) and the rising part of the signature of primordial tensor perturbations (recombination bump). LSPE will observe a large fraction of the northern sky (around the 30% of the celestial sphere) with a coarse angular resolution of about 1.5 degrees FWHM. LSPE will consist of two instruments: a balloon-based array of bolometric polarimeters, SWIPE, which will map the sky in three frequency bands centered at 140, 220 and 240 GHz [3], and a ground-based array of coherent polarimeters, STRIP, which will survey the same sky region in two frequency bands centered at 43 and 90 GHz [4]. This work is specific to the SWIPE instrument. SWIPE will be likely launched in winter 2018-2019 and will operate for around 15 days during the Arctic night (at latitude around 78 N), to exploit optimal observation conditions. SWIPE will scan the sky by spinning around the local vertical, while keeping the telescope elevation constant for long periods, in the range 35 to 55 deg. According to the current baseline, the azimuth telescope scan-speed will be set around 2 rpm, i.e. 12 deg/s. SWIPE will host a HWP polarization modulator as first optical element, followed by a 50 cm aperture refractive telescope, a beamsplitting polarizer, and finally two symmetric orthogonally-placed focal planes with an overall number of 110 detectors per frequency. The nominal HWP configuration consists of a step-and-integrate mode, repeatedly scanning the range $0-78.75$ deg at a rate of $11.25/min$. We generate sky-maps and angular power spectra from simulated observations in order to forecast the performances of the SWIPE instrument. Our flight simulator provides the coordinates (right ascension and celestial declination) and the polarization angle based on the SWIPE scanning strategy. Hence, we produce power spectra from the CAMB software [7] according to the last release of cosmological parameters from Planck [1] assuming a tensor-to-scalar ratio $r = 0.09$, that we use to generate a full sky map with the Synfast facility of the HEALPix package [8]. Finally, we scan this "input" map to produce the corresponding Time Ordered Data (TOD). We simulate the noise having a spectrum with a constant plateau at high frequencies with amplitude 15 $\mu Ks^{1/2}$, and a correlated $1/f^{\alpha}$ part at low frequencies with $\alpha = 2$. Moreover, we include a cross-correlated noise component equally shared by all the detectors of the form $1/f^2$ at low frequencies. The knee frequency is set to 0.1 Hz. We limit our simulation to 18 detectors, arranged in three triples sparcely located in each focal plane, and to 5 days of operation with the telescope elevation switching 5 deg every day from 35 to 55 deg. In our simulations we focus on the 140 GHz band, that will be the main SWIPE CMB channel, while observations at 220 and 240 GHz will be devoted to monitor the polarized thermal dust emission from the Galaxy. The maps are generated according to the nominal values of the telescope scan-speed and HWP (stepping) configuration. The angular resolution is set to 1.5 degree FWHM with a HEALPix $N_{side} = 128$. To produce the maps from the TODs we use the ROMA MPI-paralled code, an optimal map-making algorithm based on a generalized least squared approach [9]. The code has been extended to account for all the off-diagonal terms of the noise covariance matrix, that correspond to the cross-correlated noise among the detectors [10]. Hence, we estimate the power spectra by a pseudo-C_l approach, also knwon as the MASTER method [11]. In Fig. 1 we show the *hitmap* (i.e. the pixel-weighted coverage map) and the residual map (i.e. the difference between the output optimal map and the input map, where also the noise has been added) for the Q Stokes parameter. In Fig. 2 we show the average BB power spectrum estimated from 50 noise-only, signal-only and signal plus noise simulated maps (the error bars are given by the dispersion of the simulations).

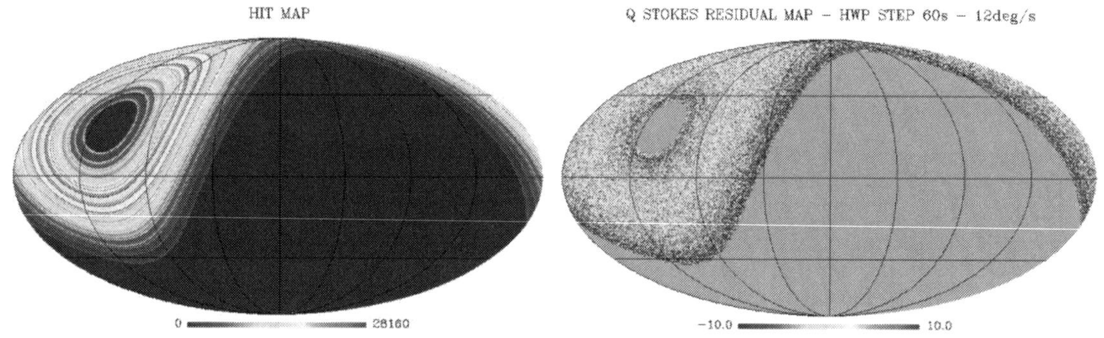

Figure 1. Maps from LSPE-SWIPE simulations. On the left: *hitmap*, i.e. the pixel-weighted coverage map. On the right: residual map (in μK), i.e. the difference between the output optimal map and the input map (where also the noise has been added) for the Q Stokes parameter. See the text for the details of the simulations.

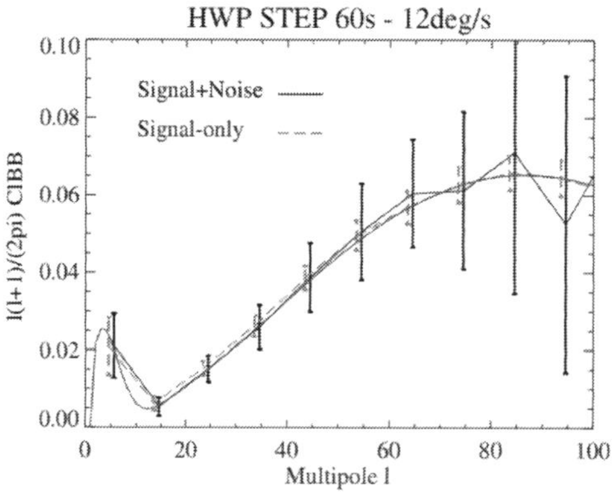

Figure 2. BB angular power spectra from LSPE-SWIPE simulations. Average spectrum and associated error bars estimated from signal plus noise simulations (in solid black line) and from signal-only simulations (in dashed red line). The input power spectrum, corresponding to $r = 0.09$, is shown for comparison in blue. Note the reionization bump at $\ell \leq 10$, and the recombination bump at $\ell \sim 80$. See the text for the details of the simulations.

4. Optimization of the polarization modulation strategy

In this Section we present preliminary forecasts on the optimization of the polarization modulation strategy for the SWIPE instrument. Wave plates are generally characterized by two orthogonal axes with different refractive indexes (the fast and the slow axis). A HWP is the particular case with a retardation of π between the phases of the two orthogonal components of the incident radiation electric field. To be effective, HWPs must be rotated during the observation, either in stepping or (continuously) spinning mode. The incident polarization is thus modulated at 4 f_r, where f_r is the rotation frequency. The use of a HWP as polarization modulator represents a powerful tool to minimize most of the instrumental systematic effects, that would otherwise severely degrade the experimental performance. A HWP allows to: i) effectively mitigate beam, calibration and other intrumental systematics, as a HWP design does not need to difference power from orthogonal polarization sensitive detectors, and additionally allows not to rotate the whole instrument; ii) reject the $1/f$ noise at the hardware level, as it moves the polarization signal to high frequencies allowing thus to sample the Stokes parameters Q and U in the white noise regime; iii) achieve a better angle coverage uniformity.

Here, we focus on two specific benefits provided by the HWP: the shift of the polarization signal to high frequencies and the improvement in the pixel angle coverage. The first effect allows to preserve cosmological information when a low-frequency cut of the data stream is performed (common practice to filter out the $1/f$ noise). The second effect helps the sky-map reconstruction since each pixel is observed from more directions. We investigate a HWP stepping every 1, 60 and 3600 s and spinning at 5, 2 and 0.5 Hz, keeping fixed the telescope azimuth scan-speed at the nominal value of 2 rpm. In Fig. 3 we show the "periodograms", i.e. the plots of temperature and polarization intensity power as function of frequency. We notice that a stepping HWP is not effective in modulating the polarization to higher frequencies; on the contrary, a (fast) spinning HWP is extremely powerful in moving the polarization away from the low-frequency part dominated by the $1/f$ noise (for instance, a HWP spinning at 5 Hz modulates the polarization signal in a narrow band around 20 Hz).

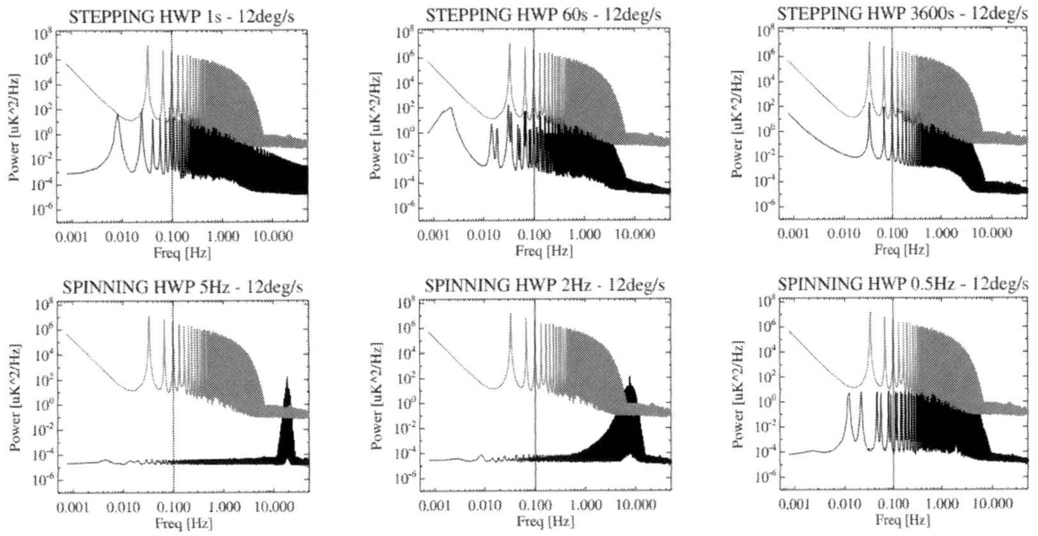

Figure 3. Periodograms, i.e. plots of temperature (in red) and polarization (in black) intensity power as function of frequency, for the HWP schemes under consideration. On the top, from left to right: stepping HWP (1, 60, 3600 s). On the bottom, from left to right: spinning HWP (5, 2, 0.5 Hz).

In Fig. 4 we show the histograms of the pixel inverse condition number, R_{cond}, that is an indicator of the angle coverage uniformity ($R_{cond} = 0.5$ in the case of perfect pixel coverage). We notice that all the HWP configurations under examination provide a very good pixel angle coverage, except the slowest stepping mode (at 3600 s).

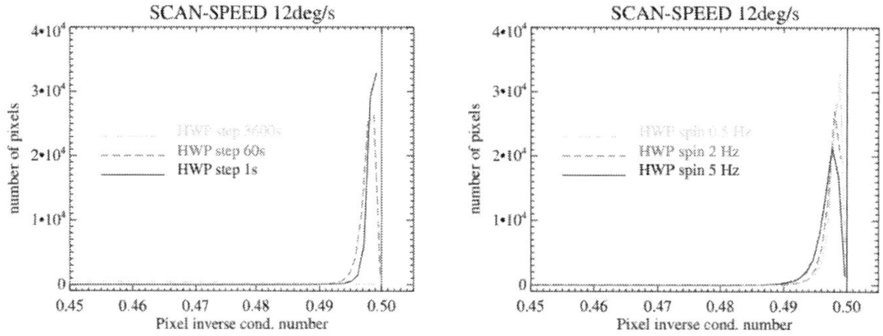

Figure 4. Histograms of the pixel inverse condition number for the HWP schemes under exam. On the left: stepping HWP (1, 60, 3600 s). On the right: spinning HWP (5, 2, 0.5 Hz).

5. Conclusions

A B-mode detection would represent an invaluable discovery in cosmology and fundamental physics since it would be a “smoking gun” evidence for inflation theory. This is due to the specific property of the curl polarization component, which is unequivocally sensitive to the tensor mode of primordial perturbations as predicted in the inflationary scenario. Among the several experiments designed to observe B-modes, we focused on the forthcoming LSPE mission. LSPE will consist of a balloon-borne bolometric polarimeter, SWIPE, and a ground-based coherent polarimeter, STRIP. The SWIPE instrument will exploit a rotating HWP to mitigate the detrimental effects of the instrumental systematics. We tested different HWP configurations, in both stepping and spinning modes, keeping fixed the telescope scan-speed. As figures of merit, we analyzed the effectiveness of the HWP design in moving the polarization signal to higher frequencies away from the $1/f$ noise and the improvement in the pixel angle coverage. We found that the polarization shift to high frequencies is typical of the (fast) spinning mode and that any HWP scheme provides a very good pixel angle coverage, except a slow stepping mode. We emphasize that our simulations do not account for the systematic effects introduced by the HWP itself, such as non-ideal rotation and internal reflections. We will face the issue of the HWP systematics of its own as well as other figures of merit aimed at the HWP design optimation in a following paper.

References

[1] Ade P A R *et al* [Planck collaboration] 2016 *A&A* **594** A13
[2] Aiola S *et al* 2012 *SPIE Conf. Series* **8446** 277 arXiv:1208.0281v1
[3] de Bernardis P *et al* 2012 *SPIE Conf. Series* **8452** 125 arXiv:1208.0282v1
[4] Bersanelli M *et al* 2012 *SPIE Conf. Series* **8446** 279 arXiv:1208.0164v1
[5] Ade P A R *et al* [BICEP2/Keck and Planck Collaborations] 2015 *Phys. Rev. Lett.* **114** 101301
[6] Kamionkowski M and Kovetz E D 2015 arXiv:1510.06042
[7] Lewis A and Bridle S 2002 *Phys. Rev. D* **66** 103511
[8] Gorski K M *et al* 2005 *Astrophys. J.* **622** 759
[9] de Gasperis G *et al* 2005 *A&A* **436** 1159-1165
[10] de Gasperis G *et al* 2016 *A&A* **593** A15
[11] Hivon E *et al* 2002 *Astrophys. J.* **567** 2

IOP Conf. Series: Journal of Physics: Conf. Series **841** (2017) 012002 doi:10.1088/1742-6596/841/1/012002

Sterile neutrinos with non-standard secret interactions imprints on Cosmic Microwave Background anisotropies

F Forastieri

Dipartimento di Fisica e Scienze della Terra, Università degli Studi di Ferrara, via Giuseppe Saragat 1, I-44122 Ferrara, Italy

Istituto Nazionale di Fisica Nucleare, Sezione di Ferrara, via Giuseppe Saragat 1, I-44122 Ferrara, Italy

E-mail: frsfnc2@unife.it

Abstract. Short baseline laboratory (SBL) anomalies have shown preference for light sterile neutrinos with eV masses. These particles, if confirmed, would be produced in the early universe and would add their contribution to the relativistic energy density basically increasing the effective number of extra relativistic species ($N_{\rm eff}$). It has been shown that when the matter potential produced by the sterile interactions becomes smaller than the vacuum oscillation frequency, sterile neutrinos are plentifully produced by the scattering effects in the sterile neutrino sector. This behaviour, however, leads to a $\Delta N_{\rm eff} \simeq 1$ which is in tension at $3-5\sigma$ with the actual constraints given by the latest Cosmic Microwave Background radiation (CMB) observations. In order to avoid the thermalization of eV sterile neutrinos in the early universe, secret interactions between the sterile and active sectors mediated by a massive vector boson ($M_X < M_W$) have been proposed. In particular, interactions mediated by a gauge boson having $M_X < 10\,{\rm MeV}$ would suppress the sterile neutrino production for $T > 0.1\,{\rm eV}$ and seem to save the cosmological constraints coming from big-bang nucleosynthesis (BBN) and mass bounds. In this framework, cosmological observations represent a powerful tool to constrain neutrino physics complementary to laboratory experiments. In particular, observations of the CMB have the potential to constrain the properties of relic neutrinos, as well as of additional light relic particles in the universe. In this work we present the effects of the strength of the interaction on the neutrino fluid perturbations and on the CMB anisotropies power spectrum.

1. Introduction

The standard cosmological picture predicts the existence of a thermal background of relic neutrinos with $T_{\nu 0} = 1.9\,{\rm K}$; present cosmological data provide a $10\,\sigma$ evidence for the presence of this background. Moreover, cosmological data are so precise that can go further, constraining several properties of the cosmic neutrino background. From a theoretical point of view the thermal history of neutrinos is well established: they are kept in equilibrium with the cosmological fluid by weak interactions until the temperature of the primordial fluid falls below $T \sim 1\,{\rm MeV}$ and, since the decoupling happens when neutrinos are ultrarelativistic, they preserve a thermal spectrum. The energy contribution of the standard three neutrino families is taken into account into the relativistic content through the effective degrees of freedom $N_{\rm eff}$ whose standard value is 3.046 [1]. The radiation energy density of the universe reads:

$$\rho_r = \rho_\gamma + \rho_\nu = \left(1 + \frac{7}{8}\left(\frac{4}{11}\right)^{\frac{4}{3}} N_{\rm eff}\right)\rho_\gamma \,. \quad (1)$$

Thus in the standard cosmological model the only free parameters are the masses of the three eigenstates. Although this simple picture is perfectly consistent with the observed data there are still some open questions: starting from the absolute scale and the hierarchy of the masses, passing through the Majorana or Dirac nature of these particles and the mass generation mechanism, investigating the reason of the existence of only three active neutrino families and so on. Luckily present-day cosmological data are good enough to allow the study of several non-standard models and to test more complicated scenarios such as non-standard interactions between neutrinos or hidden sterile eigenstates.

The existence of an extra neutrino eigenstate is hinted in short-baseline neutrino experiments [2, 3], in particular laboratory data suggest a sterile neutrino mass of the order of 1 eV and active-sterile mixing angle $\theta_{as} \simeq 0.1$. The production of this new state in the primordial universe would lead to an increase in the extra-relativistic energy content $\Delta N_{\text{eff}} \simeq 1$ in tension with the actual bounds coming from the Planck experiment [4] and with other cosmological bounds [5, 6, 7]. In order to avoid this behaviour several mechanisms have been proposed in the literature, e.g. scenarios with large primordial neutrino asymmetries [8, 9] or with low-reheating temperature [10].

An alternative approach introduces a Fermi-like non-standard interaction between active and sterile eigenstates mediated by a massive gauge boson X, with $M_X \ll M_W$ [11, 12, 13]. The strength of the non-standard secret interaction G_X drives the collisional rate and as long as the universe expands the matter potential generated by the secret coupling declines leading to a resonance in the sterile neutrino sector. This translates into a sterile production due to the combination of the resonant Mikheyev-Smirnov-Wolfenstein effect [14, 15] and non-resonant Dodelson-Widrow production [16]. Considering a coupling constant $g_X \leq 10^{-2}$ and for masses of the mediator larger or of the order of 10 eV, the sterile production would have a considerable effect on the light elements production during the Big Bang Nucleosynthesis (BBN). Assuming smaller values of the mediator mass, sterile neutrinos would still be produced at $T \ll 0.1\,\text{MeV}$ [17]. Mirizzi et al. in [18] have shown that in a certain region of the parameter space (g_X-M_X) where the mass of the mediator is smaller enough to allow the sterile production after the neutrino decoupling, the process of flavour equilibration is fast and produces a sizeable ν_s abundance. In addition this mechanism reduces the effective number of neutrinos to $N_{\text{eff}} \simeq 2.7$ at matter radiation equality (see Sec. 2 for further details) [17, 18].

The existence of a possible window for this secret interaction which allows the presence of a sterile eigenstate without adding a tension in the extra-relativistic energy content is tempting. Aim of this paper is to investigate a secret Fermi-like interaction considering a scenario with 3 active massive neutrinos, having $\sum_i m_{\nu_i} = 0.06\,\text{eV}$ and one light $m_{\nu s} \sim 1\,\text{eV}$ sterile neutrino.

2. The model

The existence of a fourth sterile neutrino eigenstate requires the extension of the standard 3 neutrino families scenario to a 3 + 1 active-sterile neutrino mixing scenario. The active-sterile flavour evolution is described in [19, 8]. Since the flavour equilibration happens when neutrinos have a temperature $T < M_X$, we can consider the low energy approximation for the non-standard interaction, i.e. the effective strength of the interaction takes the form:

$$G_X = \frac{\sqrt{2}}{8}\frac{g_X^2}{M_X^2} \,. \tag{2}$$

The collisional rate of the interaction Γ_X is calculated using the relaxation time approximation [20] and describes a strong collisional effect that would bring the system towards a quick flavour equilibrium among the different neutrino species. The effective production rate Γ is the product of the average probability of conversion between active and sterile neutrinos $\langle P(\nu_a \to \nu_s)\rangle$ with the collisional rate:

$$\Gamma_X = G_X^2 T_\nu^5 \frac{p}{\langle p \rangle}\frac{n_s}{n_a} \,, \tag{3}$$

where n_s and n_a are the densities in number of sterile and active neutrinos, while p and $\langle p \rangle$ are the momentum and the average momentum respectively.

IOP Conf. Series: Journal of Physics: Conf. Series **841** (2017) 012002 doi:10.1088/1742-6596/841/1/012002

It is possible to constrain the secret interaction using different cosmological observables. In [17] the authors perform a study on deuterium primordial abundances $^2H/H$ for a coupling constant $g_X \gtrsim 10^{-2}$ and masses of the mediator $M_X \gtrsim 10\,\mathrm{MeV}$, thus excluding a large part of the parameter space. Alteration in the production of light elements during BBN is due to a larger value of N_{eff} and to the spectral distortion of electron neutrinos when active-sterile oscillations occur close to the neutrino decoupling. Taking smaller values of the mediator mass the sterile production is suppressed before the neutrino decoupling and this choice leaves unchanged the BBN dynamics, but at temperatures less than 1 MeV sterile neutrinos are still in a collisional regime, due to their secret self-interactions. Neglecting the contribution of the resonance and considering only a pure collisional production, the average probability of conversion takes the form:

$$\langle P(\nu_a \to \nu_s)\rangle \simeq \frac{1}{2}\sin^2\theta_{as}\,. \tag{4}$$

In this framework even a small population of sterile neutrinos can generate a large scattering rate at relatively low temperatures for sufficiently large values of the coupling constant G_X. In particular if $G_X > 10^8\,G_F$, where G_F is the Fermi constant, the decoupling of the non-standard interaction would take place at redshift $z \sim 5\times 10^4$, which means a temperature of a few tens eV; this implies the following energy density equilibration. Going from an initial energy density to a final one [7]:

$$\rho_\nu^{in} = 3\cdot\frac{7}{8}\left(\frac{4}{11}\right)^{\frac{4}{3}}\rho_\gamma \;\to\; \rho_\nu^{fin} = 4\left(\frac{3}{4}\right)^{\frac{4}{3}}\cdot\frac{7}{8}\left(\frac{4}{11}\right)^{\frac{4}{3}}\rho_\gamma\,. \tag{5}$$

The consequence of this behaviour is a reduction in the energy density of the neutrino sector which translates into a slightly lower value for the effective number of neutrino families with respect to the standard value:

$$N_{\mathrm{eff}} = 4\cdot\left(\frac{3}{4}\right)^{\frac{4}{3}} = 2.7\,. \tag{6}$$

In this work, we want to study this model (we will refer to it as the $(3+1)\nu$ model) looking at the effect of the interaction on cosmological neutrino perturbations and consequently CMB temperature anisotropies. We will include the effect of the interaction in the density and pressure perturbations inside the neutrino fluid, this modifies the CMB angular power specturm (APS) by increasing the power mainly at high multipoles i.e. at small angular scales (see Sec. 3).

3. Results

In the cosmological framework neutrino perturbations are evolved through the Boltzmann equation and we adopt the same notation of [21]:

$$\frac{\partial\Psi}{\partial\tau} + i\frac{q(\vec{k}\cdot\hat{n})}{\epsilon}\Psi + \frac{dln f_0}{dlnq}\left[\dot{\phi} - i\frac{q(\vec{k}\cdot\hat{n})}{\epsilon}\Psi\right] = \frac{1}{f_0}\hat{C}[f]\,, \tag{7}$$

where τ is the conformal time, f is the distribution function of neutrinos and Ψ is the perturbation. The right hand side of the equation is the collisional term which consists in a complicated integral involving specific scattering terms for the considered processes. Its calculation would be too demanding from a computational point of view and generally excessively sophisticated for the purpose of this study. Thus, as said before, we use the relaxation time approximation where:

$$\hat{C}[f] \simeq a\Gamma\,\delta f\,. \tag{8}$$

Inside this collisional term, $2 \leftrightarrow 2$ processes are taken into account and the most generic form of the scattering rate can be written as:

$$\Gamma = \begin{bmatrix} sin^2\theta_{as} & 0 & 0 & sin\theta_{as}\,cos\theta_{as} \\ 0 & 0 & 0 & 0 \\ 0 & 0 & 0 & 0 \\ sin\theta_{as}\,cos\theta_{as} & 0 & 0 & cos^2\theta_{as} \end{bmatrix} G_X^2\,T^5\,, \tag{9}$$

where we considered the mass basis in order to evolve the perturbations inside the public code CAMB [22, 23], basically considering the sterile state as a superposition of neutrino mass eigenstates ν_1 and ν_4. In Figure 1 it is possible to appreciate the effects of the interaction on the density perturbations $\delta_\nu = \delta\rho_\nu/\rho_\nu$ (the first order of Boltzmann hierarchy) for different perturbation wave numbers.

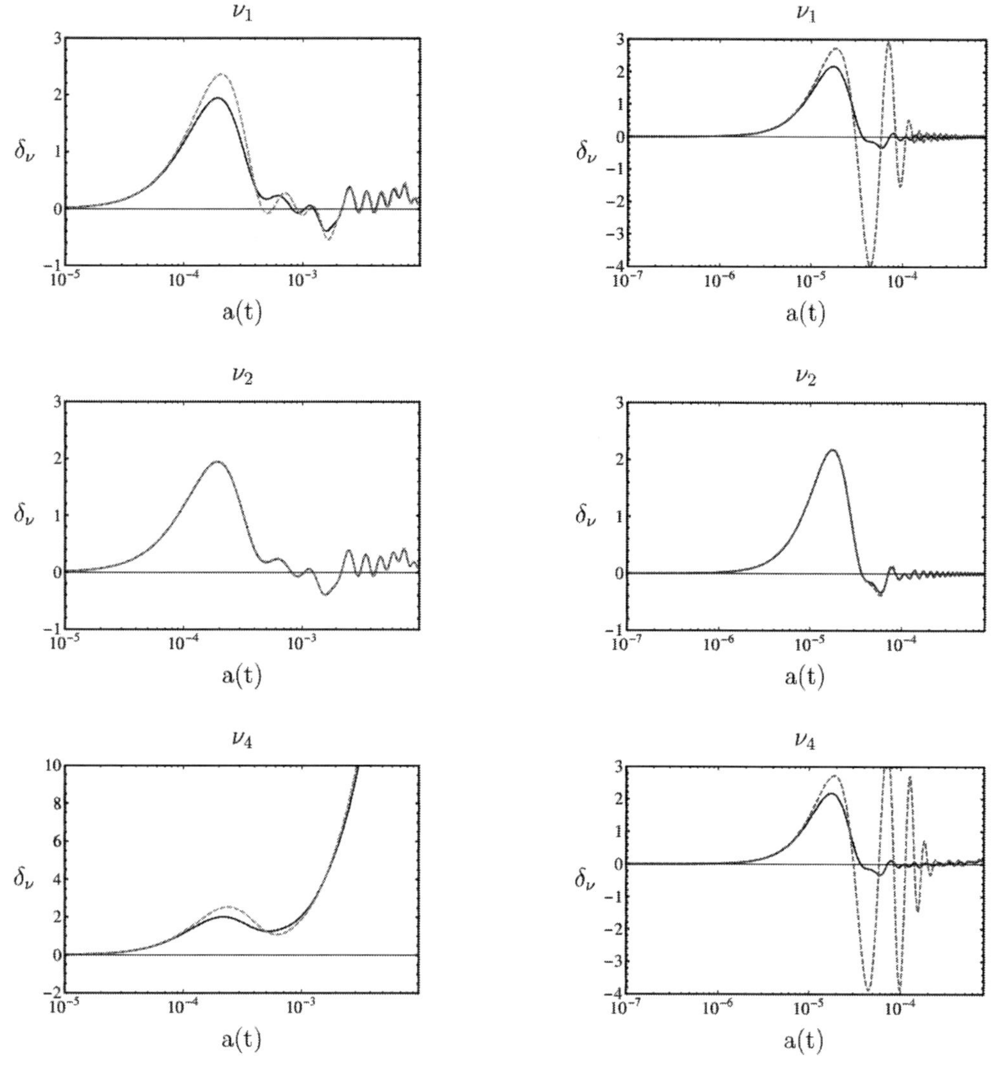

Figure 1. Density perturbations as function of the universe scale factor $a(t)$ for the $(3+1)\nu$ model for two different wave numbers, left column $k = 0.05\,\mathrm{Mpc}^{-1}$, right column $k = 0.5\,\mathrm{Mpc}^{-1}$, in case of no interaction (black solid line) and in case of interaction mediated by a coupling constant $G_X = 10^5\,\mathrm{G_F}$ (colored dashed line).

Out of the horizon the physics is unchanged and the evolution is dominated by the metric perturbation. As the mode enters the horizon it starts to oscillate and from this moment on the dynamic is influenced by micro-physics, thus the larger the magnitude of the interaction the larger the effects on the oscillations of the density perturbations. Considering the interacting cases, the blue dashed line refers to the mode $k = 0.5\,\mathrm{Mpc}^{-1}$ which crosses the horizon at approximately $z \sim$ few $\times\, 10^5$, the mode $k = 0.05\,\mathrm{Mpc}^{-1}$ (red dashed line) enters the horizon at a redshift an order of magnitude lower, this translates into a greater effect. Notice that the intensity of the effect depends also on the probability of conversion which is greater for $\nu_s - \nu_s$ due to the $cos^2\theta_{as}$ dependency. The corresponding effect on the temperature anisotropies power spectrum is an increase of the power over all the multipoles, especially on small angular scales as expected from the perturbation analysis which revealed a greater impact of the interaction on scales that enter the horizon at early times with respect the photon recombination.

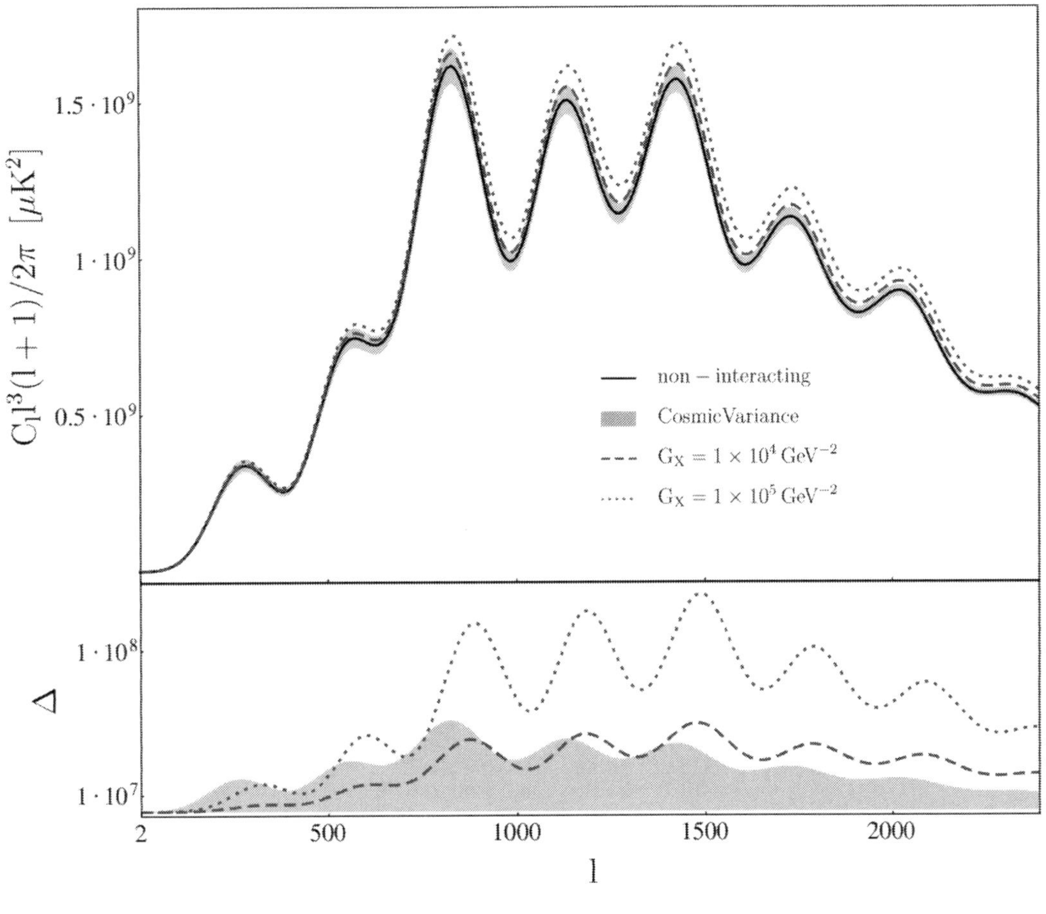

Figure 2. Upper panel: theoretical angular power spectrum for the $(3 + 1)\nu$ model and for two different values of the coupling constant G_X, the blue dashed line is obtained setting $G_X = 1 \times 10^4\,\mathrm{GeV}^{-2}$, the blue dotted one represents the case $G_X = 1 \times 10^5\,\mathrm{GeV}^{-2}$. The black solid one is the 3+1 non-interacting case ($G_X = 0$) with the corresponding cosmic variance (grey filled area). In the lower panel we show the corrensponding differences among the interacting and non interacting cases. The line-style notation is the same in the two panels. In order to underline the effect of the interaction all the spectra have been obtained starting from the same set of cosmological parameters for the best-fit [4] of the non-interacting case.

In Figure 2 it is possible to appreciate the impact of the interaction on the main observable of the CMB framework. The upper panel shows the APS for the $(3 + 1)\nu$ model in case of no interaction and for two different values of the coupling constant $G_X = 10^4\,\mathrm{GeV}^{-2}$ and $G_X = 10^5\,\mathrm{GeV}^{-2}$ respectively. In the bottom panel we present the difference of the two interacting cases with respect to the non-interacting one.

4. Conclusions

We have shown the effects of a non-standard Fermi-like interaction among three active and one light sterile neutrinos in the primordial universe framework; we studied their impact on the perturbations and, thus, on the angular power spectrum of the CMB. When the interaction is dominant, if neutrinos are tightly coupled their anisotropic stress is suppressed and eventually vanishes, thus the damping of density perturbation due to neutrino free streaming is no more effective and the neutrino fluid undergoes acoustic oscillations. Since the fluid starts to oscillate when the wavelength of a perturbation enters the horizon and the interaction is stronger at early times, the effect is greater on small scales. Considering an interaction strength $10^9 \div 10^{10}$ times greater than the standard Fermi weak interaction, the impact on the CMB APS is, over all the multipoles, larger than the theoretical uncertainty given by the cosmic variance. This translates into a quantifiable effect that can balance the presence of a sterile massive neutrino eigenstate suggested by the SBL anomalies. Therefore, the model can be tested against available CMB data in a future work.

References

[1] Mangano G, Miele G, Pastor S and Peloso M 2002 *Phys. Lett.* **B534** 8–16
[2] Abazajian K N *et al.* 2012 (*Preprint* 1204.5379)
[3] Gariazzo S, Giunti C, Laveder M, Li Y F and Zavanin E M 2016 *J. Phys.* **G43** 033001
[4] Ade P A R *et al.* (Planck) 2016 *Astron. Astrophys.* **594** A13
[5] Hamann J, Hannestad S, Raffelt G G and Wong Y Y Y 2011 *JCAP* **1109** 034
[6] Hannestad S, Tamborra I and Tram T 2012 *JCAP* **1207** 025
[7] Mirizzi A, Mangano G, Saviano N, Borriello E, Giunti C, Miele G and Pisanti O 2013 *Phys. Lett.* **B726** 8–14
[8] Mirizzi A, Saviano N, Miele G and Serpico P D 2012 *Phys. Rev.* **D86** 053009
[9] Saviano N, Mirizzi A, Pisanti O, Serpico P D, Mangano G and Miele G 2013 *Phys. Rev.* **D87** 073006
[10] Yaguna C E 2007 *JHEP* **06** 002
[11] Hannestad S, Hansen R S and Tram T 2014 *Phys. Rev. Lett.* **112** 031802
[12] Dasgupta B and Kopp J 2014 *Phys. Rev. Lett.* **112** 031803
[13] Bringmann T, Hasenkamp J and Kersten J 2014 *JCAP* **1407** 042
[14] Mikheev S P and Smirnov A I 1986 *Nuovo Cimento C Geophysics Space Physics C* **9** 17–26
[15] Wolfenstein L 1978 *Phys. Rev. D* **17**(9) 2369–2374
[16] Dodelson S and Widrow L M 1994 *Phys. Rev. Lett.* **72**(1) 17–20
[17] Saviano N, Pisanti O, Mangano G and Mirizzi A 2014 *Phys. Rev.* **D90** 113009
[18] Mirizzi A, Mangano G, Pisanti O and Saviano N 2015 *Phys. Rev.* **D91** 025019
[19] Dolgov A D, Hansen S H, Pastor S, Petcov S T, Raffelt G G and Semikoz D V 2002 *Nucl. Phys.* **B632** 363–382
[20] Hannestad S and Scherrer R J 2000 *Phys. Rev. D* **62**(4) 043522
[21] Ma C P and Bertschinger E 1995 *Astrophys. J.* **455** 7–25
[22] Lewis A, Challinor A and Lasenby A 2000 *Astrophys. J.* **538** 473–476
[23] Howlett C, Lewis A, Hall A and Challinor A 2012 *JCAP* **1204** 027

IOP Conf. Series: Journal of Physics: Conf. Series **841** (2017) 012003 doi:10.1088/1742-6596/841/1/012003

Hyperspectral camera based on a Fabry-Pérot with varying beam incidence

L Giovannelli, F Berrilli and D Del Moro

Università di Roma "Tor Vergata", Via della Ricerca Scientifica, I-00133 Roma, Italy

Abstract.
Fabry-Pérot interferometers are commonly used as tunable filters in a variety of scientific fields. We here consider the use of a single Fabry-Pérot as a universal tunable filter in hyperspectral imaging applications, when it is necessary to recover image information of the scientific target at different wavelengths.
Fabry-Pérot are commonly used to scan small portions of the spectrum, defined by suitable prefilters. This is necessary due to the multi-peak response of a single interferometer. In order to overcome this issue we propose a novel triple-pass configuration, able to provide a fast and continuous sampling of the electromagnetic spectrum in the visible-infrared range without the need for a collection of dedicated prefilters.

1. Introduction

The ideal instrument for spectroscopy in a variety of scientific fields is a filter with a tunable centerband, which could cover a wide range of the electromagnetic spectrum, sampling light intensity at different wavelengths. Fabry-Pérot Interferometers (FPI) are commonly used as tunable filters in Astronomy (in ground based observatories i.e. VLT [1], Gemini [2], GTC [3], or in space missions i.e. Dynamic Explorer [4], JWST [5], Bepi-Colombo [6]), Information and Communications Technology [7, 8], Photonics [9–11], Atmospheric Physics [12–14] and other research fields.
FPIs have some major advantages with respect to long slit spectrographs:

- Fast response: can be tuned changing the spectral position in few ms. A full scan is completed in tens of seconds in the case of an FPI whereas can last minutes in the case of a long slit spectrograph.
- High transparency: given the same collecting area and the same spectral resolving power, the flux of light exiting an FPI can be 30 times the one exiting a long slit spectrograph [15].
- Adjustable: an FPI is not only tunable in wavelength at fixed step, you can program the portion of the spectrum you want to sample jumping by where it is not necessary to register the information. Furthermore, it is possible to vary the spectral resolving power in different portions of the spectrum.

The spectral response of a single FPI is a comb of peaks along the whole spectrum, which corresponds to different interference orders. The interference comb can be opportunely fine-shifted varying the optical cavity spacing. The goal of a FPI based instrument is to select just one interference order to obtain a nearly monochromatic signal. For this reason an FPI is always used

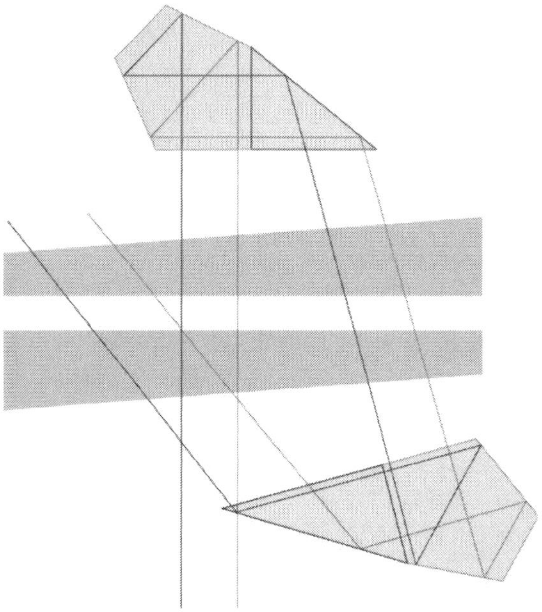

Figure 1. Optical layout. The FPI (gray) is seen from above. As usual to avoid unwanted ghost reflections from the external optical surfaces, the optical plates are wedged. The pentaprisms (light green) collect the beam and provide a 90 degree reflection. The inclination angle is set by the prisms (light blue), optically contacted with the pentaprisms. Two rays of the optical beam are shown in red and black. The superposition of the reflected beam is avoided due to a suitable arrangement in 3D.

with a prefilter to select the portion of the spectrum to be investigated. For a more comprehensive description of single FPI see e.g. [16–20]. This may be not sufficient since the bandwidth of the prefilter is often much greater than the Free Spectral Range (FSR), i.e. the distance between two subsequent peaks of the FPI comb. The solution is often to use two or more FPIs in series, with slightly different FSR to obtain the superposition of just one transparency peak in the prefilter bandwidth (see e.g. [21, 22]). We here propose for the first time a different solution using just one FPI in an triple pass configuration, with a varying beam incidence, to obtain an extended FSR (see Fig. 1). The result is a compact instrument which is here proposed as a universal tunable filter.

We report here just a few examples of scientific cases in which an FPI could be an improvement compared to nowadays instruments:

- Earth observation from satellites. Hyperspectral datasets are used, for example, to distinguish terrain characteristics and the type of cultivation. See i.e. the datasets provided by AVIRIS satellite, which uses 224 spectral channels with contiguous 10 nm sampling from 400 nm to 2500 nm [23].
- Archaeological and arts applications. Hyperspectral cameras are used e.g. to reveal hidden features in paintings and artifacts in reflectometry when X-rays and IR techniques fails [24]. In this case 1000 partially overlapped bands of 2.5 nm in the visibile and near-infrared region were used.
- Planetary surface mapping. We here focus in particular on imaging spectroscopy instruments used on different planetary missions to understand chemical composition of the

soil and the atmosphere as well as the morphological evolution of planets and minor bodies of the Solar System [25,26]. E.g. CRISM is an imaging spectrometer with a scannable field of view that can cover wavelengths from 0.362 μm to 3.92 μm at 6.55 nanometers/channel ($\sim$ 500 channels).

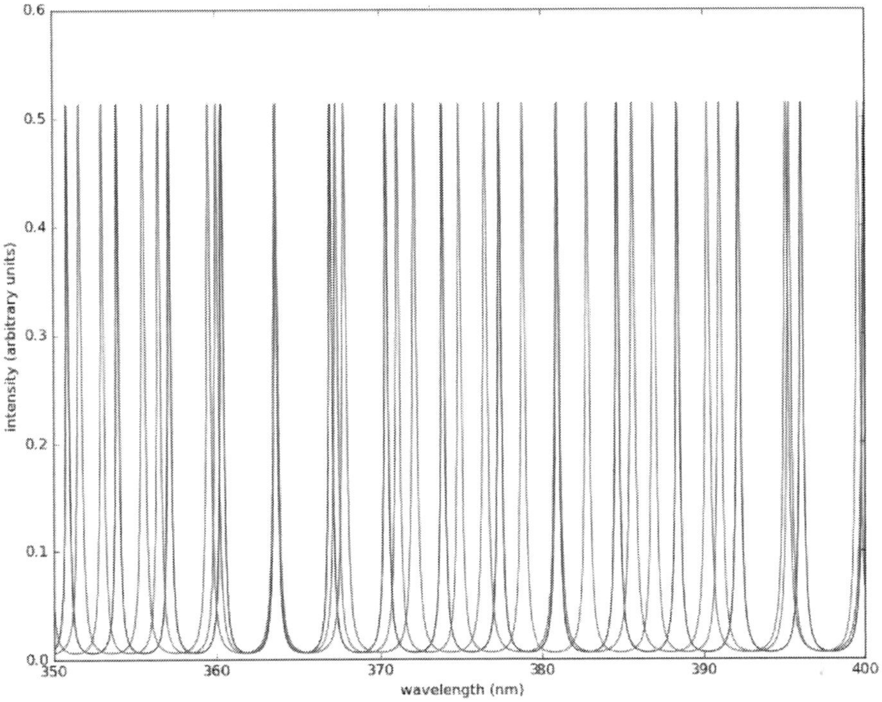

Figure 2. We show the simulation of the FPI spectral profile with three different angle of incidence. The first passage in the optical cavity (blue) has $\theta_1 \simeq 0$ to maximize the interference order and so spectral resolution. The second passage (green) has $\theta_2 \simeq 17.75$ in order to have a green peak superposed with a blue one every 20 green peaks. The third passage (red) has $\theta_3 \simeq 36.87$ in order to have a red peak superposed with a blue one every 4 red peaks. The incoming intensity is set to 1 all over the spectrum (a.u.).

2. The new design with varying incidence angle

Usually a double or triple FPI configuration is used to extend the FSR of the instrument. Using different optical cavity spacing (OCS) every FPI is characterized by a different interference order $m = \frac{2n}{\lambda}\, d\, \cos\theta$, where n is the refractive index of the air, λ is the wavelength of interest, d is the OCS and θ is the angle of incidence of the beam respect to the optical surface normal. The FSR depends on the interference order: $FSR = \lambda/m$. Combining the spectral profiles together, it is possible to extend the FSR. We here propose the same effect with a single cavity. This is possible because we vary the incidence angle of the beam instead of d. We impose FSR2 = 1.05 FSR1 and FSR3= 1.25 FSR1 in order to have one coincident peak out of 20 in the first case and one out of 4 in the second (see Fig. 2). We impose $\theta_1 = 0$ to maximize the interference order and so spectral resolution. From FSR calculation we get: $\theta_2 \simeq 17.75°$ and $\theta_3 \simeq 36.87°$. This geometric configuration is obtained using a pentaprism optically contacted with a prism with a suitable angle to fine tune the new incidence angle of the beam (see Fig. 1). This optical configuration guarantees the same inclination angle even in the case of minor misalignments of the optical elements.

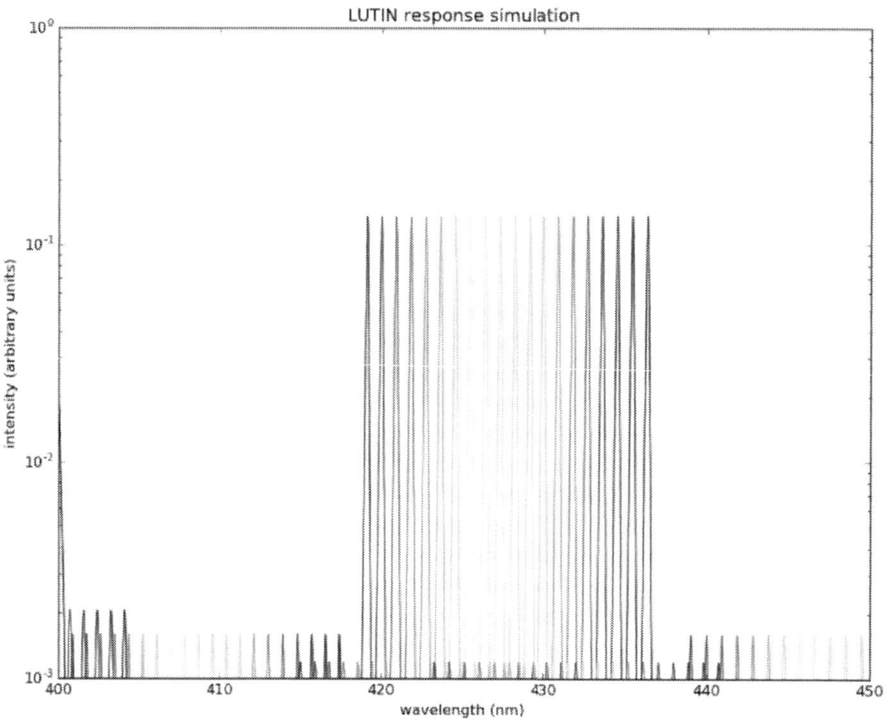

Figure 3. We show the simulation of the spectral profile after all three passages in the optical cavity. The different colours represent different steps during an FPI scan of the visible spectrum. Apart from weak secondary peaks, there is only one peak in the reported 50 nm portion of the spectrum. We set to 50 nm the variation of the optical cavity spacing, resulting in a spectral step of $\simeq$ 0.9 nm, while the FWHM $\simeq$ 0.2 nm. The incoming intensity is set to 1 all over the spectrum (a.u.).

We present the simulation for a FPI with an OCS of 24 μm in the range $400 \div 450nm$. The simulation take into account the effective finesse of a non-ideal FPI as in [17], using, as an example, the same parameters of the LUTIN prototype as in [20]. For a single pass, the resulting FWHM is $\simeq$ 0.8 nm while the FSR $\simeq$ 3.5 nm. With the triple pass the FSR around 400 nm is extended to $\sim$ 80 nm, while around 800 nm is $\sim$ 300 nm. The spectral purity of the instrument is high in the extended FSR region, as the contribution from secondary peaks is $< 4\%$. In this way it is possible to use standard broadband filters as pre-filters, like Johnson UBVRI filters, covering with a single FPI the whole VIS-NIR spectrum. We show in Fig. 3 the spectral profiles for every incidence angle, the combined spectral profile and a scan sequence with a 50 nm optical cavity scanning step. The effect of the inclination over the reflectivity of the coatings is not taken into account here, since is highly dependent on the type of coating used in FPI outer and inner optical surfaces. However a preliminary calculus suggest that the final peak transparency of the FPI in the proposed inclined configuration must be $\sim 70\%$ respect to the case of three FPIs in series. Further investigation on this point will be performed once the technique will be applied to a specific real case. We stress that the three spectral profiles remain tuned during the scan, with only one parameter to be controlled, while a triple FPI system needs three cavity gaps to be tuned in sincro during the scan. Furthermore, given that the three inclination angles are fixed, the polarization induced by the reflection on the inclined optical surfaces can be easily taken into account by usual polarization calibration routines anyhow used in spectropolarimeters based on FPIs (see i.e. [27]).

3. Conclusions

In recent years, many imaging spectroscopy instruments have flown on different planetary missions, e.g. VIMS on Cassini, CRISM on Mars Reconnaissance Orbiter (MRO), and hyperspectral remote sensors from there have provided the measures that allowed us to understand the formation and evolutionary processes planets and minor bodies have undergone, through the surface mapping and composition study and the identification of specific surface features. In fact, hyperspectral imaging instruments play a major role in most of the planetary missions. The new instrument concept introduced in this work is a change of paradigm, since it provides imaging at a wavelength of choice, with a < 1 nm bandpass, in a very wide spectral range. These capabilities can be either exploited to reduce the amount of data to be acquired, processed and transmitted back, or to increase the data acquisition rate, or to tune the spectral sampling to the observation necessities. The flexibility of this instrument is such that it can play both the high-cadence imaging camera role when telemetry allows, and, at large distances from Earth, it can limit the data volume that has to be transmitted to the bare minimum, by selecting only the most interesting wavelengths.

References

[1] Hartung M *et al.* 2004 *Society of Photo-Optical Instrumentation Engineers (SPIE) Conference Series* **5492**
[2] Scott A *et al.* 2006 *Society of Photo-Optical Instrumentation Engineers (SPIE) Conference Series* **6269**
[3] Cepa J *et al.* 2003 *Society of Photo-Optical Instrumentation Engineers (SPIE) Conference Series* **4841**
[4] Killeen T L *et al.* 1983 *Applied Optics* **22** 22
[5] Yoshikawa I *et al.* 2010 *Planetary and Space Science* **58** 1-2
[6] Doyon R *et al.* 2008 *Society of Photo-Optical Instrumentation Engineers (SPIE) Conference Series* **7010**
[7] Chan L Y *et al.* 2002 *Electronics Letters* **38** 1
[8] Vahala K J *et al.* 2003 *Nature* **424**
[9] Villatoro J *et al.* 2009 *Optics letters* **34** 16
[10] Murphy K A *et al.* 1991 *Optics letters* **16** 4
[11] Wu C *et al.* 2011 *Optics letters* **36** 3
[12] Rees D *et al.* 1996 *Journal of Atmospheric and Terrestrial Physics* **58** 16
[13] Xia H *et al.* 2007 *Applied Optics* **46** 29
[14] Arshinov Y *et al.* 2005 *Applied Optics* **44** 17
[15] Hernandez G 1988 Fabry-Perot interferometers *Cambridge University Press*
[16] Berrilli F *et al.* 2011 *Society of Photo-Optical Instrumentation Engineers (SPIE) Conference Series* **8148**
[17] Giovannelli L *et al.* 2012 *Society of Photo-Optical Instrumentation Engineers (SPIE) Conference Series* **8446**
[18] Giovannelli L *et al.* 2012 *Journal of Physics Conference Series* **383**
[19] Giovannelli L *et al.* 2014 *Society of Photo-Optical Instrumentation Engineers (SPIE) Conference Series* **9147**
[20] Giovannelli L *et al.* 2014 *Journal of Physics Conference Series* **566**
[21] Cavallini F 2006 *Solar Physics* **236** 415
[22] Tritschler A Schmidt W Langhans K & Kentischer T 2002 *Solar Physics* **211** 17
[23] Clevers J G P W 1999 *ISPRS Journal of Photogrammetry and Remote Sensing* **54** 299
[24] Snijders L Zaman T & Howell D 2016 *Journal of Archaeological Science: Reports* **9** 143
[25] Brown R H Baines K H Bellucci G *et al.* 2004 *Space Science Reviews* **115** 111

[26] Murchie S Arvidson R Bedini P *et al.* 2007 *Journal of Geophysical Research (Planets)* **112** E05S03
[27] Viticchié *et al.* 2010 *The Astrophysical Journal* **723** 1

Derivation of the Hubble parameter using galaxy clusters

A Kozmanyan[1], H Bourdin[1], G de Gasperis[1,2], P Mazzotta[1,3] and N Vittorio[1,2]
[1]Dipartimento di Fisica, Università di Roma "Tor Vergata", via della Ricerca Scientifica 1, I-00133, Roma, Italy
[2]Sezione INFN Roma 2, via della Ricerca Scientifica 1, I-00133, Roma, Italy
[3]Harvard Smithsonian Centre for Astrophysics, 60 Garden Street, Cambridge, MA 02138, USA

E-mail: arpine.kozmanyan@roma2.infn.it

Abstract. In this work we describe a possible way to use X-ray and microwave observations of nearby galaxy clusters to derive the value of the Hubble constant, that parametrises the expansion rate of the Universe. We provide a brief introduction to the Sunyaev-Zel'dovich effect that allows to detect galaxy clusters at microwave frequencies, and the method to combine it with X-ray observables. We emphasize what kind of considerations should be done when applying the method on real data and study the effect of the geometry of the clusters on the final result.

1. Introduction

Galaxy clusters are the largest gravitationally collapsed structures known in the Universe. They are dominated by dark matter, but also contain a considerable amount of hot ionised gas called Intra-Cluster Medium (ICM). This ionised medium allows the detection of galaxy clusters in at least two ways. The ICM electrons emit bremsstrahlung radiation at X-ray wavelengths. Additionally, when passing through a galaxy cluster the Cosmic Microwave Background (CMB) Radiation photons undergo Inverse Compton scattering by the ICM electrons. This introduces a modification to the original blackbody spectrum of the CMB radiation (the so called Sunyaev-Zel'dovich effect), thus making it possible to detect the clusters using the CMB as a backlight.

The two effects are interconnected, since they both depend on the distribution of the ICM electrons in the galaxy cluster. Using both of the signals a thorough investigation of matter distribution in the cluster can be done. The knowledge of the distribution of matter inside the cluster on the other hand allows to model the expected Sunyaev-Zel'dovich (SZ) and X-ray signals for a given cluster assuming a certain cosmological model. A derivation of the best-fitting parameter values can then be done comparing the modeled signals with the observed ones.

The physics of the ICM is quite complicated and our knowledge of it is limited, however, galaxy clusters could provide a complementary cosmological probe along with the classical ones such as type Ia supernovae, CMB, or baryon acoustic oscillations. The quantity that can be constrained using SZ and X-ray signals jointly is the angular diameter distance for individual clusters. Hence cosmological parameters such as the Hubble constant today, the total density of matter and the equation of state of dark energy can be potentially constrained.

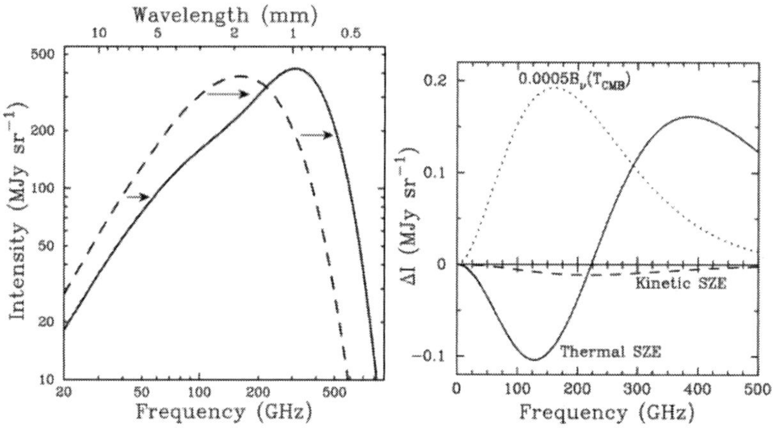

Figure 1. Left: The distortion of the blackbody CMB power spectrum (dashed line) due to the SZ effect (solid line). The amplitude of the effect is exaggerated in order to show visually the overall shift of the spectrum to higher energies. Right: The amplitude of the distortion depending on frequency for the thermal and kinematic effects. Both figures are taken from Carlstrom et al. [8].

At the moment, however, the data are not sensitive enough to be sensitive to the latter quantities. Instead, the Hubble constant can be estimated easily. Current estimates of the parameter from CMB anisotropies by the Planck Collaboration [1] $H_0 = 66.93 \pm 0.62$ km s^{-1} Mpc^{-1} seem to be in slight tension with the value derived by Riess et al. [2] using Cepheid-calibrated type Ia supernovae $H_0 = 73.24 \pm 1.74$ km s^{-1} Mpc^{-1}. Other probes such as the baryon acoustic oscillations provide recent results of $H_0 = 68.0 \pm 1.1$ km s^{-1} Mpc^{-1} by Cheng et al. [3] or in combination with CMB measurements from WMAP $H_0 = 69.6 \pm 0.7$ km s^{-1} Mpc^{-1} from Bennett et al. [4]. Using galaxy clusters, instead, Bonamente et al. [5] found a value of $H_0 = 76.9^{+3.9+10.0}_{-3.4-8.0}$ km s^{-1} Mpc^{-1} for fixed $\Omega_m = 0.3$ and $\Omega_\Lambda = 0.7$ using a sample of 38 clusters. The newly available Planck data will allow to improve the estimate derived from galaxy clusters.

Our work is structured in the following way. We start with a brief introduction to the Sunyaev-Zel'dovich effect and its characteristics in Section 2. In the same section we give a short description of the methods of observation of the SZ signal from galaxy clusters at microwave frequencies (Section 2.1). In Section 3 we describe the method to derive the Hubble parameter and discuss our preliminary results about the systematics introduced by the geometry of the clusters. Finally, we summarise our conclusions in Section 4.

2. The Sunyaev Zel'dovich effect from Galaxy Clusters

Density perturbations in the primordial Universe get amplified due to gravity. Eventually they collapse under their own mass and form smaller structures that, after series of mergers, will combine into larger structures. In this way, by a hierarchical sequence of mergers of galaxies, the largest gravitationally bound structures such as clusters of galaxies are formed. The main constituent of the galaxy clusters is the dark matter ($10^{13} - 10^{15} M_\odot$), which generates the gravitational potential felt by the luminous components. The latter are the hot ionised gas that constitutes the intra-cluster medium ($10^{12} - 10^{14} M_\odot$) and the galaxies themselves ($10^{11} - 10^{13} M_\odot$).

The variety of these components provides us with a number of ways to detect galaxy clusters. One can look at the light emitted by the stars in the member galaxies of the cluster at optical and radio wavelengths. The electrons in the ICM also emit bremsstrahlung radiation that is possible to detect at X-ray wavelengths [6]. And finally the same electrons can Inverse Compton scatter part of the CMB photons that pass through the cluster on their way to us. This shifts the energies of the scattered photons to higher frequencies [7]. As a result, the original blackbody spectrum of the CMB gets distorted (see Fig. 1).

The exact shape of the distortion depends on the nature of the scattering between the photons and the electrons. In particular, we are interested in the scattering due to the thermal motion of the ICM electrons (thermal SZ), since it provides a way to detect the galaxy clusters using the CMB as a backlight.

Since in galaxy clusters the electron temperatures are very high ($k_BT \simeq keV$), after the electron-photon scattering, the photons gain energy. As the Compton scattering conserves the photon number, the overall effect is a shift to higher energy of the initial Planck spectrum (illustrated in the left part of Fig. 1). As a result the measured CMB intensity is lower at frequencies $\nu < \bar{\nu}$ and higher at frequencies above it as shown in the right part of Fig. 1. The frequency $\bar{\nu}$ is where there is no variation in the intensity and it corresponds to $\bar{\nu} = 217$ GHz for non-relativistic thermal SZ effect. This spectral signature can be used to identify galaxy clusters in the microwaves. The relative change in the CMB intensity introduced by the thermal SZ effect is:

$$\frac{\Delta I}{I_{CMB}} = g(x) \int \sigma_T n_e \frac{k_B T_e}{m_e c^2} dl. \tag{1}$$

Here $g(x)$ is a function of dimensionless frequency $x \equiv h\nu / k_B T_{CMB}$, I_{CMB} the CMB intensity, T_e the temperature of the ICM electrons, n_e the number density of the electrons and σ_T, m_e, c are the Thomson scattering cross-section, the electron mass and the speed of light respectively. Finally, the integration on dl is done along our line of sight. The function $g(x)$ describes the spectral shape of the effect and has the following form:

$$g(x) = \frac{x^4 e^x}{(e^x - 1)^2} \left(x \frac{e^x + 1}{e^x - 1} - 4 \right). \tag{2}$$

The integral on the right-hand side of Eq. 1 is the Comptonisation parameter y:

$$y = \frac{\sigma_T}{m_e c^2} \int n_e \times k_B T_e \times dl = \frac{\sigma_T}{m_e c^2} \int P_e \times dl, \tag{3}$$

where we introduced the electron pressure profile as the product of the number density and the temperature ($P_e = n_e \times k_B T_e$). The Comptonisation parameter y is the quantity that describes the amplitude of the SZ effect and depends on the cluster properties, i.e. the thermal electron pressure of the ICM along the line of sight.

Eq. 1 has been derived in the non-relativistic approximation for clusters of temperatures $k_BT < 10\ keV$. At higher temperatures corrections need to be introduced to the above formulae that will eventually also shift the location of the characteristic frequency $\bar{\nu}$ [9, 10].

An additional effect can be caused due to the overall peculiar motion of the galaxy cluster relative to the CMB rest frame (kinematic SZ effect). In this work we will concentrate only on the thermal effect and will denote it simply as SZ.

2.1. Observations of the thermal SZ signal from galaxy clusters

There is a large number of astrophysical and cosmological components that emit at microwave frequencies. Examples are the synchrotron emission generated by the relativistic cosmic-ray electrons spiraling in the Galactic magnetic field, the free-free emission that is the bremsstrahlung radiation from electron-ion collisions in our Galaxy, the thermal emission from dust grains on the Galactic plane and of course the CMB. The signal that we measure from the sky is the sum of the contributions from all these components. In order to perform any physical investigation we need first to perform a component separation. This can be achieved by multifrequency observations making use of the different spectral signatures of the components [11, 12].

For illustration purposes let us assume to have only three components: the CMB with the SZ component embedded in it and the thermal emission from Galactic dust. The total signal at a given frequency and direction in the sky (the latter is characterised by a given pixel $\vec{p}$ in the eventual map of the sky) can be written as the linear combination of the three components:

$$s(\vec{p},\nu) = A_{CMB}(\vec{p}) + A_{SZ}(\vec{p}) * f_{SZ}(\nu) + A_{dust}(\vec{p}) * f_{dust}(\nu). \quad (4)$$

Here $A_i(\vec{p})$ are the amplitudes of the component signals at a given direction $\vec{p}$, and $f_i(\nu)$ are their spectral dependences. Note that $A_i(\vec{p})$ depend only on the direction on the sky and $f_i(\nu)$ instead depend only on the frequency and are defined by the model for the spectral dependence assumed for each component. In particular $f_{SZ}(\nu)$ would be described by Eq. 2 and the thermal dust contribution can be described by a variety of gray-body spectra [12, 13]. The total signal $s(\vec{p},\nu)$ is a function of frequency and direction in the sky. Thus, having measurements of the total signal for a given pixel at different frequencies, one can solve a set of equations for a given pixel and derive the values for $A_i(\vec{p})$.

3. Derivation of the angular diameter distance from joint SZ and X-ray data of galaxy clusters

Both SZ signal and X-ray emission from galaxy clusters are caused by the ICM electrons. The SZ signal is the line-of-sight integral of the electron pressure as shown in Eq. 3. The X-ray surface brightness emission instead is the line-of-sight integral of the electron number density squared:

$$S_X \propto \int n_e^2 dl. \quad (5)$$

This means both of the signals depend on the physical size of the cluster along the line of sight.

When we observe the cluster we measure only its angular size and the redshift. However, the real physical size of the cluster given its angular size and redshift depends on the cosmological model. The relationship between the physical size of an object and its apparent angular size is given by a quantity called angular diameter distance:

$$D_A = \frac{l}{\theta}, \quad (6)$$

where l is the physical size of the object in Mpc and θ is the corresponding angular size in radians. This quantity, on the other hand, depends on the cosmology of the Universe:

$$D_A = \frac{1}{1+z}\frac{c}{H_0} \times \begin{cases} \frac{1}{\sqrt{\Omega_k}} \sinh\left(\sqrt{\Omega_k}\int_0^z \frac{dz'}{E(z')}\right), & \text{for } \Omega_k > 0 \\ \int_0^z \frac{dz'}{E(z')}, & \text{for } \Omega_k = 0 \\ \frac{1}{\sqrt{|\Omega_k|}} \sin\left(\sqrt{|\Omega_k|}\int_0^z \frac{dz'}{E(z')}\right), & \text{for } \Omega_k < 0 \end{cases} \quad (7)$$

Here H_0 is the value of the Hubble constant, c is the speed of light in vacuum and z the redshift of the object. The parameter $\Omega_k = 1 - \Omega_m - \Omega_\Lambda$ is the curvature density of the Universe, Ω_m the total density of matter and Ω_Λ is the dark energy density. Finally, the factor $E(z)$ is defined as $E(z) = \sqrt{\Omega_m(1+z)^3 + \Omega_k(1+z)^2 + \Omega_\Lambda}$. The three cases in Eq. 7 represent the cases with the Universe being negatively curved, flat and positively curved, respectively.

The relationship in Eq. 7 can be much simplified when dealing with objects at low redshift ($z < 0.1$). It can be reduced to:

$$D_A \propto \frac{1}{H_0} \times z. \tag{8}$$

Then the SZ and X-ray signals from a cluster at a given redshift and with a given angular size depend on the value of H_0. Consequently the two signals would depend on different combinations of n_e and H_0: $y \propto n_e H_0^{-1}$ and $S_X \propto n_e^2 H_0^{-1}$. This allows one to model the signal expected from the two effects for an assumed H_0 and check its agreement with the observations of the X-ray brightness and the SZ signal strength.

One needs to be careful, however, about the assumptions done for the analysis. Let us discuss two of them.

ICM profile model. The models used for the electron number density distribution and the electron pressure distribution in the cluster could introduce biases in the final result. Bonamente et al. [14] have shown, however, that the results of the calculation do not depend strongly on the models used. In their analysis the two estimates derived using two different models resulted to be in agreement.

Deviations of cluster shape from spherical. A large uncertainty in the calculation of the Hubble parameter could be introduced by assuming the sphericity of the cluster. Deviations from spherical shape will result in underestimation or overestimation of the Hubble constant. In short this can be written through the following ratio:

$$f_H = \frac{H_{est}}{H_{real}} = \frac{r_\perp}{r_\parallel}, \tag{9}$$

where $r_\perp$ is the radius that the cluster is assumed to have (hence its extent in the plane of the sky) and $r_\parallel$ is its extent in the direction of the line of sight (LOS).

So by assuming the cluster to be spherical, we will be underestimating the Hubble constant in the case when $r_\parallel > r_\perp$ (cluster prolate in the direction of the LOS). Instead when $r_\parallel < r_\perp$ (cluster oblate in the direction of the LOS) the Hubble constant will be overestimated. According to Sulkanen et al. [15] inclusion of a large number of clusters in the analysis allows to diminish the bias introduced by the asphericity of the individual clusters. This however will not be true if the sample used is predominantly prolate or oblate along our line of sight. There could be an intrinsic bias in the cluster sample, for example, if we happened to be more likely to detect clusters that are prolate in the LOS since the signal from such clusters would be stronger.

This means that there is a need to find a way to quantify the predominance of one or the other type of clusters in the sample. According to Kawhara et al. [16] the distribution of the measured individual H_{est} values for a set of clusters can give us hints about any biases inside the sample. Predominantly prolate samples will lead to a negatively skewed distribution and predominantly oblate samples to a positively skewed distribution.

We try to do a similar investigation by theoretically simulating clusters as triaxial ellipsoids with randomly distributed axis ratios in the range from 0.65 to 1 as is done in Sulkanen et al. We then give them random orientations in the sky calculating analytically the ratio f_H from Eq. 9. We find that the claim of Kawhara et al. is true only for axisymmetric clusters (clusters with two equal axes and fixed axis ratios), which is the type of clusters Kawhara et al. investigate. They simulate a single axisymmetric ellipsoid with fixed axis ratios ($0.5 : 0.5 : 1$ or $0.5 : 1 : 1$) and orient it randomly in the sky. The result is illustrated on the right side of Fig. 2.

On the contrary, the shape of the distribution does not change as obviously for the case of 10000 triaxial ellipsoids, where all three axis sizes are assigned randomly. On the left part of Fig. 2 we show the resultant distributions for our theoretical simulation of a set of triaxial ellipsoidal clusters with random orientations in the sky. Hence the distribution of f_H for a set of

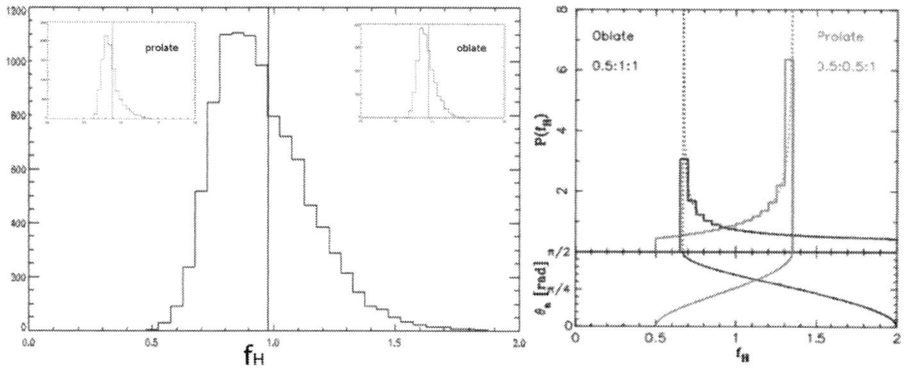

Figure 2. Distribution of the ratio between estimated and real value of the Hubble constant. On the left we show the changes in the distribution when predominantly prolate or oblate clusters are used. In this case different clusters with random axis ratios were simulated. The vertical solid line shows the average of each distribution. On the right we show the figure from [16] where axisymmetric clusters were used with a fixed axis ratio. In this case the distribution was derived by randomly orienting a single cluster with respect to the line of sight.

triaxial clusters in the sky will not always result in an obviously skewed distribution. We conclude that the skewness of the distribution is not a robust quantity. Some other characteristic of the distribution needs to be found in order to trace any cluster geometry biases in the experimental sample. A more refined analysis of those distributions is necessary with both theoretically created samples and clusters from simulations.

4. Conclusions

We have described a method to derive the Hubble constant from joint X-ray and SZ measurements of galaxy clusters. This could provide an additional insight into the discordance of the estimates of the Hubble constant from the CMB and local tracers. We have additionally investigated the possible bias introduced by the geometry of the clusters and the possibility to identify this bias from characteristics of the distribution of estimates of H_0. We have demonstrated that it is not a straightforward task as it was suggested in [16] and a further investigation is required.

References

[1] Ade P A R *et al.* (Planck Collaboration) 2016 *A&A* **596**
[2] Riess A G *et al.* 2016 *ApJ* **826**
[3] Cheng C *et al.* 2015 *Sci China-Phys Mech Astron* **58**
[4] Bennett C L *et al.* 2014 *ApJ* **794**
[5] Bonamente M *et al.* 2013 *Proc. Int. Astronomical Union, IAU Symposium* **289** 339–343
[6] Sarazin C L 1988 *X-ray emission from clusters of galaxies* (Cambridge/New York/New Rochelle/Melbourne/Sydney: Cambridge University Press)
[7] Sunyaev R A and Zeldovich Y B 1970 *Astrophysics and Space Science* **7** 3–19
[8] Carlstrom J E *et al.* 2002 *Ann.Rev.A&A* **40** 643–680
[9] Rephaeli Y *et al.* 1995 *ApJ* **445** 33–36
[10] Aghanim N *et al.* 2009 *A&A* **496** 637–644
[11] Adam R *et al.* (Planck Collaboration) 2016 *A&A* **594**
[12] Adam R *et al.* (Planck Collaboration) 2016 *A&A* **594**
[13] Meisner A *et al.* 2014 *ApJ* **798**
[14] Bonamente M *et al.* 2006 *ApJ* **647**
[15] Sulkanen M E *et al.* 2006 *ApJ* **522** 917–936
[16] Kawahara H *et al.* 2008 *ApJ* **674**

IOP Conf. Series: Journal of Physics: Conf. Series **841** (2017) 012005 doi:10.1088/1742-6596/841/1/012005

Magnetic fields in galaxy clusters in the SKA era

F Loi[1,2], F Govoni[1], M Murgia[1], V Vacca[1], H Li[3], L Feretti[4] and G Giovannini[4,5]

[1] INAF-Osservatorio Astronomico di Cagliari, via della Scienza 5, Selargius, Italy
[2] Dipartimento di Fisica, Università degli Studi di Cagliari, Strada prov.le Monserrato-Sestu Km 0.700, I-09042 Monserrato (CA), Italy
[3] Theoretical Division, Los Alamos National Laboratory, 87544 Los Alamos (NM), USA
[4] INAF – Istituto di Radioastronomia, Via P.Gobetti 101, 40129 Bologna, Italy
[5] Dipartimento di Fisica e Astronomia, Università degli Studi di Bologna, Via Ranzani 1, 40127 Bologna, Italy

E-mail: floi@oa-cagliari.inaf.it

Abstract. The study of the polarization of faint diffuse synchrotron sources, named radio halos, found in some galaxy clusters, is of paramount importance to characterize large scale magnetic fields. This is an hard task with the current radio telescopes but a next generation radio interferometer, the Square Kilometre Array (SKA), could help to shed light on the origin of cosmic magnetism. Thanks to its sensitivity, its broader bandwidth and its resolution, the SKA will allow us to perform complete and accurate studies of magnetic fields in clusters. In order to explore the potentiality of the SKA, we used state-of-art magneto-hydro-dynamical numerical simulations to produce synthetic maps of radio halos, taking into account the expected performances of the SKA1-MID in the radio band from 350 to 1050 MHz. Starting from the resulting maps, we were able to verify that radio halos could be intrinsically polarized and that SKA1-MID could detect their polarization, crucial to constraining the properties of large scale magnetic fields.

1. Introduction

Galaxy clusters are the largest gravitationally bound systems in the Universe, with typical masses of $10^{15} M_{sun}$ and volumes of about 100 Mpc^3. Their major component is Dark Matter (~80%), but they contain also an X-ray emitting thermic plasma (~15-17%), called intracluster medium (ICM), and a small percentage of luminous matter (~3-5%), due to the presence of optical and/or radio emitting galaxies.
Massive galaxy clusters form and grow through merging event, like the collision of sub-clusters. Some of the known clusters undergoing merger events show a diffuse central emission in the radio band, called radio halo. This is a faint ($\simeq 1\mu$Jy/arcsec^2 at 1.4 GHz)[1] steep-spectrum (the flux density is $S_\nu \propto \nu^{-\alpha}$, where ν is the frequency and the spectral index α >1) synchrotron source, with typical dimension of 1 Mpc, not associated with discrete radio sources and that does not show any optical counterpart (see [13] for a review). For this reason, radio halos are the direct proof of the existence of a non thermal component in the ICM, made by relativistic electrons, with density of the order of 10^{-10} particles per cm^{-3}, Lorentz factors $\gamma \gg 10^3$, and magnetic fields spread all over the cluster volume. Magnetic fields in merging clusters are turbulent

[1] 1 Jy = 10^{-26} W Hz^{-1} m^{-2}

and they have a typical strength of few μG. Our knowledge of these fields is still incomplete, but it is of paramount importance to study them in order to understand the origin and the evolution of cosmological magnetic fields. Constraining their properties is possible by analyzing the Faraday rotation effect of the polarized signal of background and cluster-embedded radio galaxies. Each time that a linearly-polarized signal pass through a foreground magnetized plasma its polarization plane rotates from its intrinsic angle Ψ_0 to the observed angle Ψ, depending on the wavelength λ squared and on the rotation measure RM:

$$\Psi = \Psi_0 + RM \times \lambda^2 \quad (1)$$

This relation holds when the radio emitting plasma is not mixed with the rotating thermal plasma. The RM is proportional to the integral of the line of sight (LOS) component of the intracluster magnetic field $B_{||}$ times the thermic plasma density n_e, performed along the distance covered by the signal across the galaxy cluster:

$$RM \propto \int_0^L B_{||} \cdot n_e \, dl \quad (2)$$

Even if this method can give good results in constraining the properties of intracluster magnetic fields, it is limited to the small areas covered by the background and cluster-embedded radio galaxies. Therefore, we can think to estimate the Faraday rotation effect of the polarized emission of radio halos, typically diffuse over the Mpc scale. Unfortunately, these objects are polarized up to a few percent and it has been possible to detect filaments of polarized emission associated with a radio halo just in three cases: in A2255 ([15], but see also [19]), in MACS J0717.5+3745 [1] and in A523 [14]. This technique is limited by the sensitivity of the current radio telescopes and to the external (beam and bandwidth) and internal depolarization. More specifically, the last case refers to the fact that the radio emitting plasma and the Faraday rotating plasma are mixed. The resulting rotation depends on the ICM depth that the linearly polarized signal crosses to reach the radio telescope: polarized signals coming from a given LOS experience a different rotation of the polarization plane and sum up incoherently, reducing the observed polarization degree.

These limits are going to be overcome with the Square Kilometre Array (SKA): the higher sensitivity and resolution, the broader bandwidth will permit to observe the polarized emission of radio halos [16, 17] and to apply sophisticated techniques to finally perform systematic and complete studies of the intracluster magnetic field. The RM Synthesis [4,6], applied to the broad bandwidth of the SKA, is a really promising technique to recover the missed polarized signal.

Making prediction on the results that the SKA will achieve is possible through numerical simulations. From cosmological magneto-hydro-dynamical simulations (MHD) one can study the evolution of magnetic fields in clusters [2,5,7,8,11,12,9,20,21,22,23], interpret observed radio and X-ray characteristics of particular systems [10,3] and produce simulated maps of radio halos [16,17].

In this work, starting from cosmological MHD simulations and numerical simulations we produced full-resolution (simulated) radio halos, in the frequency band from 350 to 1050 MHz, and their emission as it will be observed with the resolution and sensitivity of the SKA1-MID (synthetic) in the same frequency band. Starting from these simulated and synthetic radio halos, we explored the possibility that radio halos could be intrinsically polarized and we checked if their polarization could be detected with the SKA1-MID.

2. Numerical Simulations

The cosmological MHD numerical simulations studied in this work, refer to a collision between two galaxy clusters that lasts 2 Gyrs approximately. This event has been obtained as part of

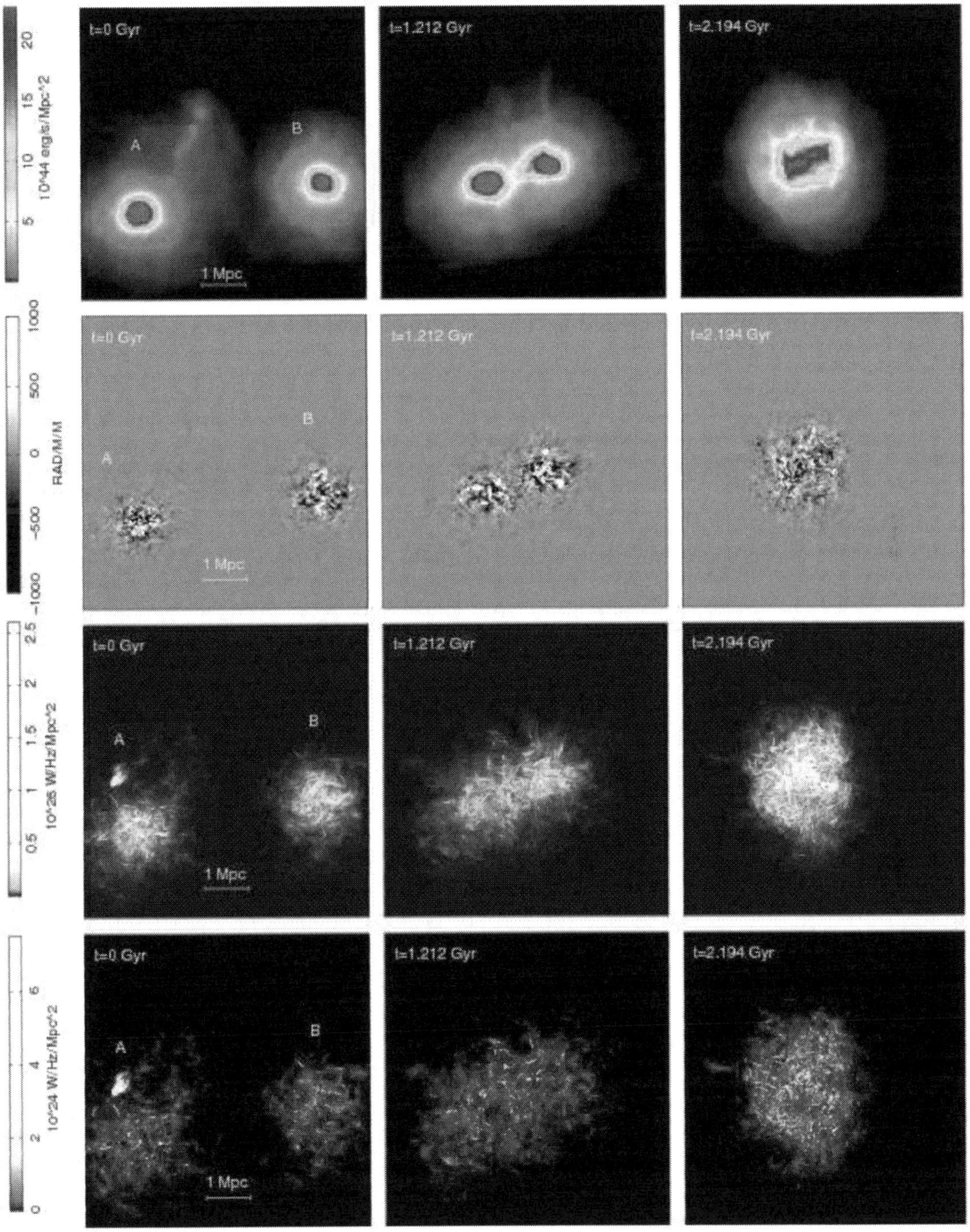

Figure 1. Simulated maps of the event: from left to right maps refer to three different stages of the collision. The first row represents the X-ray emission in the band 0.1÷2.4 keV, the spatial resolution is equal to 10.7 kpc, the field of view of the maps is equal to $(6.6\ \mathrm{Mpc})^2$. Second row shows the RM maps. Third and fourth row refer respectively to the total and linearly-polarized intensity of the radio halos in the frequency band 350÷1050 MHz.

IOP Conf. Series: Journal of Physics: Conf. Series **841** (2017) 012005 doi:10.1088/1742-6596/841/1/012005

a larger project, aimed at describing the formation of a galaxy cluster and the evolution of the intracluster magnetic field [20,21,22,23]. In this simulation the magnetic field is injected at low redshift (z$\simeq$2) from an active galactic nucleus, placed in the center of a proto-cluster of galaxies. Through the merger event the magnetic field will diffuse in the ICM and it will change its morphology and intensity, up to the μG-level. For further details on the cosmological simulation refer to [21]. Starting from the produced physical cubes of temperature, magnetic field and thermic plasma density, we simulated, thanks to the FARADAY software package [18], X-ray and radio properties of the ICM. Figure 1 showS the results obtained in three different stages of the collision (from left to right) for the emission in the X-ray between 0.1$\div$2.4 keV (first row), the RM image (second row) and the total (third row) and polarized intensity (fourth row) in the radio band from 350 to 1050 MHz, with a channel resolution of 0.5 MHz. From the last row we can infer one of the most important results of this work: radio halos could be intrinsically polarized.
In order to reproduce the images of the polarization of our radio halos as they will be observed with the SKA1-MID, we convolved the simulated maps with a Gaussian with a Full-Width-Half-Maximum of 10 arcsec and we added a noise of 1 μJy/beam [17]. In Figure 2 we report the results obtained for the radio halos in the same evolutionary stages shown in the previous figure.

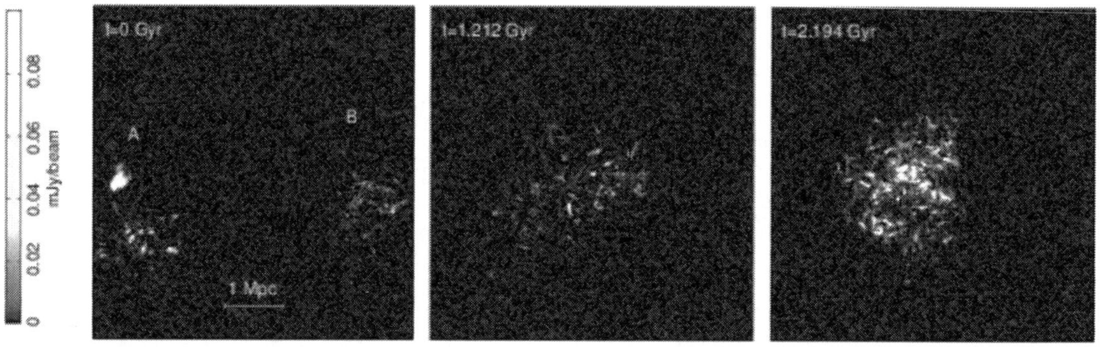

Figure 2. Polarized emission of synthetic radio halos in the frequency band 350$\div$1050 MHz, with a resolution of 10 arcsec and a noise of 1μJy/beam. The maps have been obtained after the application of the RM Synthesis technique.

From these maps, we can state that the polarization of the simulated radio halos could be detected with the SKA1-MID.

3. RM Synthesis

As already mentioned, the RM Synthesis technique has been proposed to recover the missed polarized signal of radio halos, more specifically, the one associated to the internal depolarization due to the Faraday effect. When we consider radio halos, we have to understand that the emitting and the rotating plasma are mixed, so that Eq. 1 does not hold anymore. The polarized signals emitted at different positions l along the LOS, through the cluster, experience a different rotation, called Faraday depth. It is impossible to identify the contribution of each polarized structure from the observed polarization without the application of specific techniques. The idea of the RM Synthesis is to apply different values of the Faraday depth to the polarization, in order to identify each filament and its own Faraday depth. This is done by convolving the polarization with the RM Synthesis transfer function, that represents the response that a unitary polarized signal gives, for each frequency channel of the bandwidth, assuming different values of Faraday

IOP Conf. Series: Journal of Physics: Conf. Series **841** (2017) 012005 doi:10.1088/1742-6596/841/1/012005

depth. We chose an interval of Faraday depth from -1000 rad/m^2 to 1000 rad/m^2 with a resolution of 1 rad/m^2, enough to properly cover the resolution of the RM Synthesis transfer function (6.3 rad/m^2) in the radio band from 350 to 1050 MHz. Fig. 3 shows the results for the LOS indicated in the image on the left. In the last panel on the right, we report the recovered filaments of polarized signals (red) associated to different values of Faraday depth.

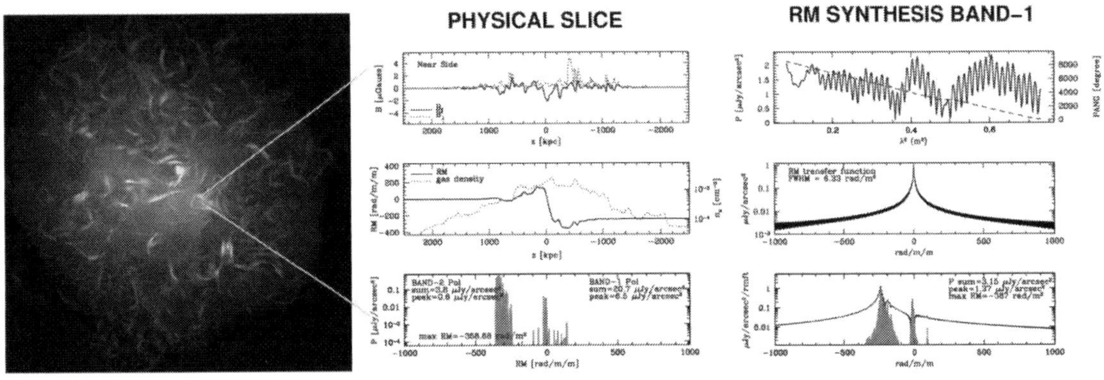

Figure 3. The image (left) shows the X-ray (red), the total (green) and polarized (blue) intensity of the cluster. The central block shows the selected LOS values of the parallel and perpendicular components of the magnetic field along the cluster length (top), the thermic plasma density and the RM values (central), for each value of polarized signal encountered along the LOS, the corresponding RM value (bottom). The right block shows the results of the RM Synthesis technique: the variation of the polarized intensity (solid line) and the polarization angle (dashed) with respect to λ^2 (top), the transfer function (central) and the recovered polarized signals along the LOS (bottom).

By using the polarization component with the maximum absolute value of the Faraday depth, we were able to estimate the RM of the cluster. In this way, we obtained a RM map (see Fig. 4) of the cluster from the polarized signal of the radio halo. The plot in Fig. 4, shows the profile of RM traced from the obtained map (full points) and the theoretical one (solid line). Assuming a simple single-scale model for the magnetic field, that consider a uniform and constant magnetic field in cells of dimension Λ_C, and a profile that scales with respect to the plasma density following a power-law trend with index η, we plotted, for comparison, the corresponding profiles (dashed lines) for three different values of the central magnetic field.

4. Conclusions

Starting from cosmological MHD simulations, by using the software package FARADAY, we reproduced the polarization properties of radio halos, during a merger event between two galaxy clusters. These radio halos show degrees of polarization as high as 10% at distances larger that 1 Mpc from the cluster center, while, toward the center, the degree is progressively lower because of the higher amount of magneto-ionic plasma. Thus, radio halos could be intrinsically polarized. From the synthetic maps of radio halos, realized by taking into account the performances of the SKA1-MID, we established that this new generation radio interferometer will have the capabilities to detect the polarization of radio halos, by applying the RM Synthesis technique to data.

The SKA will revolutionize cosmic magnetism studies. With its high resolution, sensitivity and a broader bandwidth, it will be the ideal instrument for the observation of radio halos and for the application of techniques, such as the RM Synthesis, capable to recover the polarization, fundamental to constrain intracluster magnetic fields.

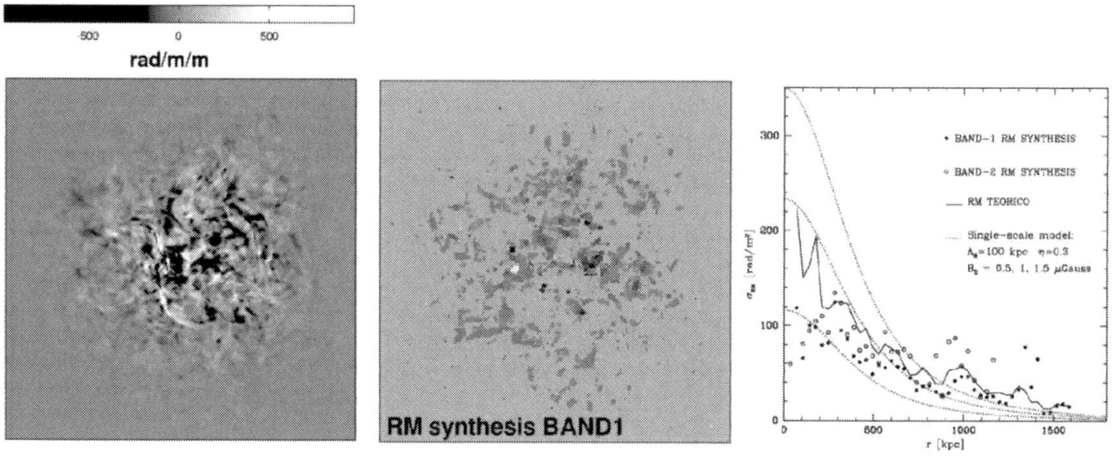

Figure 4. Theoretical RM map (left), recovered RM map from the application of the RM Synthesis (central). On the right, we plot the profile of the σ_{RM}, with respect to the cluster center, of the theoretical map (solid line), of the recovered RM map (full points) and the single-scale models with different values of the central magnetic field (dashed lines).

4.1. Acknowledgments
Francesca Loi gratefully acknowledges Sardinia Regional Government for the financial support of her PhD scholarship (P.O.R. Sardegna F.S.E. Operational Programme of the Autonomous Region of Sardinia, European Social Fund 2007-2013 - Axis IV Human Resources, Objective l.3, Line of Activity l.3.1.). This work was partially supported by PRIN-INAF 2014.

References
[1] Bonafede A, Feretti L, Giovannini G, et al. 2009 *A&A* **503** 707
[2] Bonafede A, Dolag K, Stasyszyn F, Murante G, and Borgani S 2011 *MNRAS* **418** 2234
[3] Bonafede A, Brüggen M, van Weeren R, et al. 2012 *MNRAS* **426** 40
[4] Brentjens M A, de Bruyn A G, 2005 *A&A* **441** 1217
[5] Brüggen M, Ruszkowski M, Simionescu A, Hoeft M, and Dalla Vecchia C 2005 *ApJ* **631** L21
[6] Burn B J 1966 *MNRAS* **133** 67
[7] Dolag K, Bartelmann M, and Lesch H 2002 *A&A* **387** 383
[8] Dolag K, Grasso D, Springel V, and Tkachev I 2005 *J. Cosmology Astropart. Phys.* **1** 009
[9] Donnert J, Dolag K, Lesch H, Müller E 2009 *MNRAS* **392** 1008
[10] Donnert J, Dolag K, Brunetti G, Cassano R, and Bonafede A 2010 *MNRAS* **401** 47
[11] Dubois Y, and Teyssier R 2008 *A&A* **482** L13
[12] Dubois Y, Devriendt J, Slyz A, and Silk J 2009 *MNRAS* **399** L49
[13] Feretti L, Giovannini G, Govoni F, and Murgia M 2012 *A&Ar* **20** 54
[14] Girardi M, Boschin W, Gastaldello F, et al. 2015 arXiv:1510.05951
[15] Govoni F, Murgia M, Feretti L, et al. 2005 *A&A* **430** L5
[16] Govoni F, Murgia M, Xu H, et al. 2013 *A&A* **554** A102
[17] Govoni F, Murgia M, Xu H, et al. 2015 *Advancing Astrophysics with the Square Kilometre Array* 105
[18] Murgia M, Govoni F, Feretti L, et al. 2004 *A&A* **424** 429
[19] Pizzo R F, and de Bruyn A G 2009 *A&A* **507** 639
[20] Xu H, Li H, Collins D C, Li S, and Norman M L 2009 *ApJ* **698** L14
[21] Xu H, Li H, Collins D C, Li S, and Norman M L 2010 *ApJ* **725** 2152
[22] Xu H, Li H, Collins D C, Li S, and Norman M L 2011 *ApJ* **739** 77
[23] Xu H, Govoni F, Murgia M, et al. 2012 *ApJ* **759** 40

IOP Conf. Series: Journal of Physics: Conf. Series **841** (2017) 012006 doi:10.1088/1742-6596/841/1/012006

Anisotropies in the flux of cosmic ray electrons and positrons

S Manconi[1], **M Di Mauro**[2], **F Donato**[1]

[1] Physics Department, Torino University, and Istituto Nazionale di Fisica Nucleare, Sezione di Torino, via Giuria 1, 10125 Torino, Italy

[2] W. W. Hansen Experimental Physics Laboratory, Kavli Institute for Particle Astrophysics and Cosmology, Department of Physics and SLAC National Accelerator Laboratory, Stanford University, Stanford, CA 94305, USA

E-mail: manconi@to.infn.it

Abstract. High energy cosmic ray electrons and positrons probe the local properties of our galaxy. In fact, regardless of the production mechanism, electromagnetic energy losses limit the typical propagation scale of GeV-TeV electrons and positrons to a few kpc. In the diffusion model, the presence of nearby and dominant sources may produce an observable dipole anisotropy in the cosmic ray fluxes. We present a detailed study on the role of anisotropies from nearby sources in the interpretation of present cosmic ray electron and positrons fluxes. Predictions for the dipole anisotropy from known astrophysical sources as Supernova Remnants (SNRs) and pulsars taken from the Green and the Australia Telescope National Facility (ATNF) catalogs are shown. The results are obtained from models compatible with the most recent AMS-02 data on electrons and positrons fluxes. In particular, anisotropies for single sources as well as for a distribution of catalog sources are discussed. We compare our results with current anisotropy upper limits from the Fermi-LAT and PAMELA experiments, showing that the search of anisotropy in the electron and positron fluxes represents a complementary tool to inspect the properties of close SNRs, as for example the Vela SNR.

1. Introduction

In the last years, the fluxes of electrons and positrons in Cosmic Rays (CRs) have been measured with high precision by the AMS-02 [1, 2], *Fermi*-LAT [3] and PAMELA [4] experiments. The observed fluxes can be interpreted as the emission from a variety of astrophysical sources, as for example Supernova Remnants (SNRs), Pulsar Wind Nebulae (PWNe) and the spallation of primary CRs in the interstellar medium, see *e.g.* [5]. Also, the detected fluxes have been analyzed in terms of anisotropies in their arrival directions. Searches for anisotropies in the electron plus positron flux [6], the positron to electron ratio [1] and the positron flux [7] have been presented respectively from the *Fermi*-LAT, AMS-02 and PAMELA experiments, all ending up with upper limits on the dipole anisotropy. In this work we show how the search for anisotropies in the electron and positron fluxes at GeV-TeV energies can be an interesting tool, in addition to the measured fluxes, to study the properties of near sources, as for example near SNRs.

2. Modeling the electron and positron fluxes

Electron and positron production in our galaxy is possible through different processes. *Primary* electrons can be accelerated with Fermi non relativistic shocks up to high energies in SNRs

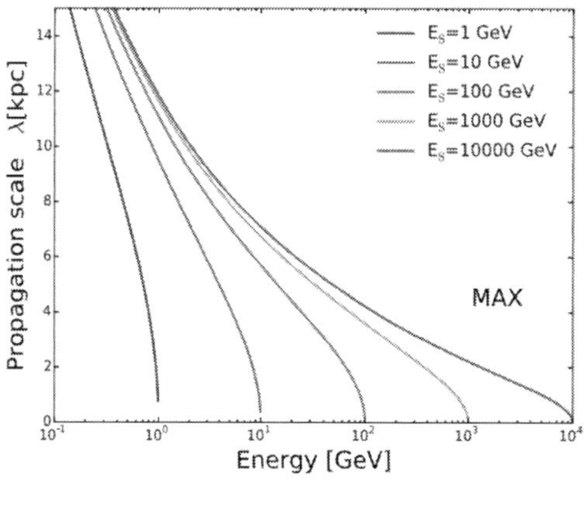

Figure 1. Propagation scale λ for electrons and positrons in our galaxy as a function of the energy at Earth, for different energies at source E_s and adopting the MAX propagation model.

[8]. In addition, both electrons and positrons are created in the strong magnetic fields that surround pulsars and then accelerated with relativistic shocks in the PWN [9]. A source of *secondary* electrons and positrons is the fragmentation of primary CR nuclei in the interstellar medium material (hydrogen, helium). Indipendent of the production mechanism, electrons and positrons propagate in the Galaxy and are affected by a number of processes. Above a few GeV, the propagation is dominated by the diffusion in the interstellar magnetic field irregularities and by energy losses, which are due to inverse Compton scattering on ambient photons and to synchroton emission. This is tipically described by a diffusion equation for the number density $\psi = \psi(E, \mathbf{x}, t) \equiv dn/dE$ per unit volume and energy:

$$\frac{\partial \psi}{\partial t} - \nabla \cdot \{K(E) \nabla \psi\} + \frac{\partial}{\partial E}\left\{\frac{dE}{dt}\psi\right\} = Q(E, \mathbf{x}, t) \tag{1}$$

where $K(E) = K_0 E^{\delta}$ is the energy dependent diffusion coefficient, the dE/dt term accounts for the energy losses and $Q(E, \mathbf{x}, t)$ includes all the possible sources. In this work a semi analytical approach is followed to solve the diffusion equation in Eq. 1 for each source, as fully detailed in [10, 11]. Within this approach, the Galaxy is modeled as a cylinder of radius $r_{\rm disc} = 20$ kpc and half height L. The parameters K_0, δ, L are usually constrained from boron over carbon ratio (B/C). In particular, in the following we show results for the MAX benchmark model derived in [12]. At high energies ($E > 10$ GeV) the electrons and positrons that we observe are probes of the local Galaxy. In fact, for leptons the energy loss timescale is smaller than the diffusion timescale. This can be visualized in Fig. 1, where the typical propagation scale λ as a function of the energy at Earth is plotted for different values of the energy at source E_s and for the MAX propagation model. We see that λ for GeV-TeV electrons and positrons is less than ~ 5 kpc for the different E_s considered. The interpretation of high energy electrons and positrons is thus connected to the inspection of local sources. The chosen modeling of electron and positron sources is functional to this aim, and is based on [11, 10, 5]. We remind here only the principal characteristics of this modeling. We include both single SNRs and PWNe, whose characteristics are taken directly from the existing catalog, and a distribution of SNRs, described by average characteristics. The secondary component is taken from [5]. The spatial distribution of SNRs is modeled with a smooth distribution of sources active beyond a radius $R_{\rm cut}$ from the Earth (*far* SNRs), and following the radial profile derived in [13]. Instead, single sources taken directly from the Green Catalog [14] are considered for $R \leq R_{\rm cut}$ (*near* SNR). To inspect the role of single near SNRs we consider $R_{\rm cut} = 0, 0.7$ kpc. The PWNe component is computed taking from the ATNF catalog [15] the sources with ages 50 kyr$< t_{\rm obs} <$ 10000 kyr. In fact, the release of the accelerated $e^{\pm}$ pairs is estimated to start at least after 40 − 50 kyr after the pulsar birth [9]. The energy injection spectrum $Q(E)$ for both SNR and PWN is

$$Q(E) = Q_0 \left(\frac{E}{E_0}\right)^{-\gamma} \exp\left(-\frac{E}{E_c}\right) \tag{2}$$

where $E_c = 5$ TeV is the cutoff energy and $E_0 = 1$ GeV. The index of the energy spectrum is expected to be different for particles accelerated in SNRs ($\sim 2 - 2.5$) and PWNe (< 2). The normalization of this spectrum is constrained using catalog quantities for single SNRs and PWNe, while using average population characteristics for the smooth SNR component. For example, the normalization for a single PWN is obtained supposing that a fraction η of the spindown energy of the pulsar W_0 is emitted in form of $e^{\pm}$ pairs. The normalization for a single SNR is constrained with radio and catalog data, as for example the radio flux, the distance and magnetic field of the remnant (see Eq. 50 in [10]). As for the smooth SNR distribution, this can be connected to average Galactic characteristics, as for example the mean energy released in electrons per century $E_{\rm tot,SNR}$.

2.1. *Anisotropy*

As shown in Fig. 1, CR electrons and positrons with observed energies in GeV-TeV range originated from relatively nearby locations in the Galaxy. This means that it could be possible that such high energy electrons originate from a highly anisotropic collection of nearby sources. Under this hypothesis, even after the diffusive propagation in the Galactic magnetic field is taken into account, a residual small dipole anisotropy should be present in the observed fluxes. In the assumption of one or few nearby sources dominating the CR flux at Earth, the dipole anisotropy is usually defined as

$$\Delta = \frac{I_{max} - I_{min}}{I_{max} + I_{min}} \tag{3}$$

being I_{max} (I_{min}) the maximum (minimum) CR intensity values. In a diffusive propagation regime this can be computed as $\Delta = \frac{3K}{c}\left|\frac{\nabla\psi}{\psi}\right|$ (see [16]), where K is the diffusion coefficient and ψ is the solution to Eq. 1. For example, the anisotropy for a single source of electrons and positrons in our model is computed as

$$\Delta(E)_{e^+ + e^-} = \frac{3K(E)}{c} \frac{2d_s}{\lambda^2(E, E_s)} \frac{\psi^s_{e^+ + e^-}(E)}{\psi^{tot}_{e^+ + e^-}(E)}, \tag{4}$$

where d_s is the distance to the source, λ is the propagation scale, $\psi^s_{e^+ + e^-}(E)$ is the source $e^+ + e^-$ number density, $\psi^{tot}_{e^+ + e^-}(E)$ is the total $e^+ + e^-$ number density given by the contributions of all the sources, both isotropic and single sources. Moreover, if a collection of electron and/or positron sources is present, the intensity of the CR flux as a function of direction in the sky $\mathbf{n}$ is (see [17])

$$I(E, \mathbf{n}) = \frac{c}{4\pi} \sum_i \psi_i(E)(1 + \Delta_i \, \mathbf{n} \cdot \mathbf{r}_i / r_i) \tag{5}$$

where the index i runs over all the sources at position $\mathbf{r}_i$ with electron and/or positron number density $\psi_i(E)$, $\Delta_i = \frac{3K(E)}{c} \frac{|\nabla\psi_i(E)|}{\psi_i(E)}$, and the total dipole anisotropy is computed directly by means of Eq. 3. To compare our prediction with the present upper limits we compute the integrated dipole anisotropies as a function of the minimum energy. We integrate fluxes in Eq. 4 up to 5 TeV.

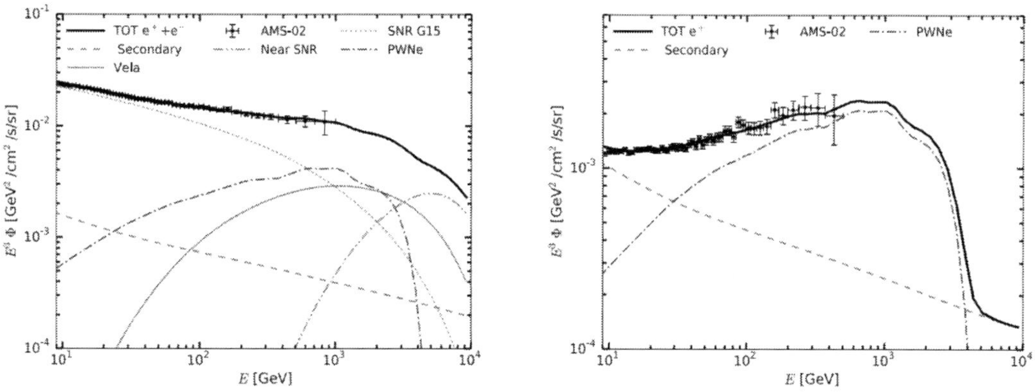

Figure 2. Best fit to $e^+ + e^-$ (left panel) and e^+ (right panel) AMS-02 data [18, 19] for the model with $R_{\rm cut} = 0.7$ kpc described in Sec. 3 and MAX propagation setup. Solid black: sum of all the plotted components; red dashed: secondary $e^+ + e^-$ (left panel) and e^+ (right panel); blue dash-dotted: $e^+ + e^-$ (left panel) and e^+ (right panel) from the collection of PWNe in the ATNF catalog; green dotted: e^- from the far SNR distribution with $R_{\rm cut} = 0.7$ kpc; black dotted: e^- for the Vela SNR; magenta double dash-dotted: e^- from the near SNRs in the Green Catalog with $R \leq 0.7$ kpc.

3. Results and Conclusions

The aim of this analysis is to provide dipole anisotropies predictions for models compatible with the observed electron and positron fluxes. Therefore, each model is fitted to the AMS-02 data on the e^+ [18] and $e^+ + e^-$ [19] fluxes. The models are fitted to the data starting from $E = 10$ GeV. This choice minimizes the effect of the solar modulation of fluxes, that is however taken into account with a modulation potential ϕ_F, according to the force field approximation. The inspected models differ mainly for the treatment of the contribution from local sources. As an example, two models and the corresponding predictions for anisotropies are discussed here.

With the first model we aim to analyze the role of single near SNRs (in particular the Vela SNR) in the high energy flux and, consequently, in the electron plus positron anisotropy. Fluxes are fitted considering a secondary component, the contribution from single PWNe in the ATNF catalog, a smooth distribution of SNRs with $R_{\rm cut} = 0.7$ kpc, the contribution from single near SNRs in the Green catalog with $R \leq R_{\rm cut}$. Among the near SNRs, Vela is treated separately from the other sources as detailed in [11]. For this analysis, its spectral index is fixed to $\gamma = 2.5$, its distance to 0.293 kpc and its age to 11.4 kyr. The results of our fit are presented in Fig. 2. A number of free parameters is used to fit our model to the data. This includes a normalization for the secondary component, a common spectral index and efficiency η for all the PWNe, a normalization for the *near* component, the magnetic field for the Vela SNR, and a spectral index and a free normalization for the smooth SNR distribution. More details on the fit parameters are given in [11]. The fit to the AMS-02 fluxes is remarkably good, with a reduced χ^2/d.o.f.= 38/89. The role of near SNR, in particular Vela, in shaping the $e^+ + e^-$ fluxes (left panel) is evident for $E \gtrsim 300$ GeV. We thus compute the corresponding dipole anisotropy for the Vela SNR and for the Cygnus Loop, which dominates the contribution of the near SNRs with $R < R_{\rm cut} = 0.7$ kpc. In Fig. 3 (left panel) the integrated $e^+ + e^-$ anisotropy as a function of the minimum energy $E_{\rm min}$ for Vela and Cygnus Loop of the model in Fig. 2 are shown. The arrows correspond to the *Fermi*-LAT upper limits in [6]. The predicted anisotropies grow with $E_{\rm min}$ and reach the maximum value of $\Delta_{e^+ + e^-} = 0.04$ for the Vela SNR and 0.15 for Cygnus Loop at TeV minimum

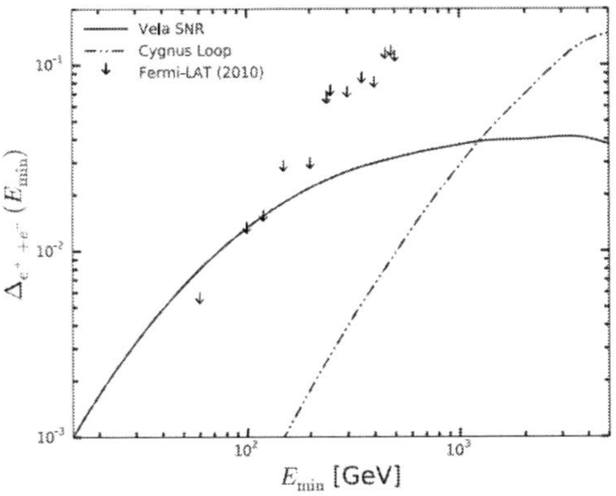

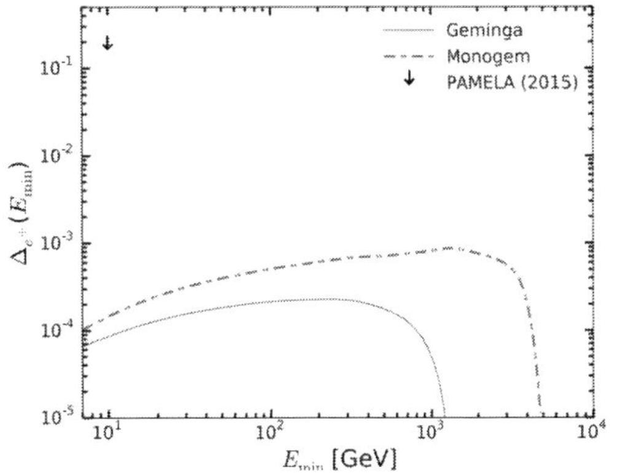

Figure 3. Left panel: predictions for the dipole anisotropies in the $e^+ + e^-$ flux for the Vela SNR (solid) and the Cygnus Loop (dot-dashed) as a function of the minimum energy for the model with $R_{\rm cut} = 0.7$ kpc. The arrows correspond to the *Fermi*-LAT upper limits [6]. Right panel: prediction for the dipole anisotropies in the e^+ flux for the Geminga (green solid) and Monogem (magenta dot-dashed) PWNe as a function of the minimum energy for the model with $R_{\rm cut} = 0$ kpc. The arrow corresponds to the PAMELA upper limit in [7].

energies. For $E_{\rm min}$ below about 150 GeV the upper limits are at the same level of the prediction for the Vela anisotropy. Thus, present *Fermi*-LAT upper limits start to test some of the models (see also [11]) for the Vela SNR that are compatible with the flux data. Future results from the full statistics of *Fermi*-LAT data, as well as ongoing experiments such as DAMPE and CALET [20], will improve the potentiality for the anisotropy to explore and eventually exclude some of the models that explains the $e^\pm$ fluxes. To explore the role of all the collection of sources in this model, we show in Fig. 4 the interstellar intensity of the $e^+ + e^-$ flux as a function of the direction in the sky in Galactic coordinates for growing minimum energies. The result obtained with Eq. 5 is shown by means of its percentage difference between the mean intensity from the entire source collection. The maximim of the intensity (yellow dot) is found to be a direction very close to Vela (black dot) for $E_{min} = 126$ and $E_{min} = 661$ GeV (top panels). At higher energies, the interplay between the Vela, Cygnus Loop and the other sources shifts the maximal intensity in direction of Cygnus Loop (bottom panels).

The second model aims to analyze the role of the most powerful PWNe among our collection of ATNF sources and, in particular, the resulting positron anisotropy. The difference with the previous model is that we consider a smooth distribution of SNRs all over the Galaxy, thus $R_{\rm cut} = 0$ kpc, and none of the SNRs in the Green catalog. Therefore, the only single sources are the PWNe. For example Geminga and Monogem PWN, for which we present Fig. 3 (right panel) the integrated dipole anisotropies in the e^+ flux, togheter with the upper limit obtained with PAMELA data in [7]. The maximum anisotropy is given by the Monogem PWN at about 1 TeV. Geminga gives a lower anisotropy due to its age (343 kyr vs. 111 kyr for Monogem, see discussion in [11]). The predicted anisotropy is more than three orders of magnitude below the Δ_{e^+} upper limit. This gap suggests that likely present or forthcoming data on positron anisotropy will not have the sensitivity to test the properties of ATNF PWNe that explains the AMS-02 data.

To conclude, in this proceeding we have shown how anisotropies in the electron and positron fluxes can be a tool to study the properties of near sources, as for example near SNRs, in addition to high precision measurements of the fluxes.

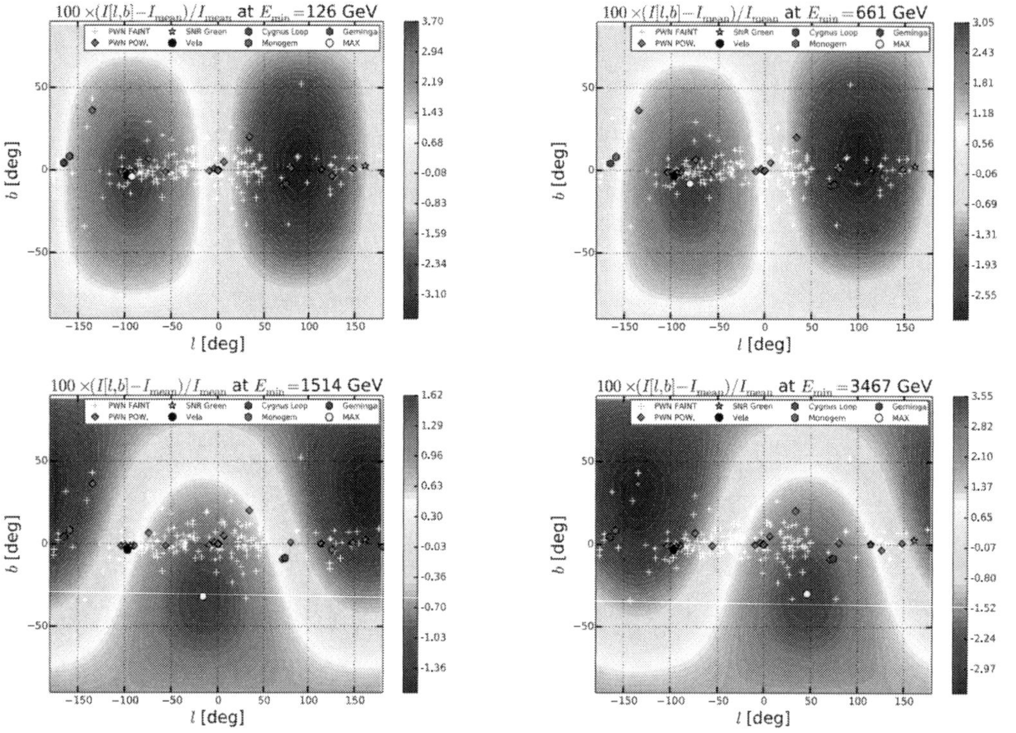

Figure 4. Countor plots of the intensity of the $e^+ + e^-$ flux as a function of the direction in the sky, obtained for all the sources in the model with $R_{\mathrm{cut}} = 0.7$ kpc. The color scale indicates the percentage relative difference between the intensity in a given direction of the sky (l, b) [deg] and the mean intensity computed from the entire source collection. The position of the maximum intensity is highlighted with a yellow dot, while the other symbols indicate the position of the sources. The four panels represent the evolution for increasing minimum energy.

References

[1] Aguilar M, Alberti G, Alpat B, Alvino *et al.* (AMS Collaboration) 2013 *Phys. Rev. Lett.* **110**(14) 141102
[2] Accardo L *et al.* (AMS Collaboration) 2014 *Phys. Rev. Lett.* **113**(12) 121101
[3] Ackermann M, Ajello *et al.* 2012 *Physical Review Letters* **108** 011103 (*Preprint* 1109.0521)
[4] Adriani O *et al.* 2009 *Nature* **458** 607–609 (*Preprint* 0810.4995)
[5] Di Mauro M, Donato F, Fornengo N, Lineros R and Vittino A 2014 *JCAP* **4** 006 (*Preprint* 1402.0321)
[6] Ackermann M *et al.* 2010 *Phy. Rev. D* **82** 092003 (*Preprint* 1008.5119)
[7] Adriani O *et al.* 2015 *Astrophys. J.* **811** 21 (*Preprint* 1509.06249)
[8] Blandford R and Eichler D 1987 *Phys. Rept.* **154** 1–75
[9] Blasi P and Amato E 2011 *Astrophysics and Space Science Proceedings* **21** 624 (*Preprint* 1007.4745)
[10] Delahaye T, Lavalle J, Lineros R, Donato F and Fornengo N 2010 *A&A* **524** A51 (*Preprint* 1002.1910)
[11] Manconi S, Di Mauro M and Donato F 2016 *ArXiv e-prints* (*Preprint* 1611.06237)
[12] Donato F, Fornengo N, Maurin D and Salati P 2004 *Phys. Rev.* **D69** 063501 (*Preprint* astro-ph/0306207)
[13] Green D A 2015 *MNRAS* **454** 1517–1524 (*Preprint* 1508.02931)
[14] Green D 2014 *Bull.Astron.Soc.India* **42** 47 (*Preprint* 1409.0637)
[15] Manchester R N, Hobbs G B, Teoh A and Hobbs M 2005 *AJ* **129** 1993–2006 (*Preprint* astro-ph/0412641)
[16] Ginzburg V L and Syrovatskii S I 1964
[17] Shen C S and Mao C Y 1971 *ApJL* **9** 169
[18] Aguilar M *et al.* 2014 *Physical Review Letters* **113** 121102
[19] Aguilar M *et al.* 2014 *Physical Review Letters* **113** 221102
[20] Torii S and CALET Collaboration 2011 *Nuclear Instruments and Methods in Physics Research A* **630** 55–57

Constraints on a k-dependent bias from galaxy clustering

E Menegoni[1]

[1] CNRS, Laboratoire Univers et Theories (LUTh), UMR 8102 CNRS, Observatoire de Paris, Université Paris Diderot; 5 Place Jules Janssen, 92190 Meudon, France

E-mail: eloisa.menegoni@gmail.com

Abstract. We forecast the future constraints on scale-dependent parametrizations of galaxy bias and their impact on the estimate of cosmological parameters from the power spectrum of galaxies: in our approach we perform a Fisher matrix analysis with two different parametrizations of scale-dependent bias. The two main results obtained from the analysis are: first, allowing for a scale-dependent bias does not significantly increase the errors on the other cosmological parameters apart from the *rms* amplitude of density fluctuations, σ_8, and the growth index γ, whose uncertainties increase by a factor up to two, depending on the bias model adopted. Second, we find that the accuracy in the linear bias parameter b_0 can be estimated to within 1-2% at various redshifts regardless of the fiducial model. The non-linear bias parameters have significantly large errors that depend on the model adopted.

1. Introduction

In the next future large experiments like DESI [1] and Euclid [2] will use galaxy clustering to simultaneously obtain information on the geometry of the Universe and the growth rate of density fluctuations by measuring the galaxy power spectrum. Since one typically observes the spatial fluctuation in the galaxy distribution, not in the mass, some independent phenomenological or theoretical insight of the mapping from one to the other is mandatory. This mapping, which is commonly referred to as *galaxy bias*, parametrises our ignorance on the physics of galaxy formation and evolution and represents perhaps the most serious source of uncertainties in the study of the large scale structure of the Universe [3].

It was shown in [4] that future galaxy redshift surveys contain enough information to break the degeneracy between the galaxy bias, the clustering amplitude and the growth factor, effectively allowing to estimate galaxy bias from the data themselves. We forecast the errors on cosmological parameters and galaxy bias parameters and we assess the robustness of our predictions against the choice of the bias and the fiducial model (for more details see [5]). We consider two different parametrizations for the scale-dependent bias: a simple power-law model and the polynomial model proposed by [6]. Both provide a reasonable good fit to mock galaxies similar to those that will be targeted by Euclid [7]. As dataset we assume a wide spectroscopic galaxy redshift survey spanning a large redshift range and consider, as a reference case, the upcoming Euclid survey [8].

2. Theoretical Setup

Following [9] we model the observed galaxy power spectrum, P_{obs} at the generic redshift z as:

$$P_{obs}(z,k) = G^2(z)b^2(k,z)\left(1+\frac{f(z)}{b(z,k)}\mu^2\right)^2 \frac{D_{Af}^2(z)H(z)}{D_A^2(z)H_r(z)}P_{0r}(k) + P_{shot}(z) \tag{1}$$

where D_A is the angular-diameter distance, $H(z)$ is the so called expansion history, i.e. the Hubble constant at redshift z, $G(z)$ is the linear growth function normalized to unity at $z=0$, $f(z) = d\log G/d\log a$ is the growth rate, $b(z,k)$ is the scale-dependent bias, $P_0(k)$ is the matter power spectrum at the present epoch, μ is the cosine angle between the wavenumber vector $\vec{k}$ and the line of sight direction, P_{shot} is the Poisson shot noise contribution to the power spectrum and the subscript r identifies the fiducial model, this one is the case of scale-independent bias ($b_0 = 1$ and $n = 0$). We parametrize the growth rate as $f = \Omega_m^\gamma$ with a constant growth index γ.

We perform the forecast using the Fisher matrix information method, we approximate the likelihood as a Gaussian in the parameters around a particular fiducial model, i.e. a value of the parameters that is assumed to approximate the 2-point clustering properties of galaxies in the real Universe. We set $h_0 = 0.7$, $\Omega_{m0} = 0.25$; $\Omega_{b0} = 0.0445$; $\Omega_{k0} = 0$; the primordial slope $n_s = 1$; and the dark energy equation of state $w_0 = -0.95$. Finally, we set $\gamma = 0.545$ and rms density fluctuation at 8 h^{-1} Mpc $\sigma_8 = 0.8$ (for more technical details see [5]).

Assuming that the fluctuation Fourier modes are Gaussian variates, the Fisher matrix at each redshift shell is [10, 11]

$$F_{ij} = 2\pi\int_{k_{min}}^{k_{max}} \frac{\partial \log P(k_n)}{\partial\theta_i}\frac{\partial \log P(k_n)}{\partial\theta_j}\cdot V_{eff}\cdot\frac{k^2}{8\pi^3}\cdot dk \tag{2}$$

where the derivatives are evaluated at the parameter values of the fiducial model. Here, the maximum frequency $k_{max}(z)$ is set by the scale at which fluctuations grow nonlinearly while $k_{min}(z)$ by the largest scale that can be observed in the given redshift shell. We set a hard small scale cut-off $k_{max} = 0.5\,h^{-1}$ Mpc at all redshifts which, together with the damping terms $P = P_{obs}e^{-k^2\mu^2\sigma_r^2}$ and $\exp\left\{-k^2\left[\frac{(1-\mu^2)\Sigma_\perp^2}{2}+\frac{\mu^2\Sigma_\parallel^2}{2}\right]\right\}$ account for non linearities [5] . On large scale we set $k_{min} = 0.001\,h^{-1}$ Mpc. $V_{\rm eff}$ indicates the effective volume of the survey defined as:

$$V_{\rm eff} \equiv \int\left[\frac{n(\vec{r})P(k,\mu)}{n(\vec{r})P(k,\mu)+1}\right]^2 d\vec{r} = \left[\frac{nP(k,\mu)}{nP(k,\mu)+1}\right]^2 V_{survey} \tag{3}$$

where $n = n(z)$ is the galaxy density at redshift z. The second equality in Equation (3) holds if the co-moving number density is constant within the volume considered. This assumption, which we adopt in our analysis, is approximately true in a sufficiently narrow range of redshifts. For this reason, we perform the Fisher matrix analysis in different, non-overlapping redshift bins, together with their mean galaxy number density. The redshift range, the size of the bin and the number density of objects roughly match the analogous quantities that are expected in the Euclid spectroscopic survey (see [5]).

2.1. Analytic models for scale-dependent bias

Our goal is to assess the impact of a scale-dependent galaxy bias $b(z,k)$ in the analysis of future galaxy surveys [10]: for this purpose we have decided to adopt two rather simple models, the Power Law and the Q-Model. The reason for choosing these models are found in their simple

Table 1. Bias parameters for the fiducial models considered in this work (see [5]).

		FM1-PL				FM1-Q		
	$n=0$	$n=1$		$n=2$				
z	b_0	b_0	b_1	b_0	b_1	b_0	Q	A
all	1	1	0	1	0	1	0	0

form allows us to compute the power spectrum derivatives in the Fisher matrix analytically. The Power Law bias model has the form [12]:

$$b(z,k) = b_0(z) + b_1(z)\left(\frac{k}{k_1}\right)^n, \tag{4}$$

where the pivot scale k_1 is introduced only to deal with dimensionless parameters. Its value does not impact our analysis and, without lack of generality, we set $k_1 = 1\ h\ \mathrm{Mpc}^{-1}$. The slope n is not treated as a free parameter but is kept fixed. We note that the Power Law model is similar to the one proposed by [13] in those $k-$ranges in which the power spectrum can be approximated by a power law. The Q-Model is also phenomenological. It has been proposed by [6] from the analysis of mock halo and galaxy catalogs extracted from the Hubble volume simulation. Its analytical expression is:

$$b(z,k) = b_0(z)\left[\frac{1+Q(z)(k/k_1)^2}{1+A(z)(k/k_1)}\right]^{1/2}. \tag{5}$$

In our analysis all three parameters b_0, Q and A are free to vary in each redshift bin. Therefore, the Q-Model has additional degrees of freedom with respect to the Power Law model. Finally, we need to specify the parameters of the fiducial model (denoted FM1-PL for the power law bias and FM1-Q for the Q-model). The parameters that identify the fiducial models are listed in Tab. 1. Note that we also consider for comparison the case of scale-independent bias (first row). This case is identical to choosing $n = 0$ in the power law model, i.e. to $b(z) = b_0(z) + b_1(z)$, so we will refer to this case as $n = 0$ fiducial. In order to evaluate the Fisher matrix we compute the derivatives of the power spectrum in Eq. (1) with respect to the parameters on the fiducial model. The 1σ error for each parameter of the model, p_i, is $\sigma_{p_i} = \sqrt{(F^{-1})_{ii}}$, where F^{-1} is the inverse Fisher matrix (for technical procedure see [5]).

3. Results

3.1. Power law case

For the power law model we have explored two cases corresponding to different choices of the power law index: $n = 1$ and $n = 2$. The parameters of the fiducial models are reported in Tab. 1. The 1σ errors on the cosmological parameters are listed in Tab. 2. For all parameters except the mass variance σ_8 and the growth index γ the errors are largely independent from n. In fact, in most cases they slightly decrease when the scale dependence is stronger. The values of σ_8 and γ show the opposite trend, although the effect is quite small in respect to other parameters (below 10 %). We conclude that allowing for a scale dependent bias has little effect on the precision in which we can measure most cosmological parameters. It is interesting to note that the accuracy of the growth rate γ is 4-5% when marginalising over all parameters, including the scale and redshift-dependent bias.

The expected 1σ errors for the bias parameters b_0 and b_1 are listed in Tab. 3 for the cases $n = 0, 1, 2$: the errors on b_1 are larger than those on b_0 and their size increase with n, while when $n = 2$ they are ten times larger than $n = 1$. On the contrary, errors on b_0 weakly depend

Table 2. 1σ errors on cosmological parameters for the fiducial models adopted in this work (see [5]).

		FM1-PL		FM1-Q
Error	$n = 0$	$n = 1$	$n = 2$	
σ_h	0.036	0.038	0.037	0.039
$\sigma_{\Omega_m h^2}$	0.015	0.016	0.015	0.016
$\sigma_{\Omega_b h^2}$	0.0034	0.0036	0.0034	0.0036
σ_{n_s}	0.036	0.042	0.036	0.044
σ_γ	0.024	0.025	0.028	0.029
σ_{σ_8}	0.0036	0.0044	0.0045	0.0047

on n. This is not surprising since b_1 is constrained by the power spectrum behaviour at high k, the larger the value of n the larger the values of k, where our analysis is less sensitive due to the damping terms and the hard k_{max} cut. A second trend is with the redshifts: errors on the bias parameters increase with the redshift, irrespectively of the value of n as shown in Fig. 1. This merely reflects the fact that the effective volume of the survey monotonically decreases when moving to high redshifts due to the smaller galaxy densities.

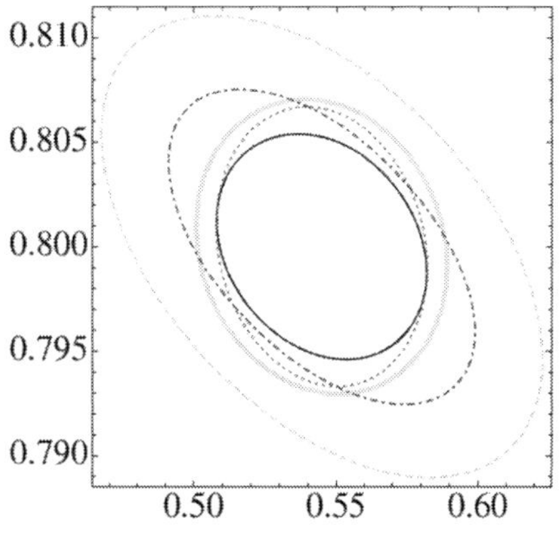

Figure 1. 68 % probability contours for σ_8 and γ. Black, continuous: standard scale-independent case, i.e. $n = 0$. Red, Dotted: FM1-PL with $n = 1$. Cyan, continuous: FM1-Q. Blue, Dot-Dashed: FM2-PL with $n = 2$. Green, dashed: FM2-Q. The FM1-PL case when $n = 1$ is only slightly larger than that corresponding to a scale-independent bias (for further details see [5]).

When compared to the results of [12], in which bias was assumed to be scale-independent, we notice that our constraints on b_0 are twice weaker than their "optimistic, internal bias" case at $z = 1.8$ and $z = 2.0$. This quantifies the effect of allowing for an additional degree of freedom, the scale dependent bias, represented by the new parameter b_1. In Fig. 2 we show the 68 % probability contours in the b_0-b_1 plane for $n = 1$ and $n = 2$, respectively. Larger ellipses refer to higher redshift bins. For the case $n = 1$ there is a strong anti-correlation between b_0 and b_1 stemming from the fact that an increase in the linear bias term b_0 can be partially compensated by reducing the amplitude of the scale-dependent term b_1. Increasing the scale dependency, i.e. setting $n = 2$ reduces the correlation between b_0 and b_1. This is due to the fact that a strong scale dependent bias has little impact on large scales ($k \ll k_1$) and therefore cannot effectively compensate a variation of the linear bias on the scales that are relevant for our analysis.

3.2. Q-model

This version of Type 1 fiducial model is characterised by three parameters, b_0, A and Q, rather than two. We have repeated the same analysis performed as in the FM1-PL case and summarised

Table 3. Errors on bias parameters for the fiducial models adopted in this work (see [5]).

		FM1-PL				FM1-Q		
	$n=0$	$n=1$		$n=2$				
z	σ_{b_0}	σ_{b_0}	σ_{b_1}	σ_{b_0}	σ_{b_1}	σ_{b_0}	σ_Q	σ_A
0.6	0.007	0.013	0.14	0.0081	1.2	0.017	3.04	0.35
0.8	0.008	0.013	0.13	0.0093	0.97	0.017	2.5	0.32
1.0	0.009	0.013	0.12	0.011	0.86	0.017	2.2	0.31
1.2	0.010	0.014	0.12	0.012	0.82	0.018	2.2	0.31
1.4	0.011	0.014	0.13	0.013	0.91	0.019	2.5	0.34
1.6	0.012	0.016	0.16	0.014	1.2	0.023	3.4	0.42
1.8	0.014	0.019	0.22	0.016	1.9	0.027	5.4	0.59
2.0	0.018	0.026	0.34	0.019	3.3	0.037	9.4	0.97

the results in Tabs. 2 and 3. The errors on the cosmological parameters are very similar to those obtained in the FM1-PL case, confirming that the accuracy in the estimate of the cosmological parameters is not much affected by the adoption of a scale-dependent bias model, even when we introduce an additional degree of freedom. Errors on the linear bias parameters b_0 (Tab. 3) are small, with a magnitude similar to that of the FM1-PL case with $n = 1$. On the contrary, the errors on A and Q are quite large, although we cannot directly compare their size to the errors on b_1. This is not entirely surprising: it is the effect of having one more parameter to marginalize over. To further investigate the possible degeneracy among the bias parameters we plot the 68 % uncertainty contours in the A-Q plane in Fig. 3. The size of the errors, and consequently the area of the corresponding ellipse increases with the redshift. They are positively correlated and the strength of the correlation also increases with the redshift.

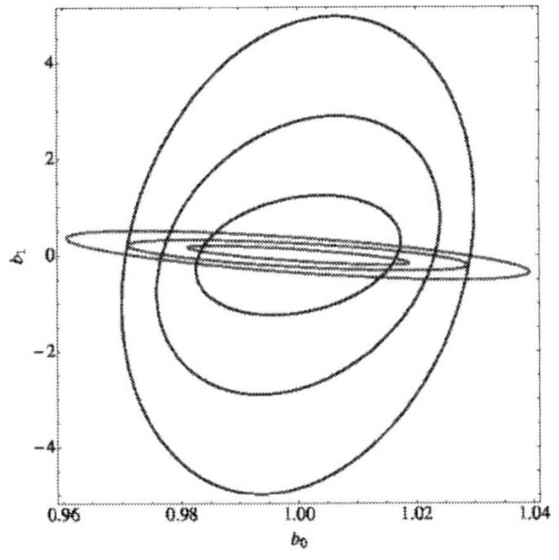

Figure 2. 68% probability contours for the parameters b_0 and b_1 of the FM1-PL model. The dotted red line is the case with $n = 1$ while the dashed line in blue shows the case $n = 2$. The redshift bins are $z = 0.6, 1.8, 2.0$ (see [5]).

4. Conclusions

We have investigated the impact of a scale-dependent galaxy bias on the results of the clustering analysis performed in next generation surveys (see [5]). Our main results can be summarised as follows:

- Allowing for a scale dependent bias does not increase significantly the errors on cosmological parameters, except for the growth index γ and the rms density fluctuation σ_8. More specifically, in our analysis we find that errors on h_0, Ω_{m0}, Ω_{b0}, Ω_{k0}, and n_s are only

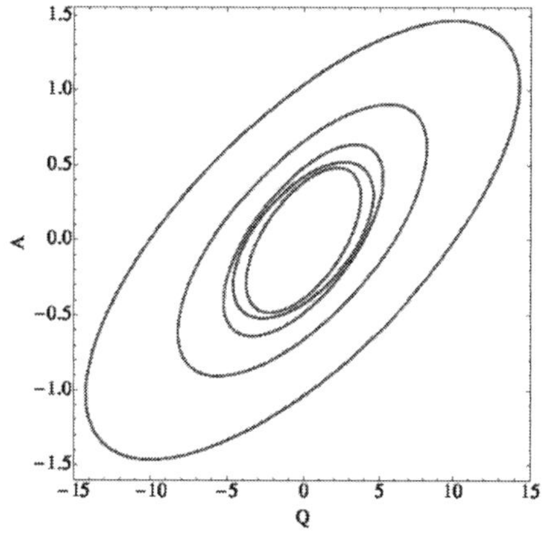

Figure 3. 68 % probability contours for the parameters A and Q of the FM1-Q model. We decided to plot only the bins $z = 0.6, 0.8, 1.6, 1.8, 2.0$, to improve clarity: larger values of redshift correspond to ellipses with larger semiaxis.

slightly larger than those expected when one assumes that bias is scale independent. In addition, γ and σ_8 are correlated. This correlation is expected since the clustering analysis constrains the amplitude of the power spectrum, proportional to the product $b(k, z)\sigma_8(z)$ and its redshift distortions, which are proportional to $f(z)/b(k, z)$.

- The linear bias parameter b_0 can be determined within a few %. The relative error is rather insensitive to the choice of the fiducial and slightly increases with the redshift. As expected, the errors increase with the number of free parameters in the model and therefore is larger in the Q-Model than in the Power Law one.
- The accuracy with which one can estimate the bias parameters that describe the scale dependency depends on the bias model and on the fiducial model. This is not surprising since in this case the scale dependence is pushed at small scales, where our Fisher matrix analysis, optimised to probe linear to mildly nonlinear scales, is less sensitive.

Acknowledgments

The research leading to these results has received funding from the European Research Council under the European Community Seventh Framework Programme ($FP7/2007 - 2013$ Grant Agreement no. 279954), therefore the financial support provided by Della Riccia Fellowship. EM acknowledges L. Amendola, M. Corsi, C. Di Porto and E. Branchini.

References

[1] Seo H J and Eisenstein D J 2007 *Astrophys. J.* **665** 14.
[2] Comparat J *et al.* 2013 *MNRAS* **433** 1146.
[3] Schlegel D, Abdalla F, Abraham T, Ahn C, Allende Prieto C, Annis J, Aubourg E and Azzaro M 2011 arXiv:1106.1706.
[4] Di Porto C, Amendola L and Branchini E 2012 *MNRAS* **423** L97-L101.
[5] Amendola L, Menegoni E, Di Porto C, Corsi M and Branchini E 2015 Accepted by *Phys. Rev. D* arXiv:1502.03994v1.
[6] Cole S *et al.* 2005 *MNRAS* **362** 505.
[7] The catalogs are publicly available at http://astro.dur.ac.uk/d40qra/lightcones/EUCLID/.
[8] Laureijs R. *et al.* 2011 arXiv:1110.3193.
[9] Seo H J and Eisenstein D J 2003 *Astrophys. J.* **598** 720.
[10] Seljak U 2001 *MNRAS* **325** 1359.
[11] Tegmark M 1997 *Phys. Rev. Lett.* **79** 3806.
[12] Di Porto C, Amendola L and Branchini E 2012 *MNRAS* **419** 985 1101.2453.
[13] Seo H.J and Eisenstein D. J. 2005 *Astrophys. J.* **633** 575.

Cosmic Microwave Background large-scale directional anomalies as seen by Planck and WMAP

L Polastri[1,2]
[1] Dipartimento di Fisica e Scienze della Terra, Università degli Studi di Ferrara, via Giuseppe Saragat 1, I-44122 Ferrara, Italy
[2] INFN, Sezione di Ferrara, via Giuseppe Saragat 1, I-44122 Ferrara, Italy

E-mail: linda.polastri@unife.it

Abstract. It has been found that large-scale anisotropies in the Cosmic Microwave Background are anomalous with respect to the predictions of the standard model of cosmology. We focused on the low multipole alignments, assuming the ΛCDM model and we confirmed that the quadrupole/octupole and the dipole/quadrupole/octupole alignments are anomalous with a significance up to 99.9%, for both WMAP and Planck data. Trying to explain the origin of this kind of anomalies we tested the dipolar model. This alternative phenomenological model explains the CMB hemispherical power asymmetry found in the WMAP and Planck data, so is possible that it can solve also other CMB directional anomalies. We show that the alignments are anomalous in the dipolar model too, roughly at the same level as in ΛCDM. We conclude that the dipolar model does not provide a better fit to the data than the ΛCDM.

1. Introduction

One of the pillars of the standard model of cosmology (ΛCDM) is the cosmological principle: the Universe is described with a good approximation as homogeneous and isotropic on large scales. The general appearance of the Universe must not depend on the observers position and on the direction of observation. In accordance with this principle the Cosmic Microwave Background (CMB) should be distributed in an isotropic way, i.e. it should not have preferred directions [1]. The data collected by the WMAP and Planck experiments are largely consistent with the ΛCDM model, but suggest that the CMB has a preferred direction. In particular, when expanding the CMB anisotropy pattern in spherical harmonics, there is an anomalous alignment for the dipole/quadrupole/octupole (ℓ=1, ℓ=2, ℓ=3) and the quadrupole/octupole, see Fig. 1.

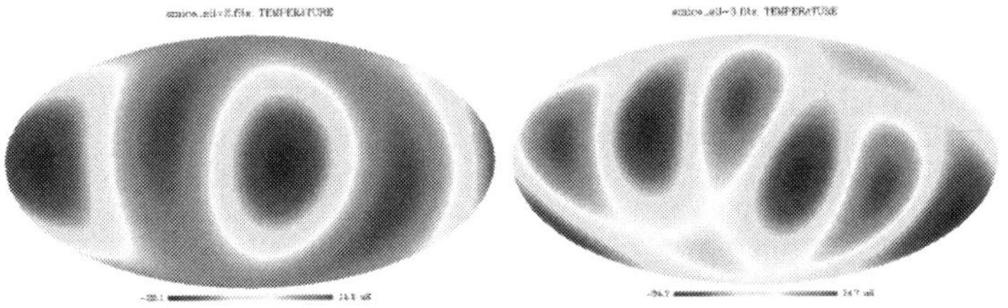

Figure 1. The quadrupole ($\ell = 2$) and the octupole ($\ell = 3$) as measured by Planck.

In literature it has been shown [2] that the quadrupole moment of the CMB is lower than expected in ΛCDM cosmology and that the octupole is planar and aligned with the quadrupole. Three independent anomalies have been identified that involve the quadrupole and the octupole: 1) the quadrupole with itself is anomalous at the 1-in-20 level by being low; 2) the octopole with itself is anomalous at the 1-in-20 level by being very planar; 3) the alignment between the quadrupole and octupole is anomalous at the 1-in-60 level. It is convenient to use a particular representation to study the directional anomalies: the Multipole Vectors (MV) formalism [3], see all details in Section 2.1. This formalism was used to study the WMAP and Planck data [4] and confirmed the anomalous alignment between quadrupole and octupole in both data sets [5–7]. At the moment, the origin of this kind of anomalies is not clear, however, it is difficult to imagine that the directional anomalies are due to systematic or instrumental problems, because two different experiments (WMAP and Planck) provided data in agreement with each other. Therefore it is hard to believe that two distinct experiments present instrumental errors capable of altering both results in the same way. Therefore it might be that these directional anomalies have a cosmological origin.

2. CMB directional anomalies

2.1. Multipole vectors

On the unit sphere, a function can be expanded as a linear combination of spherical harmonics:

$$f(\theta,\phi)=\sum_{\ell=0}^{\infty}\sum_{m=-\ell}^{\ell} a_{\ell m} Y_\ell^m(\theta,\phi), \tag{1}$$

where $a_{\ell m}$ are the coefficients and $Y_\ell^m(\theta,\phi)$ are the spherical harmonics.
We decided to use the multipole vectors formalism in order to have an easier approach to the problem. This is an alternative representation of the data on a sphere which allows us to express the information contained in each set of $a_{\ell m}$ coefficients, for any integer m = - ℓ, ..., ℓ, in unit vectors $\widehat{v}_i$ and a corresponding amplitude A^ℓ:

$$a_{\ell m} \to A^\ell, \widehat{v}_1, ..., \widehat{v}_\ell \tag{2}$$

Unfortunately no closed analytical expression for Eq. 2 is available.
The multipole vectors are invariant under transformations $a_{\ell m} \to c a_{\ell m}$, $A^\ell \to cA^\ell$ and they are used to construct scalar estimators that are invariant under rotation.
We built the estimators using the area vectors q and o_j:

$$\vec{q} = \widehat{q}_{21} \times \widehat{q}_{22}, \tag{3a}$$
$$\vec{o}_1 = \widehat{o}_{32} \times \widehat{o}_{33}, \tag{3b}$$
$$\vec{o}_2 = \widehat{o}_{33} \times \widehat{o}_{31}, \tag{3c}$$
$$\vec{o}_3 = \widehat{o}_{31} \times \widehat{o}_{32}, \tag{3d}$$

where q_{2j} (with $j = 1, 2$) represent the two MVs associated to the quadrupole and o_{3i} (with $i = 1, 2, 3$) represent the three MVs associated to the octupole.

2.2. Estimators

We provided eight different estimators belonging to two statistics "S" and "T" [3, 4, 8, 9]. They measure "distance" from a situation of complete misalignment, in terms of $\cos\theta$ resulting from the scalar product of the MVs. Orthogonality is associated to zero in both cases, whereas complete alignment, i.e. parallelism, is represented by the value 1. Notice that the "T" estimators

depend on $\cos^2\theta$, while the "S" estimators on $\cos\theta$ [10].
We considered six estimators for the quadrupole/octupole alignment:

$$S = \frac{1}{3}\sum_{j=1}^{3} |\widehat{q}\cdot o_j|, \tag{4a}$$

$$T = 1 - \frac{1}{3}\sum_{j=1}^{3}(1 - |\widehat{q}\cdot o_j|)^2, \tag{4b}$$

$$S23 = \frac{1}{3}\sum_{j=1}^{3} |q\cdot o_j|, \tag{4c}$$

$$T23 = 1 - \frac{1}{3}\sum_{j=1}^{3}(1 - |q\cdot o_j|)^2, \tag{4d}$$

$$\widehat{S}23 = \frac{1}{3}\sum_{j=1}^{3} |\widehat{q}\cdot \widehat{o}_j|, \tag{4e}$$

$$\widehat{T}23 = 1 - \frac{1}{3}\sum_{j=1}^{3}(1 - |\widehat{q}\cdot \widehat{o}_j|)^2 \tag{4f}$$

and two estimators for the dipole/quadrupole/octupole alignment:

$$DQO_S = \frac{1}{4}\left(|q\cdot d| + |o_1\cdot d| + |o_2\cdot d| + |o_3\cdot d|\right), \tag{4g}$$

$$DQO_T = 1 - \frac{1}{4}\left[(1 - |q\cdot d|)^2 + (1 - |o_1\cdot d|)^2 + (1 - |o_2\cdot d|)^2 + (1 - |o_3\cdot d|)^2\right]. \tag{4h}$$

The symbol $\hat{}$ denotes the unit vector, the vector d represents the dipole direction which reads $(l, b) = (224°, -22°) \pm 22°$ in Galactic coordinates.

2.3. Dipolar model

In order to solve the problem of hemispherical asymmetry, Gordon et al. [11] developed a particular mechanism for breaking isotropy in large-angle fluctuations of the CMB.
The pillars of this model are two: 1) a field, whose spatial fluctuations are dominated by long-wavelength contributions, 2) a non-linear response to the field by the CMB temperature fluctuations.
Statistical isotropy is preserved in the full theory, but it is spontaneously broken due to long-wavelength field fluctuations that appear as a gradient locally to the observer and carry the preferred direction into a spectrum of multipoles.
The CMB may exhibit a non linear response in two different ways: a multiplicative modulation of the anisotropy and an additive effect that is uncorrelated with the intrinsic anisotropy.
We can define for the dipolar model the temperature fluctuation field of the CMB as:

$$\left(\frac{\Delta T}{T}\right)_{mod}(\widehat{n}) = (1 + A\widehat{n}\cdot\widehat{p})\left(\frac{\Delta T}{T}\right)_{iso}(\widehat{n}), \tag{5}$$

where $\widehat{n}$ is the observed direction, $(\Delta T/T)_{mod}$ is the observed and modulated CMB temperature fluctuations, $(\Delta T/T)_{iso}$ is the usual isotropic CMB pattern, A is the amplitude of the dipole modulation and $\widehat{p}$ is a given direction. It has been found that $A = 0.07 \pm 0.02$ [6], statistically significant at $\sim 3\sigma$ and the direction $\widehat{p}$ is given by $(l, b) = (224°, -22°) \pm 22°$ in Galactic coordinates, at $\sim 3.3\,\sigma$ [12].

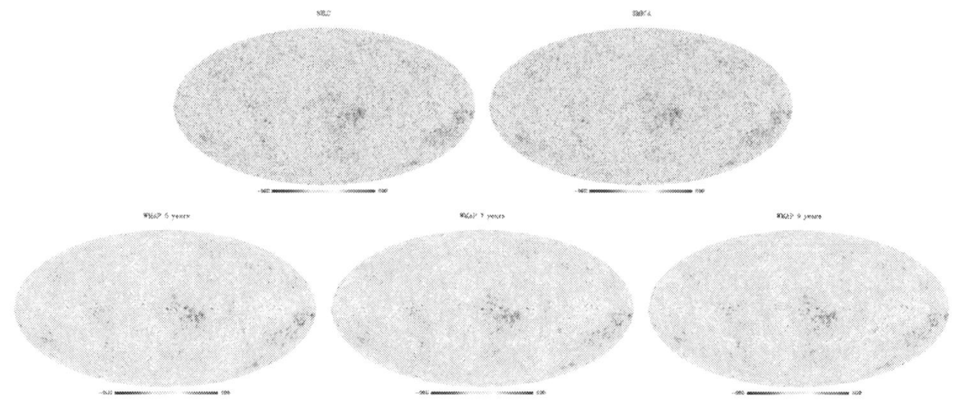

Figure 2. In the first row are shown the temperature maps by Planck: NILC (left side) and SMICA (right side). In the second row are shown the temperature maps for WMAP 5 yr, WMAP 7 yr and WMAP 9 yr respectively. In both rows the units are μK.

3. Results

The aim of this section is to show the results of the study of the CMB quadrupole and octupole anomalies in order to test the cosmological principle and verify the existence of directional anomalies for two different models: the ΛCDM and dipolar one. In the first part of the analysis, starting from the Planck 2013 fiducial model, we extracted the $a_{\ell m}$ coefficients, calculated the MV using the public code by [3], built the area vectors and consequently the estimators (presented in Section 2.2). We performed 10^5 Monte Carlo (MC) simulations to obtain a theoretical trend for the ΛCDM model. We also performed the same number of simulations based on Eq. 5 for the dipolar model.

In the second part, we extracted the estimators from the available data for two experiments: WMAP and Planck. For WMAP we analysed the Internal Linear Combination (ILC) maps for the 5, 7 and 9-year data [13–15], see bottom row of Fig. 2. The ILC method combines the available data at different frequencies to mitigate the Galactic foreground emission in the final map. Regarding the Planck mission, we consider the SMICA and NILC maps of the 2013 data release [16, 17] see top row of Fig. 2. The first one is the result of a Spectral Matching Independent Component Analysis (SMICA), which reconstructs a CMB map from the linear combination in harmonic space of several input frequency maps with weights that depend on the multipole ℓ. NILC (Needlet-ILC) instead is a method to extract the CMB by applying the ILC technique to multi-channel observations in needlet space and compute an ILC in each zone and for each scale, allowing the ILC weights to adapt naturally to the varying stretch of the other components as a function of position and multipole.

Before analysing the maps, a "boost correction" [18, 19] has been applied to the real data, because the observed dipole is affected by the motion of the satellite with respect to the CMB rest frame, see [20] for details about the boost correction.

We present our results about the estimators in Tab. 1 and in Tab. 2 where we show the corresponding probability to exceed for both ΛCDM and dipolar models [20]. The percentage of anomaly is defined as the amount of MC realizations for which the considered estimator is smaller than the observed one. The larger the percentage the more anomalous is the alignment. In Fig. 3 we show the theoretical trend for ΛCDM and dipolar models (histograms) related to the real boosted data (vertical lines). The two histograms are very close to each other and this means that the two considered models are very similar for this type of analysis. The vertical bars stand in the tail of the histograms, thus the real data confirm the anomalous alignment in

IOP Conf. Series: Journal of Physics: Conf. Series **841** (2017) 012008 doi:10.1088/1742-6596/841/1/012008

all analysed cases. Moreover, the theoretical trends of both models confirm that they are not able to explain effectively the source of the directional anomalies.

Table 1. Values of the estimators extracted from the WMAP and Planck CMB maps.

Estimator	WMAP ILC 5 yr	WMAP ILC 7 yr	WMAP ILC 9 yr	Planck SMICA	Planck NILC
S	0.799	0.804	0.807	0.794	0.804
T	0.959	0.962	0.963	0.956	0.962
S23	0.776	0.783	0.788	0.718	0.697
T23	0.949	0.953	0.955	0.919	0.908
$\widehat{\mathbf{S}}$**23**	0.869	0.877	0.884	0.859	0.877
$\widehat{\mathbf{T}}$**23**	0.982	0.984	0.986	0.979	0.985
$\mathbf{DQO_S}$	0.789	0.792	0.799	0.774	0.776
$\mathbf{DQO_T}$	0.940	0.943	0.946	0.936	0.944

Table 2. Percentage of anomaly for the quadrupole/octupole alignment and dipole/quadrupole/octupole alignment, for the WMAP and Planck data for all the analysed estimators.

	WMAP ILC 5 yr	WMAP ILC 7 yr	WMAP ILC 9 yr	Planck SMICA	Planck NILC
Estimator	ΛCDM — Dipolar	ΛCDM — Dipolar	ΛCDM — Dipolar	ΛCDM — Dipolar	ΛCDM — Dipolar
S	99.647 — 99.640	99.701 — 99.701	99.750 — 99.731	99.581 — 99.578	99.704 — 99.707
T	99.828 — 99.832	99.856 — 99.866	99.873 — 99.880	99.775 — 99.769	99.856 — 99.866
S23	99.722 — 99.724	99.793 — 99.791	99.838 — 99.830	98.649 — 98.606	97.990 — 97.951
T23	99.863 — 99.868	99.892 — 99.891	99.905 — 99.906	99.217 — 99.207	98.861 — 98.833
$\widehat{\mathbf{S}}$**23**	98.355 — 98.308	98.569 — 98.539	98.689 — 98.676	98.128 — 98.089	98.550 — 98.523
$\widehat{\mathbf{T}}$**23**	98.654 — 98.646	98.839 — 98.802	98.901 — 98.881	98.420 — 98.379	98.839 — 98.806
DQO_S	99.803 — 99.796	99.829 — 99.823	99.872 — 99.865	99.672 — 99.662	99.687 — 99.681
DQO_T	99.776 — 99.779	99.810 — 99.808	99.859 — 99.851	99.725 — 99.728	99.825 — 99.823

4. Conclusions

We report that all the data and all the estimators that we have tested exhibit anomalous alignments for both combinations of multipoles considered, typically at the 98%-99% level, and up to 99.9% in selected cases.

The consistent pattern for the alignments observed in both WMAP and Planck data strongly disfavours an origin of the effect related to unaccounted for instrumental systematics.

We have also investigated the possibility that the phenomenological dipolar model may provide a better explanation for the existence of the observed alignments with respect to ΛCDM. Although the dipolar model gathered some success in explaining some anomalies, e.g. the hemispherical asymmetry, we report negative findings regarding directional anomalies: the dipolar model does not seem to be able to accommodate for the existence of anomalies significantly better than ΛCDM.

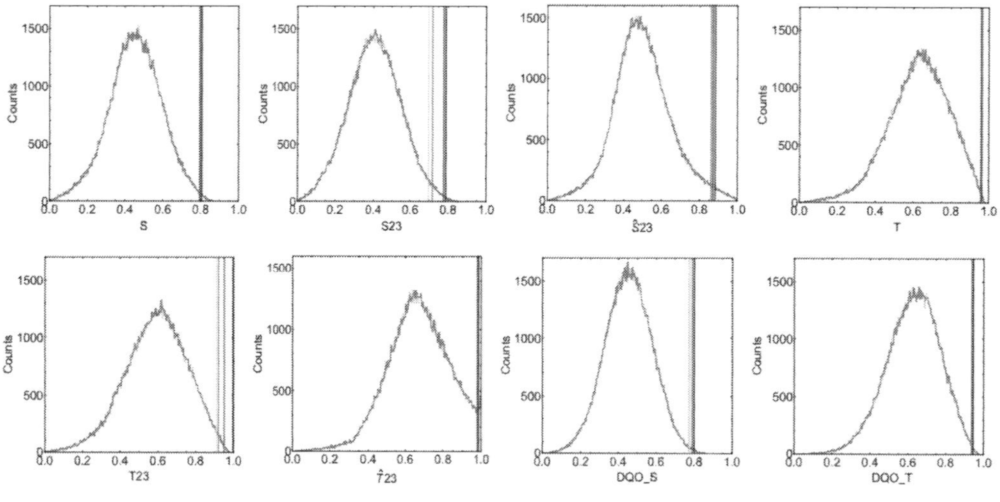

Figure 3. S statistics for the upper row and T statistics for the lower row. Green histograms for the empirical distribution of the considered estimators in ΛCDM and red for the dipolar model. Vertical lines are for the observed estimators (already boost-corrected): WMAP ILC 5 in blue, WMAP ILC 7 in pink, WMAP ILC 9 in black, Planck 2013 NILC in cyan and Planck 2013 SMICA in magenta. In each panel we show the counts in the y-axis and the estimator in the x-axis (see also [20]).

References

[1] Dodelson S 2003 *Modern Cosmology* (Amsterdam: Academic Press) ISBN 9780122191411
[2] de Oliveira-Costa A, Tegmark M, Zaldarriaga M and Hamilton A 2004 *Phys.Rev.* **D69** 063516
[3] Copi C J, Huterer D and Starkman G D 2004 *Phys. Rev.* **D70** 043515
[4] Copi C J, Huterer D, Schwarz D J and Starkman G D 2015 *Mon. Not. Roy. Astron. Soc.* **449** 3458–3470
[5] Ade P A R *et al.* (Planck) 2014 *Astron. Astrophys.* **571** A16
[6] Hoftuft J, Eriksen H K, Banday A J, Gorski K M, Hansen F K and Lilje P B 2009 *Astrophys. J.* **699** 985–989
[7] Gruppuso A and Burigana C 2009 *JCAP* **0908** 004
[8] Gruppuso A, Burigana C and Finelli F 2007 *Mon. Not. Roy. Astron. Soc.* **376** 907–918
[9] Abramo L R, Bernui A, Ferreira I S, Villela T and Wuensche C A 2006 *Phys. Rev.* **D74** 063506
[10] Copi C J, Huterer D, Schwarz D J and Starkman G D 2010 *Adv. Astron.* **2010** 847541
[11] Gordon C, Hu W, Huterer D and Crawford T M 2005 *Phys.Rev.* **D72** 103002
[12] Eriksen H K, Banday A J, Gorski K M, Hansen F K and Lilje P B 2007 *Astrophys. J.* **660** L81–L84
[13] Hinshaw G *et al.* (WMAP) 2009 *Astrophys. J. Suppl.* **180** 225–245
[14] Jarosik N *et al.* 2011 *Astrophys. J. Suppl.* **192** 14
[15] Hinshaw G *et al.* (WMAP) 2013 *Astrophys. J. Suppl.* **208** 19
[16] Ade P A R *et al.* (Planck) 2014 *Astron. Astrophys.* **571** A23
[17] Ade P A R *et al.* (Planck) 2014 *Astron. Astrophys.* **571** A15
[18] Kosowsky A and Kahniashvili T 2011 *Phys. Rev. Lett.* **106** 191301
[19] Amendola L, Catena R, Masina I, Notari A, Quartin M and Quercellini C 2011 *JCAP* **1107** 027
[20] Polastri L, Gruppuso A and Natoli P 2015 *JCAP* **1504** 018

IOP Conf. Series: Journal of Physics: Conf. Series **841** (2017) 012009 doi:10.1088/1742-6596/841/1/012009

Development and implementation of a non Gaussian model for the lateral dose prediction in a proton therapy treatment planning system

V E Bellinzona [1,2]
[1]Univeristà degli Studi di Pavia, Physics department, Via Bassi 9 27100 Pavia, Italy
[2]Ludwig-Maximilians-Universität München, Medical Physics, Am Coulombwall 1, 85748 Garching b. München Germany
E-mail: elettra.bellinzona@gmail.com

Abstract. Challenging issues in treatment planning system for hadrontherapy are the accurate and fast calculation of dose distribution, the reduction in memory space required to store the dose kernel of individual pencil beams and the shortening of computation time for dose optimization and calculation. In this framework, the prediction of lateral dose distributions is a topic of great interest because currently the double gaussian parametrization is typically used as approximation although other parameterizations are also available. The best accuracy for this kind of calculations can be obtained by Monte Carlo methods, at the expense of a long computing time. This work aims to present a flexible computational model for the calculation of the lateral profile of a pencil proton beam and the results of its implementation in a treatment planning system. The model calculation are compared with the currently used double gaussian approximation and the Monte Carlo calculations, and the tests are performed in water and in presence of inhomogeneities.

1. Introduction

Particle radiotherapy treatments constitute about the 1% of the total number of treatments for patients receiving radiotherapy worldwide [1]. Why particle therapy?
The answer is based on two main characteristic of ions and protons [2]. The *physical* advantage respect to conventional radiotherapy, is due to the characteristic depthdose profile, analytically based on the Bethe-Bloch formula described by the Bragg curve, characterized by a small entrance dose and a distinct maximum near the end of range with a sharp fall-off at the distal edge, called *Bragg peak*, that results in a customizable and precise dose deposition.
From a *radio-biological* point of view protons and heavier ions have different Radio-Biological Effectiveness (RBE)[3]; protons and photons have a similar RBE in the plateau region, and higher RBE on the peak region while heavier ions show an even higher RBE the Bragg peak, compared to photons. This is due to the high ionization density produced by the ions as they traverse a cell, resulting in more disruptive damage to the DNA double helix. The higher RBE combined with the lower Oxygen-Enhancement Ratio (OER)[3], especially in and around the Bragg peak, can be a considerable advantage in tumour cell killing, particularly for resistant tumour cells [4].
Analytically, the Bragg curve is the results of three interactions processes: stopping, scattering

IOP Conf. Series: Journal of Physics: Conf. Series **841** (2017) 012009 doi:10.1088/1742-6596/841/1/012009

and nuclear interactions. Focusing on the lateral beam shape (on the plane perpendicular to the beam axis), these effects result in a broader shape, compared with the one expected from a purely electromagnetic calculations. The angular deflection is mainly due to the *Multiple Coulomb Scattering (MCS)*. In a first approximation, the MCS angular distribution is nearly gaussian, and for this reason, is approximated with a double Gaussian function [5, 6].
To predict the exact form of the MCS angular distribution and its characteristic width as a function of proton energy as well as scattering material and thickness, we propose a flexible computational model based on Molière's theory [7, 8], that requires as input the particle energy (or the beam energy distribution as a phase-space) and the desired material depths.

2. The model

The model is composed by a purely analytical electromagnetic core based on Molière's theory and a nuclear tails part. It has been developed for a proton pencil beam in water but it can be easily extended to more realistic cases to reproduce mathematically the lateral beam scan in a given region [9].

2.1. Electromagnetic core

The Molière theory is mathematically based on the standard transport equation, whit the Bessel transforms and the small angle approximation i.e. $sin(\theta) \approx \theta$.
Physically, the theory only accounts for electromagnetic interactions and assumes the Rutherford form for the single scattering cross section:

$$\rho\sigma(\theta) = 2\chi_c^2 \frac{1}{(\theta^2 + \chi_\alpha^2)^2} , \tag{1}$$

where ρ is the number of atoms per cm^2, and the integration over ϕ is performed. χ_c and χ_α are the two crucial parameters of the model. The first one is connected to the rms scattering angle:

$$\chi_c^2 = 0.1569 \cdot 10^{-6} Z^2 z^2 \frac{x}{A} \frac{1}{p^2\beta^2} , \tag{2}$$

where z is the atomic number of the incident particle, Z and A the atomic and the mass number of the target, x is the thickness of the target (g/cm^2), p is the momentum in GeV/c and $\beta = v/c$. The second one is the parameter connected to the electron screening of the Coulomb potential:

$$\chi_\alpha^2 = \mu^2 \chi_0^2 \tag{3}$$

$$\mu^2 = \left(1.13 + 3.76 \frac{z^2 Z^2}{137^2 \beta^2}\right) \tag{4}$$

$$\chi_0^2 = \left(\frac{\hbar}{p} \frac{Z^{1/3}}{0.468 \cdot 10^{-8}(cm)}\right)^2 \tag{5}$$

When $\mu = 0$ (and consequentially $\chi_\alpha^2 = 0$) Eq. (1) is the Rutherford cross section of a point charge. The standard form for the Molière distribution is given by [10] and the projected angle θ_x (that has the same trend of θ_y) follows the distribution described in [10]. In the small angle approximation, using simple geometry consideration, the connection between the mean square of the projected angle in the transverse plane and the spatial angle is obtained:

$$\langle\theta^2\rangle = 2\langle\theta_x^2\rangle . \tag{6}$$

Therefore, the connection between the rms of the space and projected angles is:

$$\theta_{xR} = \frac{\theta_R}{\sqrt{2}} = \frac{\chi_c\sqrt{B}}{\sqrt{2}} \tag{7}$$

where B is the solution of the equation:

$$B - \ln B = b \tag{8}$$

$$b = \ln \frac{\chi_c^2}{\chi_\alpha^2} - 0.154432 = \ln \Omega_0 - 0.154432 \tag{9}$$

2.1.1. Mixture and compound The generalization of this theory to a compound or mixture is quite easy, by the modification of Eq. (2), (3) and (5) [9] and considering a proper summation over the constituents as $Z_M = \sum_i n_i Z_i$, $A_M = \sum_i n_i A_i$.
where Z_i and A_i are respectively the atomic number and atomic weights of the molecule or mixture of n_i constituents.

2.1.2. Energy loss and lateral displacement Since in realistic hadrontherapy cases the paths of protons in water is very long, the basic equations have to be modified to take into account the energy loss process. Considering that the parameters χ_c and χ_α depend on p, β, the energy loss problem can be solved if one finds the dependence of these quantities on the water thickness x traversed [9]. Including the energy loss effects, the general formula for passing from angular to spatial displacements can be derived. The rms y_M of the transverse displacement on a measuring plane at x due to a layer dt at the depth t is given by $(x-t)\theta_{xR}$, where the angle is given by Eq. (7), that presents a logarithmic dependence on χ_c, which contains the thickness x (see Eq.(2), (8)). The physical meaning is that two successive layers act in a dependent manner, since the second layer receives trajectories deflected by the first one. The following final formula can be obtained (details in [9]):

$$f_M(y) = \frac{1}{\pi \chi_c \delta} \int_0^\Gamma \cos\left(\frac{y\eta}{\chi_c \delta}\right) \exp\left[-\frac{\eta^2}{4}\left(b - \ln\frac{\eta^2}{4}\right)\right] \eta d\eta \, . \tag{10}$$

2.2. Nuclear effects

To complete the description of the overall phenomena, nuclear effects have to be properly taken into account. An additional term is added to describe the interaction of non-primary particles that affects the tails to the distribution. This is particularly relevant when big thicknesses are involved, such as in the case of hadrontherapy (it can be estimated as about 1% per cm of depth). To account for these effects, a modified Cauchy-Lorentz distribution [11, 12] is considered:

$$t(y) = \frac{1 - A\exp\left[-\frac{y^2}{2b^2\sigma^2}\right]}{\pi b\left(\frac{y^2}{b^2} + 1\right)} \tag{11}$$

where the three free parameters are the amplitude A, the half width half maximum (HWHM) b and the variance σ^2.

2.3. Complete model and validation

Now, the two parts are combined by adding a weight factor W_p [13], that express the percentage of events without nuclear interactions (primary protons), as a function of the traversed thickness [9]. The total normalized final distribution for the lateral displacement, can be formulated as:

$$f(y) = W_p f_M(y) + (1 - W_p)\frac{t(y)}{\int_{-\infty}^{+\infty} t(y)\mathrm{d}y} , \tag{12}$$

where $f_M(y)$ is the distribution of Eq. (10) and W_p is the weight factor. Both $f_M(y)$ and $t(y)$ are normalized to unit area.

The free parameters A, b and σ^2, have been determined by fitting the complete FLUKA MC lateral displacement distributions, and they are parametrised with Chebyshev polynomials. These results have been validated against FLUKA [14, 15] and HIT Heidelberg Ion Beam Therapy Center [16]. A selection of results is reported in Fig.1. For all the analysed depths and energies, the model agrees very well with the FLUKA predictions and with the measurement data.

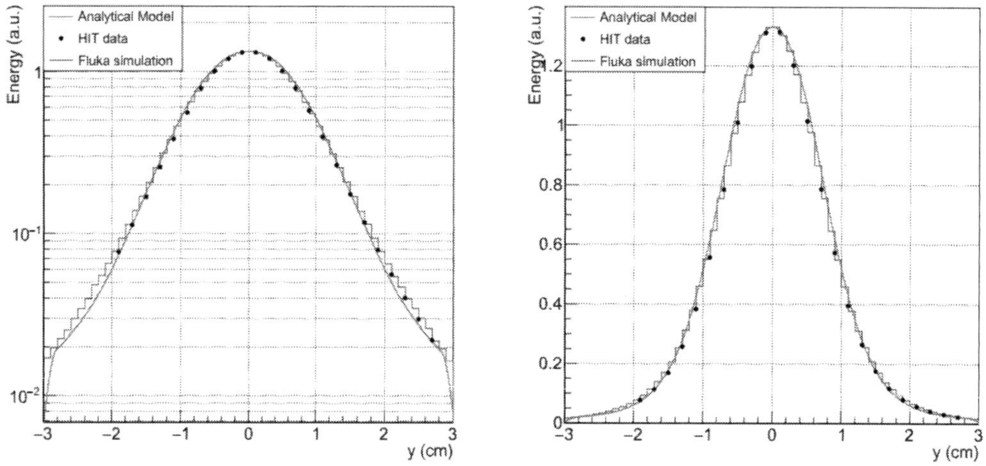

Figure 1: Comparison between analytical model, FLUKA simulation and experimental data for E=157.43 MeV in water at a thickness z=11.55 cm in logarithmic (left), to better evaluate the difference in the tail region, and linear (right) scales. The curves are normalized to the maximum of the FLUKA histogram, obtained by fitting the highest dose channels with a smooth curve. In the normalization zone the error of the data points is 2%.

3. Treatment Planning System (TPS) implementation

To test the possible clinical advantages of using our model, these results have been implemented in CERR - A Computational Environment for Radiotherapy Research TPS [17, 18] at Ludwig-Maximilians-Universität München. The rational was to compare the dose deposition obtained using our model with the results obtained using the implemented double Gaussian approximation (DGA), taking as gold standard the results of the MC method.
Several energies and depths have been considered for single beams and for a full treatment plan in water. Furthermore, plans containing inhomogeneities i.e. bone and fat, have been studied. For a more detailed explanation see [19].
The doses evaluated by the TPS, have been analysed quantitatively using statistical tests, lateral profiles comparison, residual analysis and DVH (Dose Volume Histogram) for the full plans. A selection of results is reported in Fig.2 for single beam in water and in presence of bone and fat inhomogeneities and for a full plan in Fig.3 *(Right)*. Fig.3 *(Left)* shows the direct comparison between the model DVH and the DG ones for the full plan; the delivered dose to the PTV (Planning Target Volume [3]) is almost equal in the two cases but the dose on surrounding healty tissues ((PTV3D+1)-PTV) is lower with the model.
Overall one can conclude that a plan obtained using the model for the lateral evaluation of the dose, is a promising candidate instead the DG ones in terms of treatment quality.

4. Conclusion

This study shows that the model has better accuracy than DGA method with adequate computation time in water, and it is in a good agreement with the MC for all the analysed

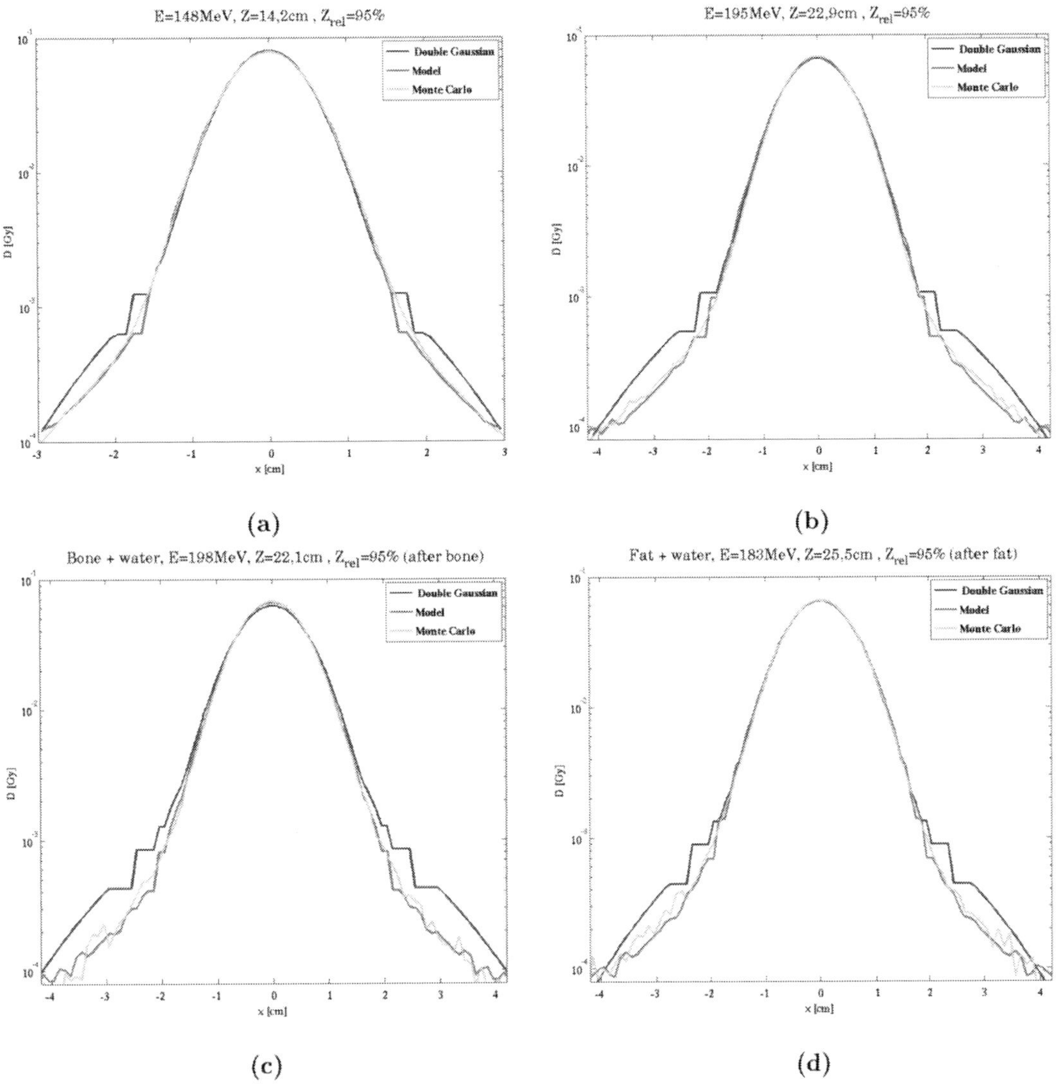

Figure 2: Lateral profile of the dose deposition evaluated by CERR TPS at a relative depth of $Z_{rel} = 95\%$, in logarithmic scale for a single beam of energy E=148 MeV **(a)** and E=195 MeV **(b)** in water, for a single beam with $E = 198$ MeV after a bone inhomogeneity region **(c)** and for a single beam with $E = 183$ MeV after a fat inhomogeneity.

cases. Since the comparison with data and MC shows very good agreement, the model can be proposed to be used instead the currently used DGA. This solution can enhance the accuracy of the lateral profile evaluation in hadrontherapy practice, that can be a critical parameter especially in the case of deep tumours near organs at risk. Since the model is much faster that the MC [20, 21] and has same accuracy in water, it can be proposed in future to be used for the online beam monitoring, and also as a powerful and flexible tool for the forward dose evaluation. A future step will be the study of the effects of different materials and different geometries, including interfaces and the use of other particles already clinically used, e.g. carbon ions, α particles.

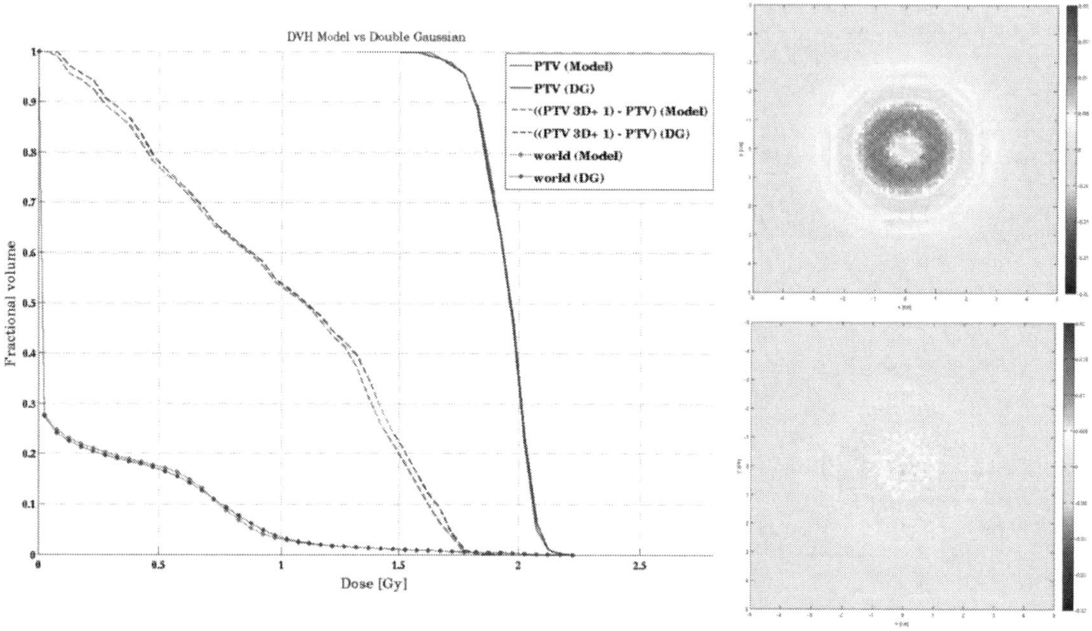

Figure 3: On the *left*: DVH calculated by CERR, for the PTV, the safe margin of PTV + 1 cm and the WORLD regions, both for the plan obtained with the DG (blue curves) and the model (red curves). On the *(right)*: Net difference between the dose deposition in Gy, obtained for the full treatment using DG lateral parametrization and the MC *(top)* and using the model evaluation compared with the MC *(bottom)*; both for the relative depth $Z_{rel} = 50\%$.

References

[1] Huan G and Bosco G 2012, *Translational Cancer Research* 1(3):**127-136**
[2] Schardt D, Elsasser T, and Schulz-Ertner D 2010, *Reviews of modern physics* 82(1):**383**
[3] Paganetti H 2011 *CRC Press*
[4] Durante M and Paganetti H 2016 *Progress in Physics* 79(9):096702
[5] Fruhwirth R and Regler M 2001 *Nuclear Instruments and Methods in Physics Research* 456(3):**369-389**
[6] Bellinzona E V et al. 2015 *Physica Medica* 31.5: **484-17492**
[7] Molière G 1948 *Mehrfach und Vielfachstreuung Z. Naturforsc.* 3a **78-97**
[8] Bethe H A 1953 *Phys. Rev.* 89 **1256-66**
[9] Bellinzona E V 2016 *Phys. Med. Biol.* 61 **102-1717117**
[10] Scott W T 1963 *Rev. Mod. Phys.* 35 **231-312**
[11] Soukup M et al. 2005 *Phys. Med. Biol.* 50, **5089-5104**
[12] Li Y et al. 2012 *Phys. Med. Biol.* 57 **983-97**
[13] Ulmer W 2007 *Rad. Phys. and Chem.* 76 **1089-107**
[14] Bohlen T T et al. 2014 *Nuclear Data Sheets* 120, **211-214**
[15] Ferrari A et al. 2005 *CERN-2005-10*, INFN/TC05/11, SLAC R 773
[16] Schwaab J et al. 2011 *Phys. Med. Biol.* 56 **7813-27**
[17] Deasy J O et al. 2003 *Medical Physics* 30.5: **979-17985**
[18] Schell S et al. 2010 *Medical Physics* 37(10):**53301740**
[19] Bellinzona E V 2016 Ph:D.thesis Univeristy of Pavia & Ludwig-Maximilians-University Munich
[20] Parodi K et al. 2010 *Phys. Med. Biol.* 55 **5169-171787**
[21] Mairani A et al. 2013 *Phys. Med. Biol.* 58 **2471**

IOP Conf. Series: Journal of Physics: Conf. Series **841** (2017) 012010 doi:10.1088/1742-6596/841/1/012010

Characterization of physicochemical and colloidal properties of hydrogel chitosan-coated iron-oxide nanoparticles for cancer therapy

E Catalano[1] and A Di Benedetto[2]

[1] Department of Clinical Molecular Biology (EpiGen), Akershus University Hospital, University of Oslo (UiO), Oslo, Norway.
[2] Dipartimento di Fisica G. Occhialini, Università Milano-Bicocca, Milano, Italia.

E-mail: enrico.catalano@medisin.uio.no

Abstract. Superparamagnetic iron oxide nanoparticles have recently been investigated for their potential to kill cancer cells with promising results, owing to their ability to be targeted and heated by magnetic fields. In this study, novel hydrogel, chitosan Fe_3O_4 magnetic nanoparticles were synthesized to induce magnetic hyperthermia, and targeted delivering of chemotherapeutics in the cancer microenvironment. The characteristic properties of synthesized bare and CS-MNPs were analyzed by various analytical methods: X-ray diffraction, Fourier transformed infrared spectroscopy, Scanning electron microscopy and Thermo-gravimetric analysis/ differential thermal analysis. Magnetic nanoparticles were successfully synthesized using the co-precipitation method. This synthesis technique resulted in nanoparticles with an average particle size of 16 nm. The pure obtained nanoparticles were then successfully encapsulated with 4-nm-thick chitosan coating. The formation of chitosan on the surface of nanoparticles was confirmed by physicochemical analyses. Heating experiments at safe magnetic field (f = 100 kHz, H =10-20 kA m^{-1}) revealed that the maximum achieved temperature of water stable chitosan-coated nanoparticles (50 mg ml^{-1}) is fully in agreement with cancer therapy and biomedical applications.

1. Introduction

Magnetic nanomaterial-based systems are exploited to develop several novel therapeutic opportunities for cancer therapy based merely upon their unique physicochemical and magnetic properties [1]. Nowadays, magnetic nanoparticles (MNPs), due to their particular physicochemical properties, have attracted growing interest in biomedicine and bioengineering such as magnetically triggered drug delivery [2], bio-separation [1], magnetic resonance imaging [1], bio-macromolecule purification, and magnetic hyperthermia [3, 4]. Magnetic hyperthermia has long been used as a treatment option for cancer [1]. In this method MNPs are introduced to the tumor tissue and then exposed to an alternating magnetic field (AMF). This method leads to heat generation by the MNPs, that can kill the cancerous cells [3, 4]. In magnetically triggered drug delivery systems, MNPs are normally coated with a thermosensitive polymer as drug carrier. When these MNPs are exposed to an AMF, due to the heat generated by magnetic core, phase transition occurs in the thermosensitive polymer and drug is released [1]. The most common MNPs used for magnetic hyperthermia or drug delivery are Fe_3O_4 or

γ-Fe_2O_3 [5]. Chitosan, as a multifunctional natural polymer containing large numbers of active amino groups together with hydroxyl groups in the backbone, can interact with mucins and open the tight junctions between epithelial cells, which are in favor of the drug transport. The chitosan-based delivery systems have shown great potential to deliver nucleic acids, therapeutic proteins, vaccinations and other bio-molecules [6].

2. Materials and methods

2.1 Synthesis of Iron Oxide Nanoparticles

Magnetic iron oxide nanoparticles (MNPs) were prepared by alkaline co-precipitation of ferrous chloride tetrahydrate, $FeCl_2 \bullet 4H_2O$ (1.34 g) and ferric chloride hexahydrate $FeCl_3 \bullet 6H_2O$ (3.40 g) at 1:2 ratio. The salts were dissolved in 150 mL deionized water within a three necked glass balloon. The glass balloon was placed in a heating mantle and stirred with a magnetic stirrer. It was vigorously stirred at 90°C in the presence of N_2 gas. Ammonium hydroxide (NH_4OH) was added to the system drop wise. The process was ended by washing with deionized water until the solution pH was 9.0. The solution was then centrifuged at 5000 rpm for 30 minutes. The precipitates were collected and dried in the incubator at 55°C. The black precipitates were then turned into brown.

2.2 Synthesis of Chitosan-coated Magnetic Iron Oxide Nanoparticles

Chitosan-coated magnetic iron oxide nanoparticles (CS-MNPs) were in situ synthesized by the co-precipitation of Fe (II) and Fe (III) salts in the presence of chitosan and trisodium phosphate molecules. Chitosan was previously prepared with degree of deacetylation 75% by titrimetric method [7]. Trisodium phosphate was used for the crosslinking of low molecular weight chitosan polymers. Chitosan (0.15 g) was dissolved in 30 ml of 1% acetic acid and the pH was adjusted to 4.8 by 10M NaOH. Iron salts (1.34 g $FeCl_2 \bullet 4H_2O$ and 3.40 g $FeCl_3 \bullet 6H_2O$) were dissolved in 30 ml of 0.5% chitosan solution. The solution was then vigorously stirred at 2000 rpm. 10 ml of 22.5% trisodium phosphate and different amounts of 32% NH_4OH (18, 20, 22, 25 mL) were added to the solution to obtain the final NH_4OH concentration of 31%, at room temperature. The ammonia solution was added very slowly to produce smaller sized nanoparticles. The resulting solution was stirred for an additional 1 hour. The colloidal chitosan coated magnetic Fe_3O_4 nanoparticles were extensively washed (3 times) with deionized water and separated by centrifugation and drying.

2.3 X-ray diffraction (XRD)

X-ray diffraction (XRD) was used to investigate the mineral phase of obtained materials. XRD measurements were performed using a D8 ADVANC diffractometer (BRUKER AXS GmbH, Germany) with Cu Kα radiation (λ = 1.54178 Å), with the operation voltage and current at 40 kV and 40 mA, respectively. The 2θ range was from 5°-90° in steps of 0.02°.

2.4 Fourier transformed infrared spectroscopy (FTIR)

The FTIR spectroscopy of the sample was taken in the region between 500-4000 cm^{-1} (with Perkin Elmer 1650, USA) on a Thermo-Nicolet Avatar 370 model FTIR in order to understand the chemical and structural nature of the chitosan coated iron-oxide nanoparticles.

2.5 Thermo-gravimetric analysis/differential thermal analysis/derivative thermo gravimetric analysis (TG/DTA/DTG)

TGA shows the change in mass with increase of temperature. This studies were carried out by TG/ DTA 6300, SII Nano Technology, Japan by heating the sample at 20°C /min in the temperature range 100 - 1000°C in nitrogen atmosphere.

2.6 Morphological analysis of the CS-MNPs
The size of the prepared sample (CS-MNPs) was related to properties such as bioavailability and surface modification for in vivo application. The particles were inspected on an FEI Quanta 200 scanning electron microscope (SEM) at an accelerating voltage of 20 kV to identify average diameter, interconnectivity, and any agglomeration.

2.7 Zeta potential measurements
The evolution of ζ potential towards pH of the nanoparticles solution was evaluated using a ζ potentiometer (Zetasizer Nano ZS90; Malvern Instruments, Malvern, UK) in order to monitor the stability of nanoparticles in aqueous media. ζ potential measurements were performed both for bare and chitosan coated Fe_3O_4 and a comparison was performed.
The nanoparticles-containing solution were diluted 100-fold with deionized water before testing. The analysis was carried out at room temperature, three measurements per sample were performed and the values reported as the average ± SD.

2.8 Calorimetric measurements
The heat generation ability of the iron-oxide nanoparticles for biomedical application was investigated using a homemade induction heating instrument that induced an alternating magnetic field in five turns (5 cm diameter) induction coil. A close loop water temperature control was made to keep the temperature of the coil at ambient temperature. The stable suspension of nanoparticles (50 mg ml^{-1}) in 1.5 ml plastic microtube, thermally insulated using styrofoam, was placed in the centre of coil and then irradiated by magnetic field (f = 100 kHz, H = 10-20 kA m^{-1}). The temperature of the sample was monitored by an alcohol thermometer. The value of the specific absorption rate (SAR) of the sample was calculated according to [Eq. (1)]

$$\mathrm{SAR} = \left(\frac{C_{\mathrm{suspension}}}{X_{\mathrm{NP}}}\right)\left(\frac{dT}{dt}\right) \quad (1)$$

where $C_{suspension}$ is the specific heat of the suspension [8], C indicates precisely the heat capacity of the fluid per unit mass of nanoparticles and their suspension in solution, X_{NP} is the weight fraction of nanoparticles (X_{NP} = 0.05) and dT /dt is the initial slope of the temperature curve vs. time.

3. Results and discussion

The characteristic properties of synthesized bare and CS-MNPs were analyzed by various analytical methods. Crystal structures of synthesized MNPs were analyzed by XRD. The chemical groups and chemical interactions involved in synthesized MNPs were identified using the FTIR methods. The sizes of magnetic core and morphological properties were observed through SEM. The qualitative and quantitative information about the volatile compounds of CS-MNPs were performed using TG/DTA.

3.1. Structural and component analysis of CS-MNP nanocarriers

The crystal structures of Fe_3O_4 MNPs and CS-Fe_3O_4 MNPs were characterized by X-ray diffraction (XRD), see in Fig. 1(a). No obvious diffraction peak is detected, but only a broad hump at around 2θ = 20.5° in the XRD pattern of chitosan, indicating that raw chitosan has an amorphous structure. Fig. 1a shows the XRD pattern of synthesized crystalline iron oxide (Fe_3O_4) nanoparticle. Fe_3O_4 exhibits a strong and sharp peak in the diffractogram. The diffraction peaks are obtained at (220), (311), (400), (422), (511), and (440) reflections, which are the characteristic peaks of magnetite (Fe_3O_4). The peaks in the XRD patterns of iron oxide are compared with the standards, (JCPDS card file no 77-1545), appeared in both bare and chitosan coated iron oxide nanoparticle. The diffractograms of chitosan are showed in Fig. 1(a)-(c), which revealed the characteristic peak between 9°-20° with comparable degree of crystallinity (Fig. 1a) [9]. The XRD spectrum of chitosan is reported in various works [10, 11] which show peaks of low intensity values around 2θ=10°, and a broad peak at around 2. XRD result exposed a broader peak in chitosan coated iron-oxide nanoparticle (CS-MNPs), which indicates the smaller crystallite size of the produced CS-MNPs than the one of the starting iron oxide nanoparticles. This may be due to the method of preparation of composite that causes the coating of chitosan on iron oxide nanoparticle. The crystallite size of both iron oxide and produced composite (CS-MNPs) were calculated from peaks at 2θ = 35.55° corresponding to iron oxide phase using Scherrer [Eq. (2)]

$$D_p = \frac{0.94\lambda}{\beta_{1/2} cos\theta} \quad (2)$$

where β is the FWHM of diffraction peak, λ is the wave length of X-ray (0.154 nm), L is the crystallite size, and θ is the Bragg peak position. The crystallite size was found to be 13.4 nm and 11.6 nm for MNPs and CS-MNPs respectively.

3.2 FT-IR analysis of CS-MNPs

FT-IR spectral analysis was employed to confirm the introduction of chitosan in Fe_3O_4 MNPs and the results are shown in Fig. 1(d). From the spectrum of chitosan, distinctive adsorption peaks appear at 1080 cm^{-1}, 1653 cm^{-1} (skeletal vibration involving the C–O stretching) and 2848 cm^{-1} (asymmetric stretching of the C–H bridge) [12]. The absorption peak appears at 3432 cm^{-1} assigned to the axial stretching vibration of O–H, superimposed to the N–H stretching band and inner hydrogen bonds of the polysaccharide, indicating the remaining amino and hydroxyl groups providing the coordination ability with metal [13]. The characteristic absorption peak appears at 580 cm^{-1} attributed to the Fe–O bond, confirming the existence of Fe_3O_4 [14]. In comparison with the chitosan spectrum, all the significant peaks of chitosan in the range of 900-1200 cm^{-1} are present in the CS-MNPs with a small shift. Moreover, one main characteristic peak appears at 1635 cm^{-1} due to the bands of COOM (M represents metal ions) groups, indicating the COOH groups of reaction of chitosan with the surface OH groups of Fe_3O_4 particles and resulting in the formation of the iron carboxylate [15]. In terms of CS-MNPs, the stretching vibrations of C=N bond at 1640 cm^{-1} could be observed. This peak clearly indicates the formation of the Schiff's base as a result of the reaction between the carbonyl group of glutaraldehyde and the amine group of chitosan chains. Additionally, the C–O absorption peak of secondary hydroxyl group becomes stronger and moves from 1080 cm^{-1} to 1060 cm^{-1}. Finally, the absorption of Fe–O bond of Fe_3O_4 still exists in the CS-MNPs, which further demonstrates that Fe_3O_4 nanoparticles have been wrapped by crosslinked chitosan, and the results are in good agreement with the conclusions of XRD analyses.

3.3 Scanning Electron Microscopy (SEM) of CS-MNPs

The typical SEM micrograph for the CS-MNP nanoparticles was shown in Fig. 2(a). From Fig. 2(a), it is clear that Fe_3O_4 nanoparticles bonded with chitosan have well-shaped spherical form with smooth surface, and its average particle size is approximately 16±3 nm with a narrow size distribution. The diameters of CS-MNPs are slightly bigger than the sizes of bare MNPs, due to the chitosan polymer coating on the surface of Fe_3O_4 particles (thickness of chitosan layer ~ 4 nm). The SEM image can be considered as the persuasive evidence that CS-functionalized MNPs are fabricated successfully. Meanwhile, it displays a slight reunion due to its smaller particle size, high surface energy and surface activity, thus, composite nanoparticles are prone to attract each other to formulate a stable state.

3.4 Thermal Gravimetric Analysis

The TGA-FTIR analysis of bare and chitosan coated Fe_3O_4 nanoparticles provides qualitative and quantitative information about the volatile components of the nanoparticles. The TGA curve shows that the weight loss of bare MNPs over the temperature range from 50°C to 800°C is about 7%. This might be because of the loss of residual water in the sample. The CS-MNPs gave their distinctive TGA curves, which can provide indications of the content of chitosan polymers [16]. The chains of chitosan degrade at about 210°C and the temperature of decomposition was around 600°C (Fig. 3).

The results of TGA-DTGA demonstrated that most of the organic layer, chitosan, was removed as CO_2 at high temperatures. TGA result indicated that the chitosan content increased at higher temperature and maximum amount is found about 29% by weight.

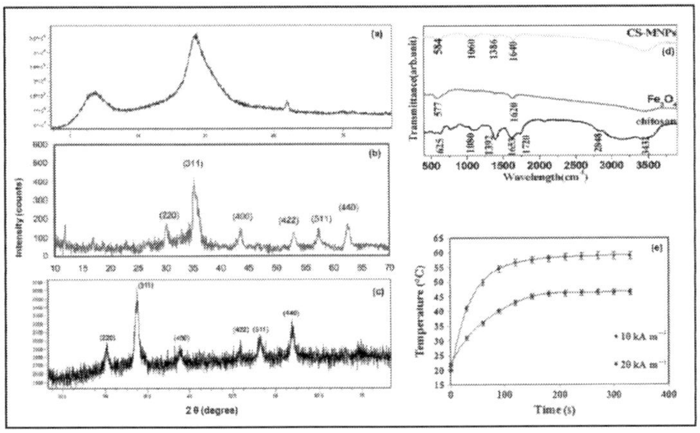

Figure 1. XRD pattern of synthesized crystalline iron oxide (Fe_3O_4) nanoparticle (a), chitosan (b), Chitosan-coated MNPs (c) and FT-IR spectra of CS-MNP nanoparticles (d), temperature vs. time for aqueous suspension of iron-oxide nanoparticles at different magnetic fields (e).

Figure 2. SEM images of CS-MNPs (a) and (b) Zeta potentials of bare Fe_3O_4 nanoparticles and CS-MNPs at different pH values.

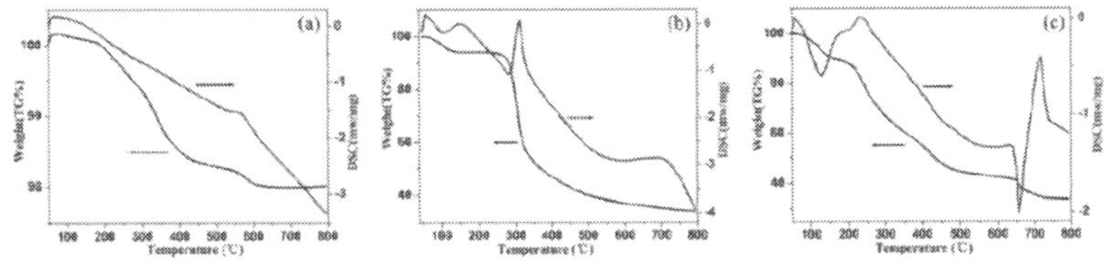

Figure 3. Thermogravimetric analysis curves of (a) Fe_3O_4 magnetic nanoparticles, (b) chitosan, and (c) CS-MNPs.

3.5 Zeta Potential In order to further prove the influence of chitosan polymers on the property of Fe_3O_4 nanoparticles, zeta potentials of bare MNPs and CS-MNPs were collected at different pH values

(Fig. 2(b)). The zeta potentials, as a measurement of repulsive or attractive forces between particles, play an important role to find not only the isoelectric point (pI), but also the colloidal dispersion stability. Zeta potentials of uncoated and chitosan polymer coated magnetic nanoparticles (0.1 mg·L^{-1}) were measured in 10^3 M NaCl aqueous solution at different pH values. The solution pH was adjusted by NaOH or HCl. Results show that PI values of unmodified and CS modified MNPs are 6.7 and 4.8, respectively. Upon surface modification with CS polymer containing multiple carboxyl groups, the PI shifts to a lower pH value because of the introduction of acidic surface groups.

3.6 Heating ability of CS-MNPs

The induction heating instrument used for calorimetric measurement is shown in figure 1(e). The temperature vs. heating time for aqueous suspension of chitosan-coated iron-oxide nanoparticles (50 mg ml^{-1}) exposed to the alternating magnetic field with different intensities (H = 10 and 20 kA m^{-1}, f = 100 kHz) are shown in figure 1. As shown in figure 1(e), after a short time (less than 3 min), the temperature of the suspension is saturated at T_{max} = 46 and 59°C for H = 10 and 20 kA m^{-1}, respectively. This saturation is due to the decreasing of the magnetization near the T_c and the balance of heat exchange with surrounding medium. Also, the heating curves show that the maximum temperature of the magnetic suspension falls within the desired temperature range (40-60°C) which suggests that the prepared samples are suitable for biomedical applications.

4. Conclusions

Magnetic nanoparticles were successfully synthesized by the co-precipitation method. This synthesis technique resulted in nanoparticles with an average size of 16 nm. The pure obtained nanoparticles were then successfully encapsulated with 4-nm-thick chitosan coating. The formation of chitosan on the surface of nanoparticles was confirmed by XRD, FTIR and SEM analyses. Cytotoxicity of bare and chitosan-coated nanoparticles was evaluated by MTT assay with MCF-10A and MCF-7 cell lines (data not shown). The results revealed that the both samples have negligible toxicity (below 500 μg ml^{-1}). Heating experiments at safe magnetic field (f=100 kHz, H=10-20 kA m^{-1}) revealed that the maximum achieved temperature of water stable chitosan-coated nanoparticles (50 mg ml^{-1}) is fully in agreement with cancer therapy and biomedical applications.

Acknowledgments: The research leading to these results has received funding from the European Union Seventh Framework Programme (FP7-PEOPLE-2013-COFUND) under grant agreement n° 609020 - Scientia Fellows.

References

1. Catalano E *Advanced Magnetic and Optical Materials* 2016 WILEY-Scrivener Publishing, USA 5 141-164
2. Mc Bain S C, Yiu H H and Dobson J 2008 *Int. J. Nanomed.* 3 169
3. Jordan A, Scholz R, Wust P, Fahling H and Felix R 1999 *J. Magn. Magn. Mater.* 201 413
4. Jordan A *et al.* 1999 *J. Magn. Magn. Mater.* 194 185
5. Majeed J, Pradhan L, Ningthoujam R S, Vatsa R K, *et al.* 2014 *Colloid Surf. B* 122 396
6. Subbiah R *et al.* 2012 *Carbohydr. Polym.* 89 (4) 1289-1297
7. Varshosaz J *et al.* 2016 *Int J Biol Macromol.* 93(Pt A):1192-1205.
8. Kim D, Zink B L, Hellman F and Coey J M D 1994 *Phys. Rev. B* 65 214424
9. Jiang Y *et al.* 2008 *Nanotechnology* 19 075714.
10. Rinaudo M 2006 *Progress in Polymer Science* 31(7) 603-632.
11. Kumar S *et al.* 2012 *Int. J. Mol. Sci.* 13(5), 6102-6116.
12. Li P *et al.* 2008 *Ind. Eng. Chem. Res.* 47 7700-7706
13. Zhang HF 2007 *React. Funct. Polym.* 67 322-328
14. Li GY *et al.* 2008 *Colloids Surf. A* 20 11-18
15. Badruddoza AZM *et al.* 2011 *J. Hazard. Mater.* 185 1177-1186
16. Peniche H et al. 2005 *J. Appl. Polym. Sci.* 98 651-657.

The INSIDE project: on-line monitoring and simulation validation with the in-beam PET scanner

V Ferrero[1,2] on behalf of the INSIDE collaboration

[1] Department of Physics, University of Torino, Via Pietro Giuria 1, Torino, Italy
[2] Istituto Nazionale di Fisica Nucleare INFN, Sezione di Torino, Via Pietro Giuria 1, Torino, Italy

E-mail: vferrero@to.infn.it

Abstract. The quality assurance of particle therapy treatment is a fundamental issue that can be addressed by developing reliable monitoring techniques and indicators of the treatment plan accuracy. Monitoring using Position Emission Tomography (PET) systems is the only *in-vivo* non invasive technique employed clinically and has been carried out in particle therapy since 1997. However, the PET monitoring of β^+ emitter isotopes is typically done after the treatment, resulting in a large fraction of lost data because of the isotopes rapid physical decay. The INSIDE collaboration has recently installed an in-beam PET scanner at the Italian National Center of Oncologic Hadrontherapy in Pavia, Italy. Here, there is an ongoing project in order to start testing the method on patients. This work focuses on the online performances of the scanner with clinical beams.

1. Introduction

Particle range verification is one of the key issues for treatment quality assessment in hadrontherapy, and requires the development of reliable monitoring techniques and indicators of the treatment plan accuracy. Charged particles release all of their energy inside the body, with a sharp ionization peak at the end of their range called the Bragg Peak (BP), so the only non-invasive way to verify the treatment accuracy is to collect the secondary radiation generated by the beam-tissue interactions. Positron Emission Tomography (PET) has been employed clinically since 1997 [1], and offers the possibility to directly monitor the spatial distribution of positron (β^+) emitter isotopes created during irradiation. Specifically, the emitted positron annihilates with an electron in the tissue, producing two back-to-back 511 keV photons, which in turn can interact and release part (i.e., Compton scattering) or all (photoelectric effect) of their energy in the detection crystals. Two detectors triggering in time coincidence allow the definition of the line of response associated to a certain annihilation event. To correctly identify the lines of response the scattered events need to be discarded, so only the 511 keV events (correlated to the photoelectric effect) are selected. The reconstructed isotopes activity distribution inside the patient thus gives an insight about the delivered treatment.

Defining the particle range from the activity measurement is not straightforward, because it has a non-trivial dependency on the ion species, the energy, and the tissue composition. However, from the activity distribution it is possible to infer the BP position since its profile along the beam path has a characteristic fall-off a few millimeters before the BP itself [2, 3, 4].

PET monitoring is typically performed after the treatment session, which leads to a large fraction of lost treatment data. The β^+ emitter isotopes produced have a rapid physical decay (their half lives last from few ns to a maximum of 20 min for the ^{11}C isotope) and can undergo functional pathways such as perfusion and diffusion. Moreover, the radioactivity produced during the treatment is of a maximum of few kBq/ml/Gy, almost two orders of magnitude below the typical activity concentrations in nuclear medicine (about 50 kBq/ml in hot spots), and the loss of short half time nuclides combined with the physiological washout makes after-treatment PET not optimal for *in-vivo* monitoring. Dedicated PET systems integrated within the treatment room are being investigated, so as to detect both the long and short half time nuclides radioactivity signal, and to also minimize biological washout.

The INSIDE collaboration has recently completed the construction of an in-beam PET detector, now under commissioning at the Italian National Center of Oncologic Hadrontherapy (CNAO) synchrotron facility in Pavia. The system characterization using proton and carbon beams is ongoing, and first measurements showed the capability of the in-beam PET to operate during the irradiation delivery and to reconstruct the beam-induced activity map in real-time.

2. The INSIDE in-beam PET

The INSIDE in-beam PET mechanical structure has been optimized to fit the nozzle of the CNAO treatment room, and can be easily placed in working position by aligning the markers on the PET heads with the positioning lasers of the room.

The system features two planar heads of 10×25 cm^2 area, mounted on a mobile support and placed at a 25 cm distance from the isocenter. Each head is made of 2×5 detection modules based on 16×16 matrices of segmented Lutetium Fine Silicate (LFS) scintillating crystals coupled to Hamamatsu silicon photomultipliers [5]. A detection module is read out by 4 TOFPET ASICs [6], which give information about the energy and the detection time of each event.

A data processing system based on field programmable gate array (FPGA) selects the 511 keV events and transmits them to a custom data acquisition system implemented on a server (32 Hyper Thread cores and 128 GB RAM) with a real-time high performance software that applies on-line data sorting and writes time-tagged coincidence data.

The PET images are reconstructed through a maximum likelihood expectation maximization algorithm [7] with 5 iterations. The Field Of View (FOV) of the reconstructed images is $16 \times 10.4 \times 26.4$ cm^3, with a voxel size of $1.6 \times 1.6 \times 1.6$ mm^3.

3. Experimental set-up

The INSIDE in-beam PET was tested at the CNAO facility with different irradiation configurations. The first characterization tests were performed with single pencil beams on both homogeneous and inhomogeneous polymethyl methacrylate (PMMA) phantoms [8, 9, 10], showing the system capability to acquire and reconstruct data in real-time.

The INSIDE in-beam PET experimental set-up is shown in Fig. 1. The PET scanner is positioned in the treatment room so that the center of the FOV is at the isocenter. The working distance between the heads is set at 50 cm, symmetrical with respect to the isocenter.

The aim of the test was to perform a first evaluation of the system online performances for a real treatment plan (i.e. actually delivered in clinical treatments) delivery, in view of future applications with patients. A $15 \times 15 \times 20$ cm^3 homogeneous PMMA phantom was irradiated with proton beams. Two real Treatment Plans (TP A and TP B), were delivered twice each, respectively called TP A1 and TP A2, TP B1 and TP B2. The phantom was changed after each acquisition to avoid residual activity from the previous irradiation. Each phantom was positioned with the long side along the beam direction and it was centered at the isocentre.

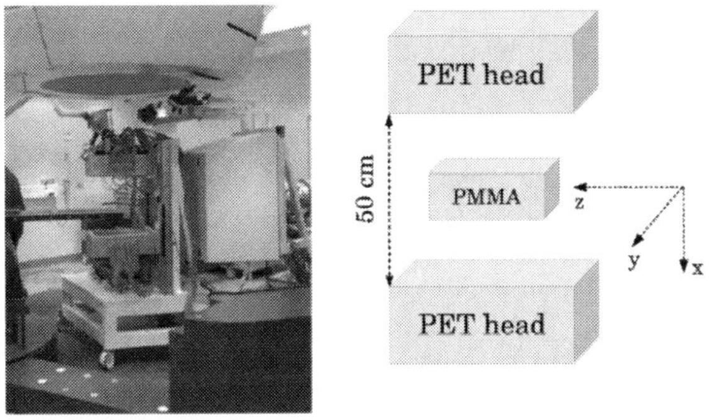

Figure 1. Left: the INSIDE in-beam PET installed in one of the CNAO treatment rooms. Right: schematic picture of the experimental set-up. The beam direction is parallel to the positive z axis.

The total number of particles delivered per treatment was 38×10^9, resulting in about 1 Gy of dose. The energy ranges for each plan are reported in Tab. 1. The maximum single event rate in the photopeak window was 13 MHz, with a coincidence time window set at 2 ns. The irradiation lasted about 4 min, but data were also acquired in the after-treatment for about another 6 min.

Table 1. Range energies for two real Treatment Plans (TP), protons.

TP	E_{min} (MeV/u)	E_{max} (MeV/u)
A	83	150
B	62	129

4. Simulations

Since the distributions of β^+ radioactivity and dose are not directly comparable, the quality assessment of charged particle therapy requires the prediction of the β^+ activity distribution from the treatment planning.

A dedicated simulation has been developed in FLUKA [11] and it comprises two steps. First, the primary beam is simulated with reduced statistics (about 1/100 of the primaries released in the whole treatment, thus reducing the simulation time of about 70x), and the 3D β^+ emitter distribution is scored. Then, a full statistic Monte Carlo simulation is performed by extracting the emission time of all positrons.

The simulation accounts for both the INSIDE in-beam PET characteristics and Figs. of merit and the delivery informations coming from the CNAO Dose Delivery System (DDS) [12]. The DDS is integrated in the CNAO synchrotron and monitors the space and time structure of the beam for each delivery. The average beam structure in time is 1 s beam on (in-spill) followed by 3-4 s beam off (inter-spill). The CNAO beam pipe and nozzle have a known geometry and have been simulated in previous works [13].

A C++ tool is developed for post-processing the FLUKA events and obtaining a time-based signal. The simulated and experimental data are analyzed with the same parameters (e.g. equal coincidence time window, energy window, and acquisition time).

The simulation plays a central role for in-beam PET clinical use, because the simulated image represents the prior to compare with the experimental image.

5. Activity comparison

The acquired data activity distribution was reconstructed by considering inter-spill and after-treatment only for a total of 700 s in all cases. The irradiation settings described previously for TP A and TP B were reproduced in the FLUKA simulations and the activity was reconstructed with the same parameters as the data.

Fig. 2 shows the activity profiles for the experimental data and the simulation for all the TPs under consideration. The experimental activity image is also encapsulated within the plots. The beam is directed along the positive z axis as shown in Fig. 1.

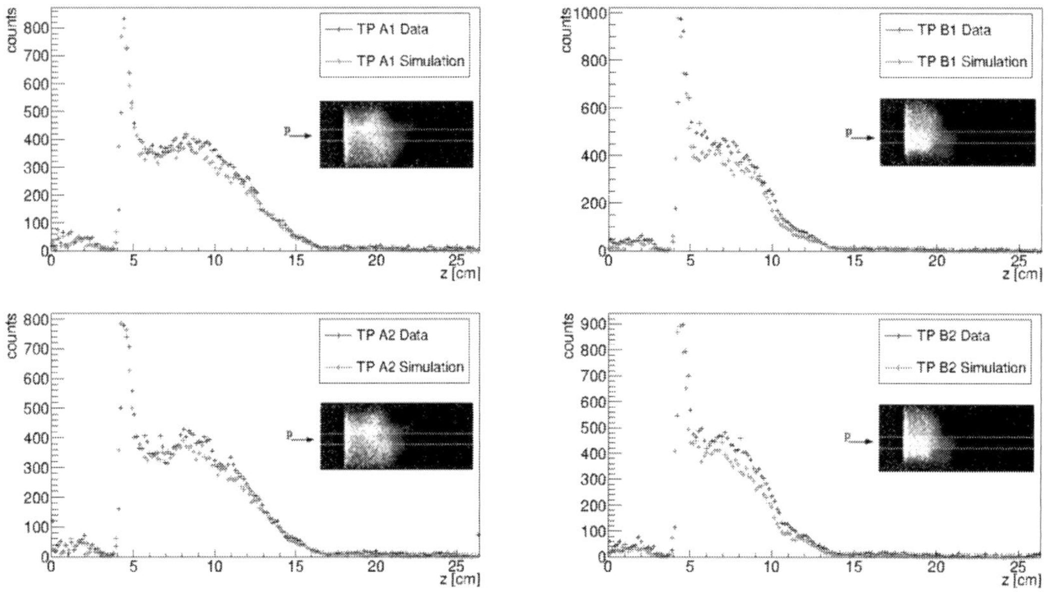

Figure 2. Experimental (black curves) and simulated (red curves) activity profiles of real treatment plans (TP). The energy ranges are reported in Tab. 1. Each TP configuration was delivered twice. The central section parallel to the PET heads of the reconstructed experimental activity image is also shown. The beam enters from the left, in the direction of the positive z coordinate.

The profiles are calculated along the beam axis, in a square area of 0.96 × 0.96 cm^2 in the transverse plane from the center of the image. The images, and thus the activity profiles, are not normalized. The plots in Fig. 2 show a satisfactory agreement between data and simulations for all the cases under consideration. The shape of these activity profiles is influenced by the chosen treatment delivery modality, in which there is a Spread Out Bragg Peak (SOBP) that depends on the energy ranges used and the material traversed. In previous works [8 - 10] the activity measurement was validated by evaluating the distal fall off of the activity profile and comparing it to the expected Bragg peak position. In the case of a SOBP though, the distal fall off is not as clearly defined, and a more complex analysis is needed. For this reason, a 3D comparison was implemented with ITK [14] libraries, applying a median filter with a 5 × 5 × 5 kernel to the whole FOV to reduce the noise, followed by a threshold filter to select only the voxels with values higher than 10% with respect to the maximum relative intensity. To further eliminate any remaining background spots, smoothing, erosion and dilation filters are applied. The DDS information about the proton beam position in the transverse plane is also applied to select the most meaningful volume of the image where the TP was actually delivered and ignore eventual artifacts from the reconstruction algorithm. Through this process a selection of the activity is made for all the images. The image activity range is then calculated by considering the depth along the z axis contained in the selected volume. The resulting distribution of the activity range of one image is then subtracted to the distribution of the image to be compared. The activity depth difference histogram is shown in Fig. 3 for the two repeated acquisitions of the TP A (TP A1 and A2), and the simulation. The corresponding activity images are also reported with the contours of the selected volume (TP A1 in white, TP A2 in green, simulation in red).

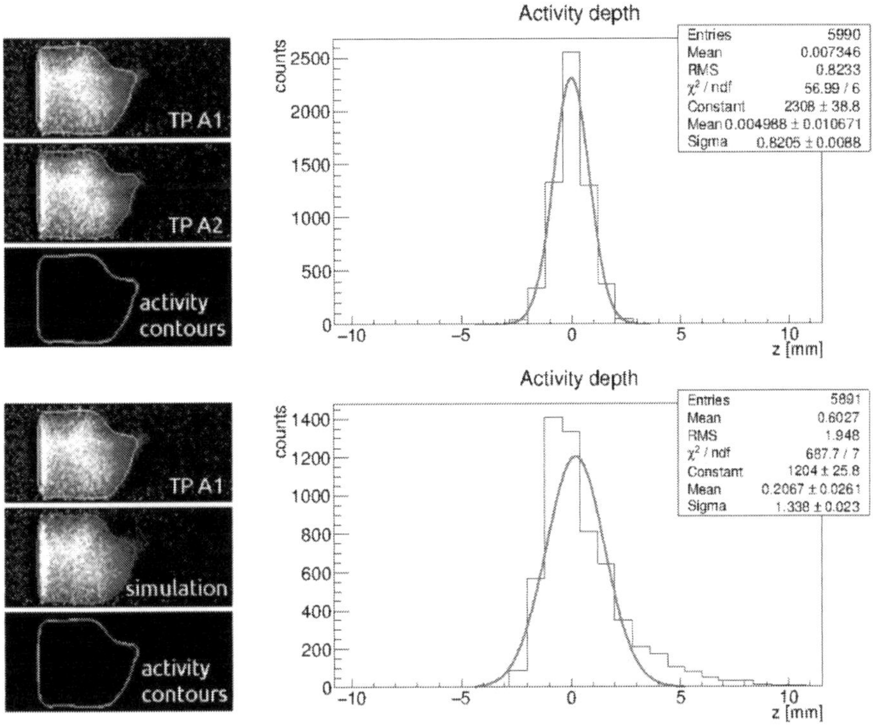

Figure 3. Example of activity depth difference histogram for the two experimental acquisitions of the TP A (TP A1 and TP A2, upper panel) and the TP A1 with the simulation (lower panel). The activity images in the plane parallel to the PET heads are shown with the contour of the selected volume: TP A1, white; TP A2, green; simulation, red. The superimposed contours of the compared images are also shown.

The agreement between TP A1 and TP A2 3D activity distributions results to be of (0.82 $\pm$ 0.01) mm, while the agreement of the TP A1 with the simulation is of (1.34 $\pm$ 0.02) mm. The activity depth difference histogram between data and simulation is not symmetric, with a tail on the right side. That is because the activity distribution of the simulation is slightly shorter than that of the data. Further studies will be carried out to improve the simulation, and to better characterize the activity comparison. Nonetheless, an agreement within 1.5 mm between the expected and experimental activity distribution is way below the safety margins (3 mm). The same analysis carried out comparing the experimental TP B and the correspondent simulation showed similar consistency.

These results are a milestone for the INSIDE project, because they prove the capability of the INSIDE scanner to acquire real-time images with a precision within 1 mm. Moreover, they validate the simulations with which the experimental data must be compared. Knowing the expected activity distribution, the on-line validation of the treatment deliver becomes possible. These results pose very promising prospects for in-beam PET monitoring and quality assurance in particle therapy.

6. Conclusions

Particle range monitoring is a key issue to improve quality assurance in particle therapy. In this work the INSIDE in-beam PET scanner response was tested for repeated irradiations on PMMA phantoms with clinical proton beams. Comparing the detected 3D activity distributions of two different experiments, we found them to be in agreement within 1 mm, while, when comparing the experiments with the simulated analysis, we found them to be consistent within 1.5 mm. A satisfactory qualitative agreement was also found for the activity profiles along the beam direction.

The detected 3D activity distribution was found to be in agreement within 1 mm for the experimental data comparison, and within 1.5 mm for the experimental and simulated analysis.

In the near future the INSIDE in-beam PET will be further characterized in order to start testing the method on patients at CNAO.

Acknowledgments

The INSIDE project was funded by the Ministry of Education, Universities and Research under the Italian national research program PRIN MIUR 20102011 2010P98A75.

References

[1] Enghardt W *et al.* 1999 *Nuc. Phys. A* **654**
[2] Pawelke J *et al.* 2007 *Proc. IBIBAM (Heidelberg)* 97–105
[3] Kraan A C *et al.* 2014 *Phys. Medica* **30** 559–69
[4] Parodi K 2015 *Med. Phys.* **42** 7153–68
[5] https://www.hamamatsu.com
[6] Rolo M D *et al.* 2013 *J. Instrum.* **8.02** C02050
[7] Camarlinghi N 2014 *J. Instrum.* **9.04** C04005
[8] Fiorina E *et al.* 2015 *Proc. IEEE (Strasburgo)*
[9] Piliero M A *et al.* 2016 *J. Instrum.* **11(12)** C12011
[10] Bisogni M *et al.* 2017 *J. Med. Imaging* **4(1)** 011005
[11] Bohlen T *et al.* 2014 *Nucl. Data Sheets* **120** 211–14
[12] Giordanengo S *et al.* 2015 *Med. Phys.* **42(1)** 263
[13] Molinelli S *et al.* 2013 *Phys. Med. Biol.* **58** 3837–47
[14] https://itk.org

IOP Conf. Series: Journal of Physics: Conf. Series **841** (2017) 012012 doi:10.1088/1742-6596/841/1/012012

Optimal configuration of a low-dose breast-specific gamma camera based on semiconductor CdZnTe pixelated detectors

B Genocchi, O Pickford Scienti and DG Darambara

Joint department of Physics, Division of Radiotherapy and Imaging, The Institute of Cancer Research and the Royal Marsden NHS Foundation Trust, London UK

E-mail: barbara.genocchi@icr.ac.uk

Abstract. Breast cancer is one of the most frequent tumours in women. During the '90s, the introduction of screening programmes allowed the detection of cancer before the palpable stage, reducing its mortality up to 50%. About 50% of the women aged between 30 and 50 years present dense breast parenchyma. This percentage decreases to 30% for women between 50 to 80 years. In these women, mammography has a sensitivity of around 30%, and small tumours are covered by the dense parenchyma and missed in the mammogram. Interestingly, breast-specific gamma-cameras based on semiconductor CdZnTe detectors have shown to be of great interest to early diagnosis. Infact, due to the high energy, spatial resolution, and high sensitivity of CdZnTe, molecular breast imaging has been shown to have a sensitivity of about 90% independently of the breast parenchyma. The aim of this work is to determine the optimal combination of the detector pixel size, hole shape, and collimator material in a low dose dual head breast specific gamma camera based on a CdZnTe pixelated detector at 140 keV, in order to achieve high count rate, and the best possible image spatial resolution. The optimal combination has been studied by modeling the system using the Monte Carlo code GATE. Six different pixel sizes from 0.85 mm to 1.6 mm, two hole shapes, hexagonal and square, and two different collimator materials, lead and tungsten were considered. It was demonstrated that the camera achieved higher count rates, and better signal-to-noise ratio when equipped with square hole, and large pixels (> 1.3 mm). In these configurations, the spatial resolution was worse than using small pixel sizes (< 1.3 mm), but remained under 3.6 mm in all cases.

1. Introduction

During the last decade, molecular breast imaging has been shown to be one of the most powerful techniques for improving early diagnosis and detection of lesions, especially in dense breast tissue, where both sensitivity and specificity decrease. Breast-specific gamma-cameras for SPECT (Single Photon Emission Computed Tomography) imaging, based on semiconductor CdZnTe (CZT) detectors, have already been introduced into clinical practice and have been investigated in various works [1 - 4]. However, there are many issues still to consider in particular in the case of aggressive breast cancers, for which earlier detection, and diagnosis are associated with a higher probability of successful intervention. A major requirement to allow for earlier detection and reliable diagnosis of a breast lesion is the development of detectors with high spatial and energy resolution. The main factors that determine the spatial resolution of a system are the collimation used (a thicker collimator produce a greater breast - detector separation), the detector geometry (pixel size, pitch and thickness), and intrinsic detector properties (intrinsic spatial and energy resolutions). CZT is a material of particular interest in semiconductor detector field due to its very high energy resolution, which leads to high spatial resolution, and high detection efficiency. A high energy resolution, indeed, allows to apply a narrow energy window around the photopeak, leading to a better rejection of scattered photons, and therefore to an increased image contrast, enhancing the detection of smaller tumours. The aim of this work is to determine the optimal combination of the detector pixel size, the hole shape, and collimator material in a low dose dual head breast-specific gamma-camera, based on a CZT pixelated detector at 140 keV in order to achieve high count-rate and the best possible image spatial resolution. The optimal combination was studied by using the state-of-the-art Monte Carlo code Geant4 Application for Emission Tomography (GATE) [5] to simulate the 140 keV photons in order to obtain count rates, energy resolution, and spatial resolution of the different detector-collimator configurations considered in this study. Further, initial simulations were carried out to investigate the induced signal in CZT detectors using the Finite Element method (COMSOL Multiphysics [6]), in order to take into account the effect of the electric field in the pixels, and in the inter-pixel separation on the charge induction efficiency (CIE) with the aim to incorporating the CIE into the GATE simulations in the next stage of the project [7].

2. Methods

Two different types of Molecular Breast Imaging (MBI) systems are currently used in clinical trials [3]. One is the *GE Healthcare Discovery NM750b* which utilises two opposite 80 x 80 array of CZT elements with dimensions of 2.5 x 2.5 x 5 mm^3, giving a field of view (FOV) of 20 x 20 cm^2. These detectors are equipped with a square-hole general purpose lead collimator with hole dimensions of 2.3 x 2.3 x 3.5 mm^3 and septal thickness of 0.2 mm. The energy resolution of this system is 7.8% at the photopeak of ^{99m}Tc (140 keV). The second MBI system is the *Gamma Medica LumaGem*, where each detector is made of an array of 96 x 128 CZT elements with dimensions of 1.6 x 1.6 x 5 mm^3, giving a 15 x 20 cm^2 FOV. The LumaGem detectors are equipped with high-sensitivity hexagonal hole lead collimators with a parallel-to-parallel hole diameter of 2.54 mm, hole length of 25 mm, and septal thickness of 0.3 mm. This system has an energy resolution of 3.9% at 140 keV. We have thus simulated six different pixel sizes from 0.85 mm to 1.6 mm with an overall FOV of about 20.5 x 16.5 cm^2. For each pixel size two different collimator materials, lead ($\rho = 11.4\ g/cm^3$) and tungsten ($\rho = 19.4\ g/cm^3$), and two different hole shapes, hexagonal and square, were investigated. In this work each hole is aligned with each pixel as shown in Fig. 1c and Fig. 1d. The simulated system has two opposite heads, each equipped with a CZT detector (yellow) and a collimator (white), as shown in Fig. 1a and Fig. 1b. A point source of photons at 140 keV has been modelled, with an activity of 1.5 MBq. The simulated scanning time was 5 min. To obtain the optimal configuration, detection efficiency, photopeak-to-total ratio, energy resolution, and system spatial resolution have been studied. Detection efficiency has been obtained dividing the total number of detected photons by the activity and the simulated scanning time; the photopeak-to-total ratio has been obtained dividing the number of photons in the photopeak by the total number of detected photons. The energy resolution has been obtained from the full width half maximum (FWHM) of the energy spectrum, and the system spatial resolution has been obtained from the FWHM of the distribution of the source position.

Figure 1: Mediolateral view (ML) (a) and side view (b) of the gamma camera modelled with GATE, and view of the hexagonal hole (c) and square hole (d) centered with the pixel (outer square).

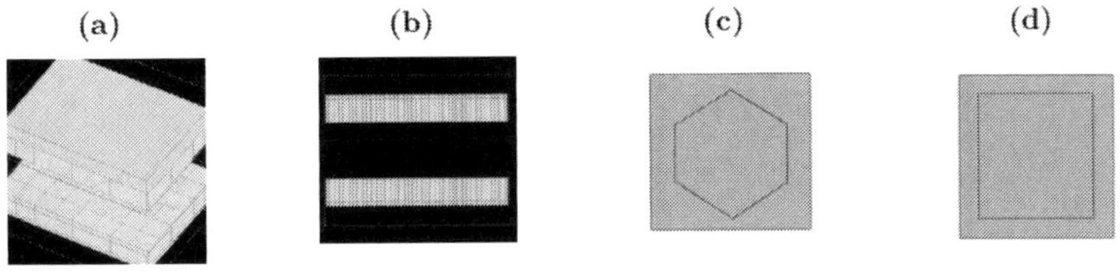

Figure 2: 3D map of the charge induction efficiency of the central pixel. The anode is on the left of the map.

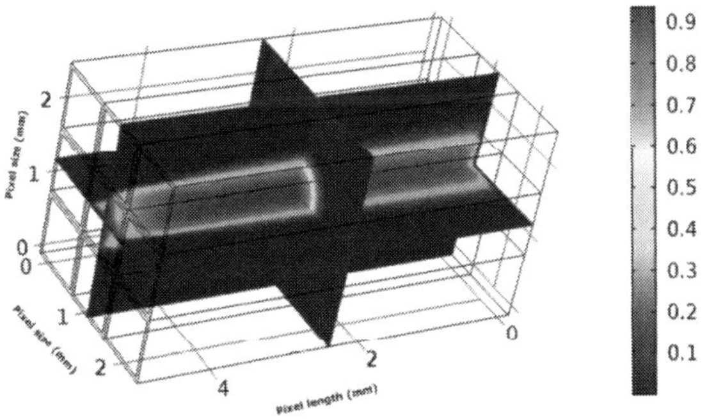

3. Results

In order to develop a complete model of the system, the transport of generated charges in the detector by drift and diffusion, and the collection of charges at the contacts and noise associated with carrier production and trapping need to be included in the simulations. These processes were modeled with COMSOL Multiphysics, a finite element method code.
A block of 3x3 pixels was modeled with a voltage of −300 V applied to the pixels. Then the adjoint electron continuity equation [8] was mapped as in Eq. 1:

$$\frac{dn^+}{dt} = \mu_n \nabla\varphi \cdot \nabla n^+ + \nabla \cdot (D_n \nabla n^+) - \frac{n^+}{\tau_n} + \mu_n \nabla\varphi \cdot \nabla\varphi_k \qquad (1)$$

where n^+ is the carrier concentration, $\mu_n \nabla\varphi \cdot \nabla\varphi_k$ is the electron generation term, $D_n \nabla n^+$ is the diffusion term, $\mu_n \nabla\varphi \cdot \nabla n^+$ is the drift term, $\frac{n^+}{\tau_n}$ is the trapping term. The map in Fig. 2 shows a high dependency of the charge induction efficiency (CIE) on the position in the pixel: as expected the CIE is higher at the centre of the pixel where there is enough space for a good charge collection. As shown in the left part of Fig.2, if a gamma-ray interacts near the boundaries between pixels or near the anode, there is a high probability that the products of the interaction escape into the neighbors pixels or escape the detector altogether, decreasing the charge induction efficiency.

Running simulations with GATE code, the *detection efficiency* has been investigated in term of *counts per second per MBq (cps/MBq)*. It was shown that the detection efficiency increases as

Table 1: The detection efficiency, expressed in term of counts per second per MBq, for square and hexagonal collimator holes depends on the pixel size.

Detection Efficiency (cps/MBq)

	Tungsten		Lead	
Pixel size (mm)	Hexagonal hole	Square hole	Hexagonal hole	Square hole
0.85	19.77	30.22	19.61	31.61
1	29.50	56.44	30.35	56.98
1.15	42.58	86.80	42.60	88.16
1.3	63.48	134.26	63.48	134.36
1.45	83.17	186.73	84.42	188.98
1.6	105.65	250. 91	107.49	257.83

Table 2: Photopeak to total counts ratio for the four configurations considered in this study as a function of the pixel size.

Photopeak To Total Counts Ratio (%)

	Tungsten		Lead	
Pixel size (mm)	Hexagonal hole	Square hole	Hexagonal hole	Square hole
0.85	44.80	59.48	44.48	60.45
1	60.56	72.52	60.70	72.43
1.15	69.70	78.00	69.76	77.81
1.3	76.29	81.90	76.29	81.94
1.45	79.27	83.67	79.78	83.83
1.6	81.42	84.99	81.26	85.10

Figure 3: Detection Efficiency (left) and Photopeak to Total Counts ratio (right) for the four configurations considered in this study as a function of the pixel size.

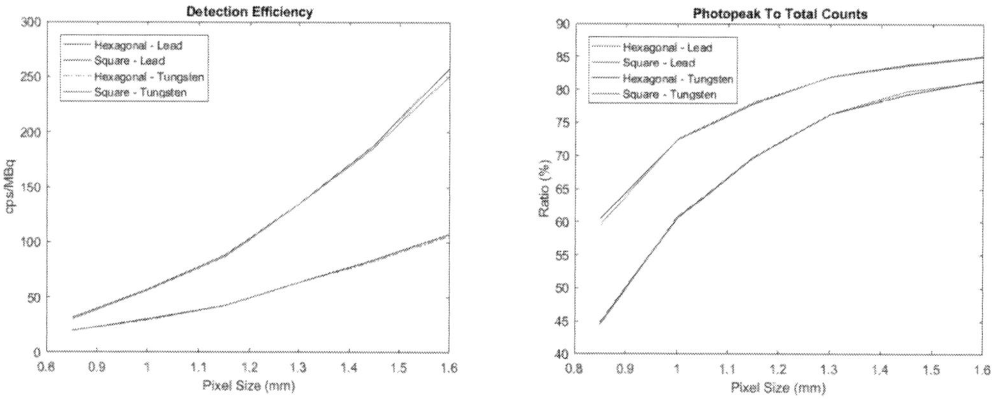

the pixel size increases, and moreover it is higher when square holes are used, as can be seen in Tab. 1 and Fig. 3.

The final image will be reconstructed only from those photons that have released all their energy in the same pixel, and therefore the *photopeak counts to total ratio* is an important parameter for the image reconstruction to be considered. Also in this case the photopeak-to-total ratio increases with the pixel size and it is shown to be higher when square holes are used, as shown in Tab. 2 and Fig. 3.

The *energy resolution* has also been investigated for the four different system configurations. We recall that a high energy resolution means that a narrow energy window around the photopeak can be applied leading to the reduction of the noise coming from Compton scattering. It was shown that the energy resolution increases with increasing pixel size, as can be seen in Tab. 3 and Fig. 4.

The *system spatial resolution* is also an important parameter to take into account as it affects the capability to resolve small lesions. It was shown that the system spatial resolution increases

Table 3: Energy resolution for the four configurations considered in this study as a function of the pixel size.

Energy Resolution (%)

	Tungsten		Lead	
Pixel size (mm)	Hexagonal hole	Square hole	Hexagonal hole	Square hole
0.85	1.33	1.34	1.34	1.32
1	1.36	1.37	1.36	1.37
1.15	1.50	1.49	1.50	1.49
1.3	1.63	1.64	1.63	1.63
1.45	1.66	1.67	1.67	1.67
1.6	1.71	1.70	1.71	1.71

Table 4: System spatial resolution for the four configurations considered in this study as a function of the pixel size.

System Spatial Resolution (mm)

	Tungsten		Lead	
Pixel size (mm)	Hexagonal hole	Square hole	Hexagonal hole	Square hole
0.85	2.07	2.16	2.04	2.13
1	2.21	2.22	2.23	2.22
1.15	2.49	3.05	2.52	3.04
1.3	2.54	2.95	2.55	2.95
1.45	2.81	3.39	2.82	3.39
1.6	2.80	3.51	2.83	3.53

Figure 4: Energy resolution (left) and system spatial resolution (right) for the four configurations considered in this study as a function of the pixel size.

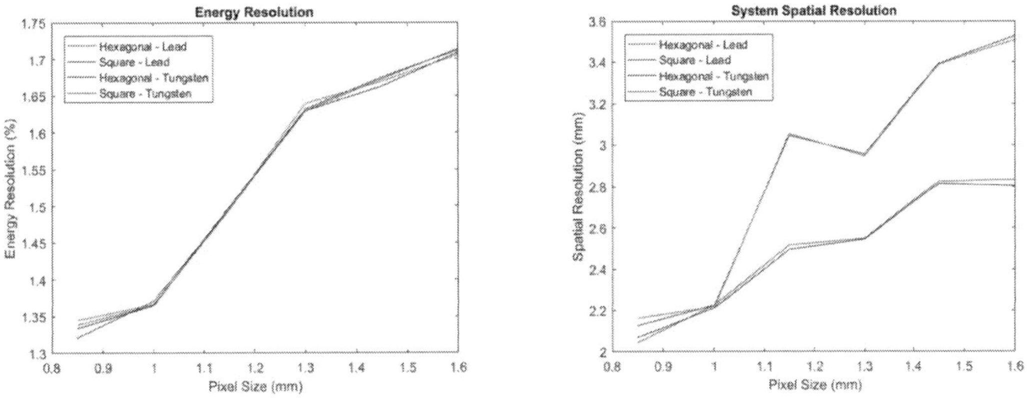

with increasing pixel size, as well as it is higher with hexagonal holes, as can be seen in Tab. 4 and Fig. 4.

4. Conclusions

As it was expected, small pixels (< 1.3 mm) have shown fewer photopeak counts and higher spatial resolution for both collimator materials, lead and tungsten, and both hole shapes, hexagonal and square. The best system spatial resolution achieved is 2.04 ± 0.03 mm with 0.85 mm detector pixel size and a lead collimator with hexagonal holes. However, for this configuration the photopeak to total ratio is low, 45 ± 3 %; this means having a very good image spatial resolution degraded by a low signal-to-noise ratio. For larger pixels (> 1.3 mm) the signal-to-noise ratio is about 80% and the system spatial resolution is $2.5 - 3$ mm depending on the pixel size. The system energy resolution achievable with CZT detectors is in the range $1.33 - 1.70$ % depending on the pixel size. For the level of activity of ^{99m}Tc used (1.5 MBq) and the simulated time (5 minutes) implemented in the current simulations, it was shown that

there is no clear difference between using tungsten or lead as collimator material. Therefore, further simulations are required to collect more statistics, and to explore any possible difference in the imaging performance. It is expected, in fact, that a system equipped with tungsten would have better imaging performance since tungsten is denser than lead and hence a larger number of photons, which do not pass straight through the holes, should be stopped, giving a better system spatial resolution.
It has to be noted that all the results obtained from the Monte Carlo simulations did not take into account the impact of the voltage applied, the point of interaction, trapping and detrapping of electrons and holes on the charge induction efficiency. In the near future the Multimodality Molecular Imaging (MMI) Team at The Institute of Cancer Research will continue working on a program to combine the results obtained from COMSOL with the ones obtained from the Monte Carlo modeling tool GATE. Moreover, in further work, an anatomically detailed anthropomorphic breast phantom (DeBRA) (Fig. 5) [9], developed by the MMI Team, will be inserted in the GATE simulations of the gamma camera to evaluate the image spatial resolution depending on the type and shape of tumour or on the tumour position in the breast. The DeBRA phantom has all the anatomical details of the breast: glands, lactiferous ducts, different shapes of tumours in different positions and pectoral muscle. Using DeBRA, it will be possible to evaluate the dose to the breast as a function of both the activity of the ^{99m}Tc injected to the patient and the time of breast scan. This will allow to optimise the system configuration trying to decrease the dose to the patients while keeping the image spatial resolution good enough to detect small lesions.

Figure 5: Lateral view (left) and cranio-caudal view (right) of the DeBRA phantom, developed by the Multimodality Molecular Imaging Team at ICR/RMH [9].

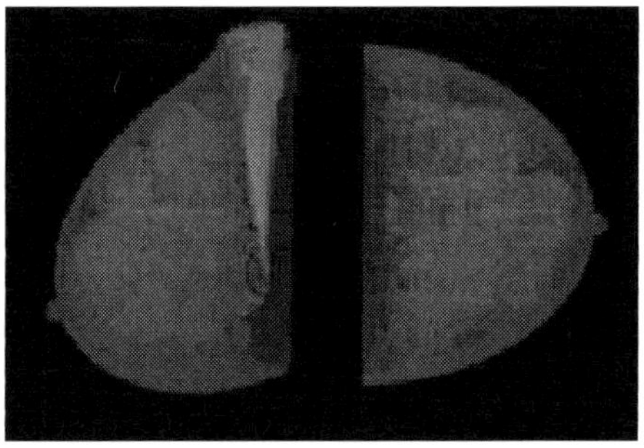

References

[1] Myronakis M, Zvelebil M and Darambara DG 2013 *IEEE NSS/MIC Conference Proceedings* 1-3.
[2] Myronakis M and Darambara DG 2011 *Med. Phys.* **38** (1) 455-467
[3] Weinmann AL, Hruska CB and O'Connor MK 2009 *Med. Phys.* **36** (3) 845-856
[4] Levin CS 2003 *Nucl. Inst. & Meth. in Phys. Res. A* **497** 60-74
[5] Jan S et al. 2004 *Phys. Med. Biol.* **49** 4543-4561
[6] https://www.comsol.com/
[7] Myronakis M, Zvelebil M, Darambara DG 2012 *JINST* **7**
[8] Prettyman TH 1998 *Journal of Radioanalytical and Nuclear Chemistry* **233** 257-264
[9] Ma AKW, Gunn S and Darambara DG 2009 *Phys.Med.Biol.* **54** 4533-4545

IOP Conf. Series: Journal of Physics: Conf. Series **841** (2017) 012013 doi:10.1088/1742-6596/841/1/012013

Computer-aided detection systems to improve lung cancer early diagnosis: state-of-the-art and challenges

A Traverso[1,2], E Lopez Torres[2], M E Fantacci[3] and P Cerello[2]

[1]Department of Applied Science and Technology, Polytechnic University of Turin, Turin, Italy
[2]Turin Section of INFN, Turin, Italy
[3]Pisa Section of INFN, Pisa, Italy

E-mail: alberto.traverso@polito.it

Abstract. Lung cancer is one of the most lethal types of cancer, because its early diagnosis is not good enough. In fact, the detection of pulmonary nodule, potential lung cancers, in Computed Tomography scans is a very challenging and time-consuming task for radiologists. To support radiologists, researchers have developed Computer-Aided Diagnosis (CAD) systems for the automated detection of pulmonary nodules in chest Computed Tomography scans. Despite the high level of technological developments and the proved benefits on the overall detection performance, the usage of Computer-Aided Diagnosis in clinical practice is far from being a common procedure. In this paper we investigate the causes underlying this discrepancy and present a solution to tackle it: the M5L WEB- and Cloud-based on-demand Computer-Aided Diagnosis. In addition, we prove how the combination of traditional imaging processing techniques with state-of-art advanced classification algorithms allows to build a system whose performance could be much larger than any Computer-Aided Diagnosis developed so far. This outcome opens the possibility to use the CAD as clinical decision support for radiologists.

1. Introduction

Lung cancer still represents a health issue in developed countries. In fact, it is the leading cause of cancer-related deaths both in Europe, and United States [1, 2], and it has a very low average 5-year survival rate, around 10-17%. Most of lung cancers are diagnosed in the late stages when the survival rate is very low if compared to the diagnosis in the preliminary stage (survival rate 45-50%). When lung cancer is detected in the early stage, treatment is very successful leading in most of the cases to the complete healing. However, preliminary stages of the disease are usually asymptomatic, making the early diagnosis a real clinical challenge. Relevant technological and clinical developments have been achieved in the last decades in order to improve the early diagnosis. In particular, screening lung cancer high-risk subjects (*e.g.* heavy smokers) with low dose Computed Tomography (CT) has been proved to reduce cancer mortality up to 20% in comparison to annual screening with chest radiography [3]. However, the detection of pulmonary nodules, the early manifestations of lung cancers, is a very challenging task. In fact, the identification of pathological Regions Of Interest (ROIs) in low-dose high resolution CT scans is a difficult and time consuming task for radiologists, mainly due to the high number of slices (on average 200/300 per patient) to be read. This task is even harder during a screening campaign, when millions of slices need to be analyzed as fast as possible, without missing any clinically relevant nodule. This situation represents a real

IOP Conf. Series: Journal of Physics: Conf. Series **841** (2017) 012013 doi:10.1088/1742-6596/841/1/012013

burden for radiologists and it has been called *data explosion* [4]. In addition, it has been proved [5] that radiologists' concentration decreases during the day with a substantial impact on the overall detection sensitivity. Supporting radiologists with tools for the automated detection of pulmonary nodules in chest CT scans can produce benefits in the overall detection sensitivity. For this reason, researchers have started developing Computer-Aided Diagnosis (CAD) systems. Several studies proved an increase of radiologist's detection sensitivity when supported by those systems [6, 7]. Despite these prominent results there is a big discrepancy between the level of developments achieved and the spread of CAD systems in clinical practice, which are rarely or not used in clinical routine. In this paper we highlight the main issues currently limiting the usage of CADs in clinical practice. In Sec. 2 we present M5L, our Web- and Cloud-based CAD system for the automated detection of pulmonary nodules in chest CT scans as possible solution to presented issues. In Sec. 3 we investigate state-of-the-art CAD systems presenting the results of the medical imaging challenge we organized to investigate new imaging processing techniques to improve the overall detection sensitivity.

2. The M5L on-demand lung CAD

2.1. Motivation

The standard approach to make CAD algorithms available in clinical routine of health facilities consists of deploying stand-alone workstations, together with a vendor-dependent Graphic User Interface (GUI). However, this solution presents several drawbacks: high fixed costs of the software licenses and the dedicated hardware and the rapid obsolesce of both. The software is usually protected by copyright and therefore it is difficult to access the real performances of the algorithms because no scientific publications are available. Additional costs could also be charged for the integration of mentioned workstations with, for example, the PACS (Picture Archiving Communication Software) already present in the hospital. Furthermore, the computational power required increases as a function of algorithm's complexity, often requiring very expensive hardware to be bought. Finally, if a CAD system has to be used during a massive screening campaign, where thousands of scans need to be analyzed as fast a possible, the flexibility in the computation resources becomes a mandatory requirement. Since screening campaigns are usually concentrated within a limited period and considering that the number of CTs which need to be analyzed is much larger than the workload in clinical routine, buying new powerful hardware for screening would not represent a cost-effective solution. For this reason, there have been some attempts to handle the analysis of medical images on distributed environments based on a GRID infrastructure [8]. Conversely, GRID computing is not suitable for most of the medical physics projects due to its rigidity and complexity of the infrastructure. Furthermore, the management of a GRID infrastructure requires dedicated man power for the service, which is usually not available and not cost-effective in a clinical structure. Finally, guaranteeing the possibility to share CAD results and more in general medical annotations between radiologists can improve the effectiveness of the diagnosis [9]. Unfortunately, the usage of stand-alone workstations makes almost impossible the possibility to allow multipanel annotations. In addition, small clinical facilities, which usually do not have a pool of resident expert radiologists could benefit from on-line consultation from radiologists of other institutions. The M5L collaboration tried to tackle all the previous issues introducing a new paradigm of CADs in clinical practice: the M5L on-demand CAD, which does not require to the hospital any additional software or hardware installation.

2.2. Material and Methods

The M5L on-demand CAD is based on three functional blocks:

- two algorithms providing automated nodule detection;

(a) The submission module. (b) The annotation module. (c) The review module

Figure 1: Screenshots of the M5L front-end.

- a web front-end;
- a cloud-back end.

M5L is a combination of two independent CAD systems: lungCAM [10] and VBNA [11] which have been extensively validated in [12]. The web front-end has been conceived to make M5L results available to radiologists without requiring any software installation. In addition, it allows to directly insert the medical annotations of the submitted cases. The web interface can be accessed at http:/m5l.to.infn.it using any browser by desktop, tablets, and mobile devices. The system has been built using the open source content manager DRUPAL [13]. Three contributed modules have been developed for CT submission, online insertion of medical reports by radiologists (not necessarily belonging to the same institution), and CAD results access or review. The system allows to use the CAD both as *concurrent-reader* or *second-reader* [14]. To tackle the issue of making CAD algorithms available without requiring any hardware installation we set up a dedicated Cloud back-end, within the INFN Computing Center in Turin, Italy. Cloud Computing is a model for enabling ubiquitous, convenient, on-demand network access to a shared pool of configurable computing resources (*e.g.* networks, servers, storage, applications, and services) that can be rapidly provisioned and released with minimal management effort or service provider interaction [15]. Our solution combines two basic concepts of Cloud Computing: Software as a Service (SaaS) and Infrastructure as a Service (IaaS). In fact, we provide to the users our software (M5L CAD), and give them the infrastructure (virtual machines) for CAD computations with the advantage that they do not have to take care of the maintenance.

2.3. Results

Presently, we can deploy up to 18 Virtual Machines, and a total number 48 of cores (processing units). The average processing time for each case is around 15 minutes. In order to have the possibility to scale resources, an elastic cluster was created based on CERNVM Online system [16]. Dedicated developed open-source tools [17] [18] guarantee a real-time monitoring and optimization of computing resources, so-that the user always has an up-to-date responsive system. In order to clinically validate the M5L CAD we set up a collaboration with the Institute for Cancer Research and Care (IRCCS) in Candiolo, Turin (Italy) [19]. Preliminary results show that around 20% of nodules, originally missed by radiologists during the first unassisted reading, are added by our CAD with a remarkable benefit on the overall detection sensitivity.

3. Comparing and combining algorithms: the LUNA16 challenge

A large number of CAD algorithms for the automated detection of pulmonary nodules have been discussed in the literature [20]. However, the direct comparison of the performances of those systems is not a trivial tasks. In fact, the sensitivities can vary tremendously depending on the datasets used for training and validation. A step toward an objective evaluation of

CAD systems was the ANODE09 study [21]. We organized Lung Nodule Analysis (LUNA16) an international medical challenge in order to provide an objective evaluation framework for automatic nodule detection algorithms using the largest publicly available reference database, and to investigate the combination of traditional imaging processing techniques with recent machine learning algorithms.

3.1. Material and Methods
The dataset used for LUNA16 is formed by 888 scans taken from the LIDC-IDRI database [22], the biggest publicly available collection of clinical and screening annotated chest CT scans. In this database, several scanner manufacturers with different reconstruction parameters are represented, the dataset suits perfectly for CAD validation purposes. We considered as reference standard all the nodules greater than 3 mm in diameter, and annotated by at least three out of four radiologists, leading to a total of 1186 nodules. A wide range of pulmonary nodules is represented, making the automated detection a high-profile challenge. Other findings (nodules annotated by 1 or 2 radiologists, nodules smaller than 3 mm, and not nodules) were considered as irrelevant findings and not considered in the performance evaluation. We decided to divide the challenge into two tracks:

- NDET (Nodule Detection) track: participants were required to develop a complete CAD system. The input for the participants to this track are the only raw CT images;
- FPRED (False Positive Reduction) track: participants were required to propose solutions for false positive reduction stage. They were given a set of candidates, including both true nodules and false positives, and they had to provide the classification.

The FPRED track has been organized to attract machine learning groups which have not been working directly on medical imaging projects, but with the expertise on most recent machine learning techniques. Participants were able to download the full dataset, together with the list of annotations from the website of our challenge (luna16.grand-challenge.org). To avoid biases due to an open challenge (no separate training / testing datasets), they were all required to perform a 10-folds cross validation [23]. We asked participants to submit their results as a comma separated value file, which includes the list of CAD marks. For each CAD mark the position (x,y and z) coordinates, the reference to the image and the corresponding nodule probability are provided. The evaluation of true positives (TP), false positives (FP) and irrelevant findings has been performed as explained in [21]. The results are evaluated in terms of FROC (Free Response Operating Characteristic) analysis [24], where the sensitivity is plotted as a function of the average number of false positives per scan. 7 values of sensitivities at seven predefined FP rates are extracted. The overall score, called CPM (Competition Performance Metric) is then the average of previous scores, ranging from 0 (worst possible system) to 1 (best possible system).

3.2. Results
When multiple systems focus on different techniques to approach a problem, a proper combination could produce a better performance than each stand-alone system taking part in the combination. Finding a method to combine CADs without accessing their internal properties (i.e the classifier) is far from being a trivial task. Since we could only access the list of CAD marks we defined a method which makes use only of those information, similar to [21]. In addition, in order to make a fair combination, systems with high performances are weighted more than systems with a lower performance. Since the output from candidate detectors is the input for the false positive reduction systems, we combined the candidate detectors from the top performing CAD systems in NDET. We asked then to the participants to the FPRED track to run their algorithms on this combined list, and we evaluated the top performing systems

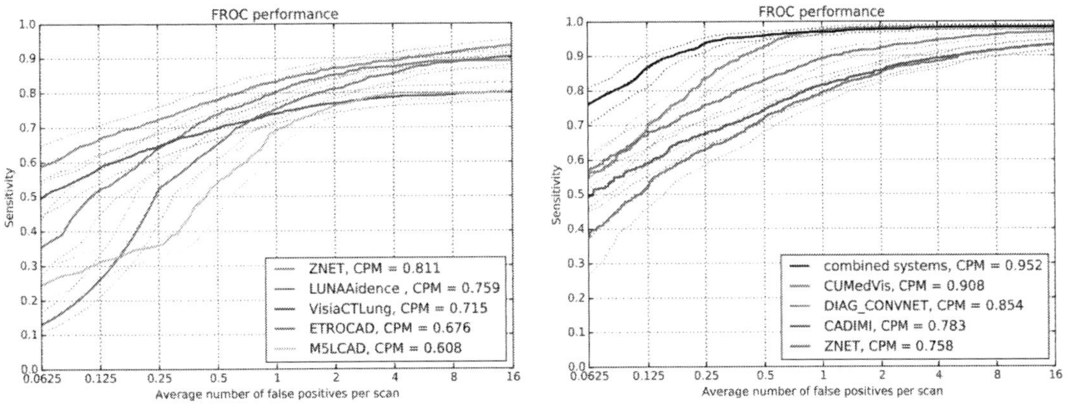

(a) (b)

Figure 2: FROC curves of the systems in (a) nodule detection track and (b) false positive reduction track. Dashed curves show the 95% confidence interval estimated using bootstrapping.

and their combination. The overall CAD is then a system which has a candidate detector, formed by a combination of different candidate detector algorithms, and a classifier, which is the combination of different false positive reduction systems. Fig. 2a shows the top performing algorithms in the NDET track. The best algorithm has a CPM score of 0.811. There are large differences in terms of sensitivity between the different systems at low false positives rates per scan. The detection of nodules with a very low FP rate is a very demanding challenge. It is worth to notice that, apart from the best system, all the other systems make use of traditional medical imaging analysis techniques. Conversely, the best system makes use of Convolutional Neural Networks (CNNs) [25] for both candidate detection, and false positive reduction. Fig. 2b shows the top performing systems in the FPRED track. The best system has a CPM score of 0.908. CNNs are used as the prediction model for all systems, showing the recent trend to apply deep learning in the medical image analysis domain. As the underlying method is similar, one could hypothesize that there could be little or no benefit when these systems are combined. On the contrary, the combination (black curve in Fig. 2b) approaches the maximum achievable sensitivity of 98.3% and saturates already starting from two false positive per scans. Although all the methods are based on CNNs, the differences in network parameters, such as the selected architectures and the input patches, makes these systems complementary in terms of the prediction.

4. Discussion

To improve the usage of CADs in clinical practice we developed M5L, a lung CAD on-demand based on the SaaS and IaaS concepts. The proposed approach solves the issue of making CAD functionality available to users without requiring any software installation or dedicated hardware, and, if properly scaled, provides the necessary amount of computational power for a large scale service. Its clinical validation on oncological patients undergoing staging or re-staging shows a positive impact of our system in the radiologists' detection sensitivity. In addition, we investigated how the combination of traditional imaging processing techniques with state-of-the-art classification algorithms could improve the overall detection sensitivity. Results show how a CAD based on this combination performs with a very high sensitivity together with a very high specificity (92% at 0.25 FP findings per scan). This solution opens the possibility to use the

CAD as *independent reader* in clinical practice as diagnosis decision support. Finally, all the combined features presented in Sec. 2 provide an innovative approach, that could be extended to other applications, allowing the possibility to combine different algorithms (as shown in Sec. 3) within the same infrastructure with a limited effort.

Acknowledgments
The authors would like to thank the technical staff of the INFN Computer Centre in Torino, for their contribution in keeping the infrastructure functional at all times, the Radiology Department of the IRCCS in Candiolo for their contribution to the clinical validation and the Diagnostic Image Analysis Group of the Radboud UMC in Nijmegen for their contribution to the organization of LUNA16.

References
[1] Malvezzi M, Bertuccio P, Rosso T, Rota M, Levi F, La Vecchia C and Negri E 2015 *Ann Oncol* **26** 779–786
[2] Siegel R L, Miller K D and Jemal A 2015 *CA: a cancer journal for clinicians* **65** 5–29
[3] Kramer B S, Berg C D, Aberle D R and Prorok P C 2011 *Journal of medical screening* **18** 109–111
[4] Rubin G D 2000 *European journal of radiology* **36** 74–80
[5] Brown M S, Goldin J G, Rogers S, Kim H J, Suh R D, McNitt-Gray M F, Shah S K, Truong D, Brown K, Sayre J W *et al.* 2005 *Academic radiology* **12** 681–686
[6] Kobayashi T, Xu X W, MacMahon H, Metz C E and Doi K 1996 *Radiology* **199** 843–848
[7] Das M, Mhlenbruch G, Mahnken A H, Flohr T G, Gndel L, Stanzel S, Kraus T, Gnther R W and Wildberger J E 2006 *Radiology* **241** 564–571
[8] Bellotti R, Cerello P, Tangaro S, Bevilacqua V, Castellano M, Mastronardi G, De Carlo F, Bagnasco S, Bottigli U, Cataldo R *et al.* 2007 *Future Generation Computer Systems* **23** 475–484
[9] Petrick N, Gallas B D, Samuelson F W, Wagner R F and Myers K J 2005 *Medical Imaging* (International Society for Optics and Photonics) pp 49–57
[10] Cerello P, Cheran S C, Bagagli F, Bagnasco S, Bellotti R, Bolanos L, Catanzariti E, De Nunzio G, Fiorina E, Gargano G *et al.* 2008 *2008 IEEE Nuclear Science Symposium Conference Record* (IEEE) pp 3147–3152
[11] Retico A, Delogu P, Fantacci M E, Gori I and Martinez A P 2008 *Computers in biology and medicine* **38** 525–534
[12] Torres E L, Fiorina E, Pennazio F, Peroni C, Saletta M, Camarlinghi N, Fantacci M and Cerello P 2015 *Medical physics* **42** 1477–1489
[13] Coombs K 2009 *Library journal* **134** 30–32
[14] Beyer F, Zierott L, Fallenberg E, Juergens K, Stoeckel J, Heindel W and Wormanns D 2007 *European radiology* **17** 2941–2947
[15] Mell P and Grance T 2011
[16] Buncic P, Sanchez C A, Blomer J, Franco L, Harutyunian A, Mato P and Yao Y 2010 *Journal of Physics: Conference Series* vol 219 (IOP Publishing) p 042003
[17] Milojičić D, Llorente I M and Montero R S 2011 *IEEE Internet Computing* 11–14
[18] Team C *HYPERLINK" http://research. cs. wisc. edu/htcondor/htc. html" http://research. cs. wisc. edu/htcondor/htc. html.[Links]*
[19] Traverso A and oth 2016 *Physica Medica: European Journal of Medical Physics* **32** 94
[20] Sluimer I, Schilham A, Prokop M and Van Ginneken B 2006 *Medical Imaging, IEEE Transactions on* **25** 385–405
[21] van Ginneken B, Armato S G, de Hoop B, van Amelsvoort-van de Vorst S, Duindam T, Niemeijer M, Murphy K, Schilham A, Retico A, Fantacci M E *et al.* 2010 *Medical image analysis* **14** 707–722
[22] Armato III S G, McLennan G, Bidaut L, McNitt-Gray M F, Meyer C R, Reeves A P, Zhao B, Aberle D R, Henschke C I, Hoffman E A *et al.* 2011 *Medical physics* **38** 915–931
[23] Efron B and Tibshirani R 1997 *Journal of the American Statistical Association* **92** 548–560
[24] Hanley J A *et al.* 1989 *Crit Rev Diagn Imaging* **29** 307–335
[25] LeCun Y, Bengio Y and Hinton G 2015 *Nature* **521** 436–444

IOP Conf. Series: Journal of Physics: Conf. Series **841** (2017) 012014 doi:10.1088/1742-6596/841/1/012014

Thermal fatigue life evaluation of SnAgCu solder joints in a multi-chip power module

C Barbagallo[1], G L Malgioglio[2], G Petrone[1] and G Cammarata[1]

[1]University of Catania, Department of Industrial Engineering, Viale A. Doria 6, 95125, Catania, Italy
[2]STMicroelectronics, IPD R&D, Stradale Primosole 50, 95121, Catania, Italy

E-mail: gpetrone@dii.unict.it

Abstract. For power devices, the reliability of thermal fatigue induced by thermal cycling has been prioritized as an important concern. The main target of this work is to apply a numerical procedure to assess the fatigue life for lead-free solder joints, that represent, in general, the weakest part of the electronic modules. Starting from a real multi-chip power module, FE-based models were built-up by considering different conditions in model implementation in order to simulate, from one hand, the worst working condition for the module and, from another one, the module standing into a climatic test room performing thermal cycles. Simulations were carried-out both in steady and transient conditions in order to estimate the module thermal maps, the stress-strain distributions, the effective plastic strain distributions and finally to assess the number of cycles to failure of the constitutive solder layers.

1. Introduction

Recently, various electrified vehicles are emerging. Power (inverter/converter) modules, based on IGBTs and diodes, are one of the core components for such vehicles and convert battery or fuel-cell power to drive motors. Power modules equipped in electrified vehicles often operate at harsh conditions. Thermal stress is known to be one of the major factors resulting performance degradations or damages in solder joints in power modules [1]. For that reason, demand of high power density electronic converters calls for numerical methods to evaluate their reliability at the design stage. The reliability of lead-free solder joints is one of the key factors in the reliability of the overall converter, since they represent mechanical, electrical and often thermal connections between the electronic components and the board [2]. Combined with the thermal expansion mismatch between the different materials of the assembly, cyclic thermal loadings result in stress reversals and potential accumulation of inelastic strain in the solder joint. This inelastic strain accumulates with repeated cycling and ultimately causes solder joint cracking and interconnect failure [3]. Many authors [3,7,8] have studied the fatigue behaviour of lead-free solders. There is a difference between a fatigue life evaluation that uses power cycling and one that uses temperature cycling. During thermal cycling, the solder layer reaches a uniform temperature. In case of the power cycle becomes short, this state is different from the real active state of the power device, in fact the real temperature distribution on the solder layer is not uniform. [4]. The damage of solder joints, which falls in the mechanical fatigue behavior of materials, is a Low-Cycle-Fatigue (LCF) process. A characteristic of LCF is that significant plastic strains occur on the macroscopic scale during each load cycle. Because of the demand for high power density electronic converters, coupled thermo-mechanical analyses oriented to the solder joint fatigue life prediction are becoming a topic of very high interest. In this framework, a numerical analysis on thermo-mechanical behaviour and fatigue life prediction for a power module is proposed.

2. Model implementation

Models were implemented and solved by a FE-based software devoted to multi-physical applications. In our study, we numerically analysed thermal state, thermal-induced stress-strain and we carried-out a LCF life prediction for a multi-chip power module. Model implementation and numerical solution are presented in the following paragraphs.

2.1. Geometry of the numerical model

The geometry of the studied module and the related stratigraphy are shown in Fig. 1. The module is mainly made by a copper baseplate where three power islands are arranged on. Each power island is composed by several layers, such as lead-free solders, a packed 3-layers Direct Bond Copper (DBC), the dies (IGBT and diode) equipped with front metals.

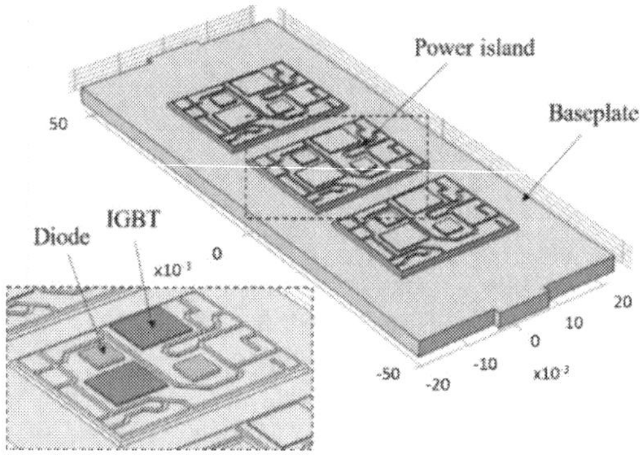

(a)

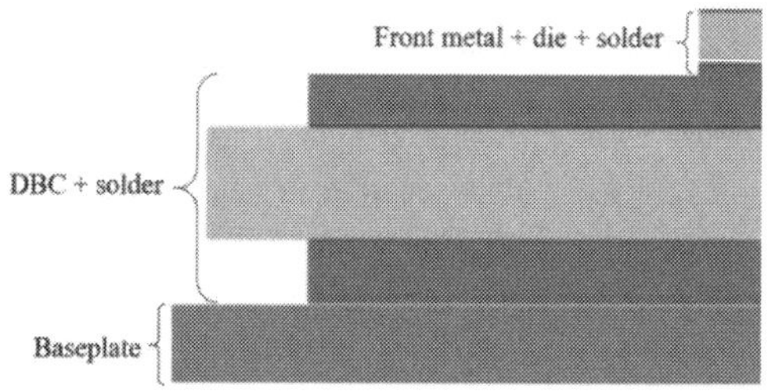

(b)

Figure 1. Geometry of the numerical model (a) and related stratigraphy (b). Baseplate is shown in grey, solders in magenta, DBC ceramic in cyan, DBC copper in red, dies in green and front metals in blue.

The physical properties of materials considered in our study are reported in Tab. 1.

Table 1. The physical properties of materials (ρ density, k thermal conductivity, C_p specific heat at constant pressure, α coefficient of volumetric thermal expansion, E Young's modulus, ν Poisson's ratio, σ_y Yield strength).

	ρ [kg/m^3]	k [W/(m·K)]	C_p [J/(kg·K)]	α [K^{-1}]	E [MPa]	ν [-]	σ_Y [MPa]
Copper	8900	360	393	1.6E-5	1.2E5	0.34	48
Solder (SnAgCu)	7370	75	220	2.2E-5	5.0E4	0.36	26
Ceramic	3260	150	734	4.5E-6	3.3E5	0.24	32
Silicon	2340	135	735	2.5E-6	1.8E5	0.22	40
Aluminium	2710	202	871	2.3E-5	7.0E4	0.33	11

2.2. *Thermo-mechanical model*

In the first step of our investigation, labelled from now Case A, a thermo-mechanical simulation was carried-out for the device, both in steady and transient conditions. It consisted in solving the following governing equations:

$$\rho C_p \frac{\partial T}{\partial t} = \nabla \cdot (k\nabla T) + Q \tag{1}$$

$$\sigma - \sigma_0 = C : (\varepsilon - \varepsilon_0 - \alpha\theta) \tag{2}$$

Eq. (1) represents the energy equation, where T is the temperature and Q is the volumetric heat generation. Eq. (2) is the Hooke's law relating the stress-strain of material, where σ is the stress tensor, σ_0 and ε_0 are initial stress and strain, C is the fourth order elasticity tensor, ":" stands for the double-dot tensor product and ε represents the total strain tensor, defined as:

$$\varepsilon = \frac{1}{2}\left(\nabla s + (\nabla s)^{\mathrm{T}}\right) \tag{3}$$

where s is the displacement. Other symbols in eqns. (1) and (2) are referred to the caption of Tab. 1. Eqns. (2) and (3) were solved in order to detect the stress-strain thermally induced on the constitutive materials. The term θ appearing in eq. (2) denotes the difference between the local temperature T and the strain reference one T_{ref}, that corresponds to the baseplate temperature. This analysis was performed considering the worst working conditions in order to compute the maximum temperature and related stress-strain during the module functioning. To this goal, the bottom surface of the baseplate was held at constant temperature T_{ref} and a volumetric heat source was applied to all devices. Values of thermal heat sources correspond to the maximum power supplied to the devices during the module functioning. On all other boundaries, an insulation condition was applied. From the mechanical point of view, a fixed constraint was chosen for the bottom baseplate surface, maintaining free all other boundaries of the model. The structural load was related to the thermal computed state only.

2.3. *Thermal fatigue model*

In a second step of our investigation, labelled as Case B, a thermal fatigue life prediction was carried-out for the module. The target of this analysis was to assess the fatigue life of solder joints subjected to thermal cycles. The model proposed for predicting the fatigue life of solder layers is plastic strain based. As usually done when applying a plastic strain approach, time-independent plastic strain conditions are applied and the resulting stresses within a component are analysed.

2.3.1. *Thermal cycle test*

The computational model was built considering a thermal cycle between two environmental fixed temperatures. To this goal, and differently from what previously described for model implementation, we solved the energy equation in "passive" conditions. That means we applied different external conditions as thermal load to the device. In particular, no internal heat source was considered, while we applied two different thermal states (T_c = -40 °C and T_h = 125 °C) to the entire device external boundaries. This procedure simulates the module standing inside a climatic test room, held at two different constant control temperatures (cold temperature, T_c, and hot temperature, T_h). By means of this approach, the plastic strain range, $\Delta\varepsilon_p$, was evaluated by using the achieved thermal states. The ε_p is defined as the effective plastic strain, computed as a component of the total strain.

Then, in analogy to the Basquin equation, we applied the Coffin-Manson equation [5] to compute the number of cycles to failure, reading as follows:

$$\frac{\Delta\varepsilon_{\mathrm{p}}}{2} = \varepsilon_{\mathrm{f}}' \,(2N_{\mathrm{f}})^{\mathrm{c}} \tag{4}$$

IOP Conf. Series: Journal of Physics: Conf. Series **841** (2017) 012014 doi:10.1088/1742-6596/841/1/012014

where ε'_f is the fatigue ductility coefficient and c is the fatigue ductility exponent of the solder material and finally N_f is the number of cycles to failure.

3. Numerical solution

Continuous equations were spatially discretized by a Finite Element approach based on the Galerkin method on non-uniform and non-structured computational grids made of tetrahedral Lagrange elements of order 2. Influence of spatial discretization was preliminary studied in order to assure mesh-independent results. A computational grid made of about 1230500 elements was retained for computations, giving 7835500 degrees of freedom. This mesh determines a maximum relative error of 4% with respect to the finer grid used for the preliminary mesh test made for this analysis. A snapshot of the used mesh is presented in Fig. 2. Steady solutions of discretized equations were carried-out by applying an iterative dumped Newton-Raphson scheme [6], classically based on the discretized PDE linearization by a first-order Taylor expansion. Algebraic systems of equations coming from differential operators discretization have been solved by an Unsymmetrical Multi-Frontal package, a direct solver particularly efficient in order to solve unsymmetrical sparse matrixes by a LU decomposition method. For time-marching simulations we adopted an Implicit Differential-Algebraic (IDA) solver based on a variable-order and variable-step-size Backward Differentiation Formulas (BDF) [6]. Computations were carried-out on computational node made of 2 x 64 bit quad-core processors speeding up to 3.3 GHz of frequency and handling 512 GB of RAM.

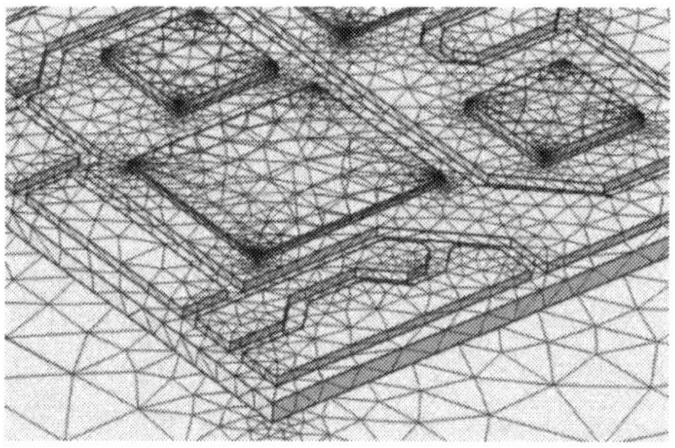

Figure 2. Mesh of the numerical model.

4. Results

Results were carried-out for different geometrical and constitutive configurations of the power module. In particular, different layer thickness were investigated and different physical properties were considered in solving the governing equations. In the following we present an extract of those results.

4.1. Thermo-mechanical analysis results

The thermal distribution obtained for a steady simulation of the module, when all devices are switched on and the maximum power (50 W for each IGBT and 10 W for each diode) is supplied through chips front metals, is shown in Fig. 3a (left side). In the enlargement of the picture, we report the maximum junction temperature value (T_{jc}(MAX) = 41.6 °C) detected by simulation on the front metal of IGBT. The von Mises stress distribution, caused by the thermal load in Case A configuration, is shown in Fig. 3b (right side). The maximum value of von Mises stress computed for each module layer was compared to the ultimate strength of the constitutive materials. From comparison, tensile resistance results widely verified in spite of the worst thermal case considered.

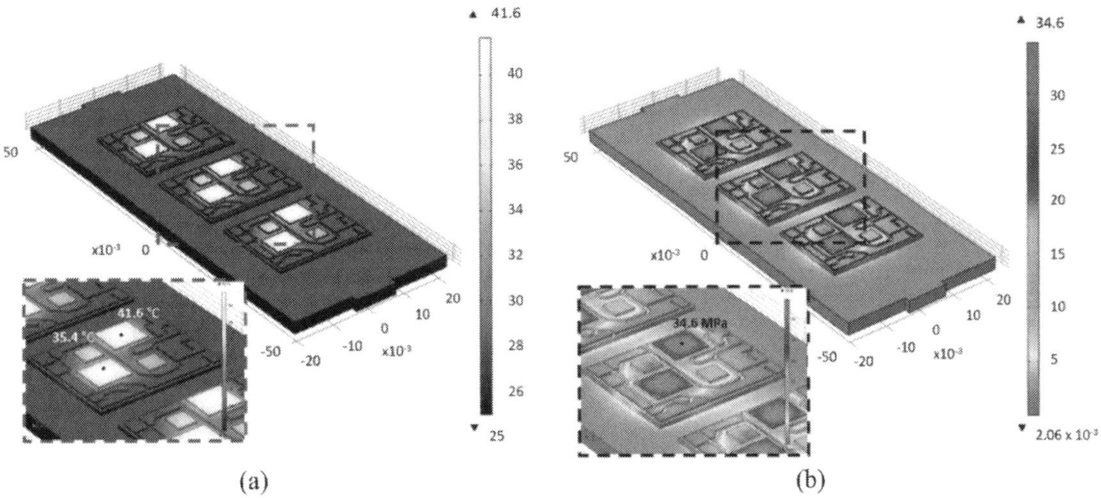

Figure 3. Thermal map [°C] (a) and stress distribution [MPa] (b) on the device and enlargements close to the central power island (Case A).

4.2. Thermal cycle test analysis results

Thermal states computed in Case B configuration were exploited for estimating the effective plastic strain distribution. The maximum value of ε_p was detected on the solder layer, joining the baseplate and the packed DBC. In particular, the most critical value of the effective plastic strain was detected in correspondence of the solder DBC corners. This is maybe due to the geometrical singularity.

Fig. 4a (left side) shows the effective plastic strain distribution in proximity of one of the solder DBC corner, where the highest value of ε_p was found in both environmental thermal conditions (T_c and T_h). This result well agrees with experimental evidences obtained by experimental activities developed on the power module. Fig. 4b (right side) shows two front-views of the module taken by a Scanning Acoustic Microscope (SAM). Pictures were taken at different time instants during a "thermal cycle" testing activity on the module. Environmental conditions applied in the climatic rooms during the experimental tests correspond to thermal levels used in computations. From experimental data reported in Fig. 4b it is possible to observe the damage produced on the module by the cyclic thermal load, in particular on the DBC solder. White regions (highlighted by the red arrows in the Fig. 4b) represent portions of the module where contact between layers fails. As attended, those regions were found in correspondence of the solder DBC corners. This establishment well agrees with some other findings of literature [7, 8].

As previously mentioned, from plastic strain evaluation it is possible to numerically assess the potential number of cycles to failure. In consequence, we were able to estimate the expected number of cycles to failure for SnAgCu solder layers, that are well-known the weakest part of the module from this point of view. From our investigation, the fatigue life prediction is approximatively 1632 cycles for the solder die and 383 cycles for the solder DBC. Those results are in agreement with experimental evidences (solder DBC damaged between 300 and 600 cycles, solder die damaged after 1100 cycles). Successively a comparison study between SnAgCu and SnPb solder joints fatigue life was performed. From results, the fatigue life prediction for SnPb solder die is approximately 162672 cycles and 3071 for solder DBC. Those results well agree with evidences in literature showing that the fatigue life of SnAgCu solders is considerably lower than that of SnPb solder layers [9]. Experimental results that have been found in ref. [9] in similar tests indicate that at high board level strains SnPb solder out performs SnAgCu solder; in particular, the characteristic life was 19500 cycles for SnPb solder and 7730 cycles for SnAgCu solder.

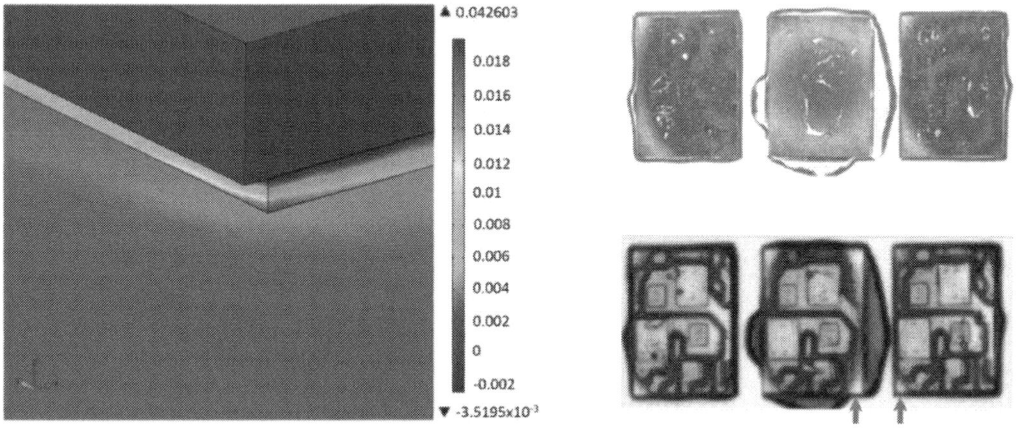

(a) (b)

Figure 4. Effective plastic strain for one of the solder DBC corner, evaluated for T_c=125 °C (Case B) (a). SAM front-view of the module during a "thermal cycle" testing (b): pictures are reported for initial conditions (picture on top) and after 300 cycles (bottom picture). Red arrows indicate some detected damaged regions.

5. Conclusions

This work highlights the opportunity to exploit a numerical approach in order to simulate, from one hand, the thermo-mechanical behaviour of power module and, from another one, the fatigue life of module performing thermal cycles a two environmental fixed temperature inside a climatic test room. From results, it appears that thermal and thermo-mechanical levels are lower than potential critical values for constitutive materials, even in the worst working conditions. The fatigue life prediction results in good agreement with literature evidences and experimental findings. Module portions subjected by the highest plastic strain are regions closed to solder DBC corners. This is maybe due to the geometrical singularity. The fatigue life of SnAgCu solders is considerably lower than that of SnPb solders: for solder DBC it is one order of magnitude shorter, and two orders shorter for solder die.

References

[1] Kim M and Yoon S W 2015 *Proc. EVS 28 (Goyang, Korea)*
[2] Delmonte N, Giuliani F and Cova P 2013 *Microelectron. Rel.* **53** pp 1611-16
[3] Sun Z, Benabou L and Dahoo P R *En. Fract. Mech.* **107** pp 48-60
[4] Shinohara K and Yu Q 2011 *Int. J. Fatigue* **33** pp 1221-34
[5] Manson S S 1966 (New York: McGraw-Hill)
[6] Hindmarsh A C, Brown P N, Grant K E, Lee S L, Serban R, Shumaker D E and Woodward C S 2005 *ACM Trans. Math. Software* **31** pp 363-396
[7] Mi J, Li Y F, Yang Y J, Peng W and Huang H Z 2014 *Sci. World J.* **1** pp 1-11
[8] Lee W W, Nguyen L T and Selvaduray G S 2000 *Microelectron. Rel.* **40** pp 231-244
[9] Blattau N and Hillman C 2005 *Proc. Int. Conf. on SMTA (Chicago, USA)*

IOP Conf. Series: Journal of Physics: Conf. Series **841** (2017) 012015 doi:10.1088/1742-6596/841/1/012015

Optical lattice clocks towards the redefinition of the second

F Bregolin[1,2], G Milani[1,2], M Pizzocaro[1], B Rauf[1,2], P Thoumany[1], F Levi[1] and D Calonico[1]

[1] Istituto Nazionale di Ricerca Metrologica (INRIM), Physical Metrology Division, Strada delle Cacce 91, 10135 Torino, Italy

[2] Politecnico di Torino, Department of Electronics and Telecommunications, Corso Duca degli Abruzzi 24, 10125 Torino, Italy

E-mail: f.bregolin@inrim.it

Abstract. Nowadays atomic optical lattice clocks can perform frequency measurements with a fractional uncertainty at the 10^{-18} level in few hours of measurement, outperforming the best caesium (Cs) standards operated in the world. Since the definition of the unit of time is based on Cs, a worldwide debate about the need to promote the redefinition of the second on a optical reference is under-way. At INRIM (Istituto Nazionale di Ricerca Metrologica) we developed an optical lattice clock based on ytterbium atoms and compared it against a Cs fountain. These results are an important contribution to the debate.

1. Introduction

The time unit, i.e. the second, is one of the base units in the International System of Units (SI). Its definition is based on the atomic transition between two hyperfine levels of the caesium (Cs) atom ground state [1]. Cs frequency standards [2], or Cs clocks, are in charge of the second realization: they generate a microwave whose frequency matches the second definition requirements, being able to excite Cs atoms on the mentioned atomic transition. This microwave can thus be used to calibrate all other oscillators.

Atomic clocks can also be based on other atomic transitions. Nowadays, among the best performing clocks there are optical lattice clocks, which use optical atomic transitions as frequency reference. World best optical lattice clocks currently provide frequency measurements with ultimate fractional uncertainties at the level of 10^{-18} after 3 hours of measurement [3, 4]. Best performing Cs standards are instead showing fractional uncertainties of about 2.3×10^{-16} after 9 days of continuous measurement [5].

The interest in developing optical frequency standards is due to their high quality factor Q, with respect to the one of Cs standards. The quality factor is defined as $Q = f/\Delta f$, where f is the frequency of the reference atomic transition (clock transition) and Δf is the clock transition linewidth. Since Δf is similar in both microwave and optical standards, the five order of magnitude higher frequency of an optical radiation with respect to a microwave lets perform frequency measurements at higher resolution.

Optical clocks need to be calibrated by measuring their oscillator frequency against a Cs standard. This comparison is currently possible through the use of the optical frequency comb

IOP Conf. Series: Journal of Physics: Conf. Series **841** (2017) 012015 doi:10.1088/1742-6596/841/1/012015

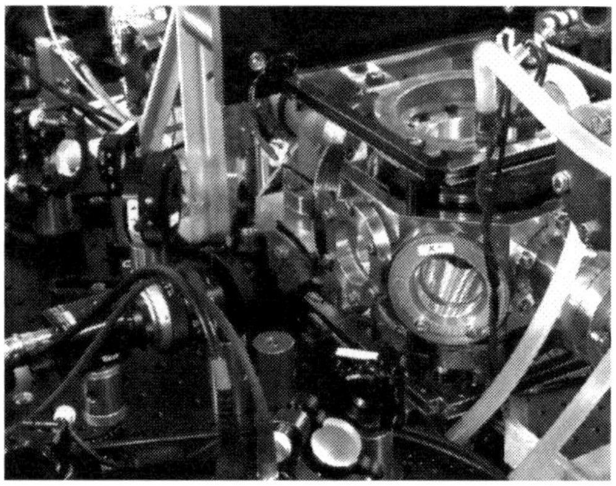

Figure 1. Picture of the INRIM ^{171}Yb optical lattice clock vacuum chamber.

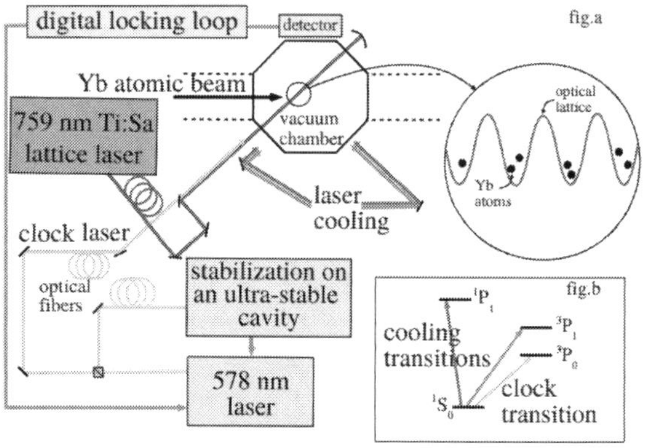

Figure 2. (a) Scheme of the INRIM ^{171}Yb optical lattice clock experimental set-up (b) Yb atomic transitions relevant for the clock operation.

[6], which bridges the frequency gap between the two oscillator frequencies. This process leads on one hand to the use of optical clocks as a secondary representation of the second, but on the other end the comparison is limited by the Cs clocks uncertainty budget. The large discrepancy between the performances of optical and microwave standards has thus opened the way for a worldwide ongoing discussion about the need to change the SI definition of the second [7].

Beyond the metrological purpose of the optical clocks research field, the availability of accurate optical frequency standards has an important impact on fundamental physics research. For instance, radioastronomy has significantly benefited from the possibility to accurately synchronize radio-antennas for very long baseline interferometry [8]. Relativistic geodesy has also been affected by the availability of accurate frequency standards, which are sensitive detectors of gravity variations [9]. Finally, optical clocks can be used to investigate the variation of fundamental constants, such as the fine structure constant [10].

At the Istituto Nazionale di Ricerca Metrologica (INRIM), in Italy, we developed and fully characterized an ^{171}Yb optical lattice clock [11]. An absolute frequency measurement has been performed against the INRIM Cs fountain clock, which is the Italian primary frequency standard.

2. The Ytterbium-171 optical lattice clock

The ^{171}Yb optical lattice clock uses as atomic frequency reference the transition between the $^1S_0$ (ground state) and the $^3P_0$ states at 578 nm. The local oscillator is a yellow laser, whose linewidth is required to be 1 Hz for probing the narrow clock transition. This is accomplished thanks to a frequency stabilization scheme based on an ultra-stable optical Fabry-Pérot cavity (see Fig. 2) [12].

The clock transition is a good frequency reference if it is completely unperturbed or if frequency shifts of the transition due to perturbations are well controlled, i.e. are measured accurately. In order to realize such controlled environment, atoms need to be wisely prepared for the interrogation process. Firstly, a cloud of Yb atoms inside a vacuum chamber (see Fig. 1) are laser cooled down to microkelvin temperatures [13], exploiting radiation pressure from laser beams interacting with blue and green Yb transitions (see Fig. 2). Atoms have to be cold in order to be afterwords successfully loaded in the potential wells of an optical one-dimensional lattice, generated as a laser standing wave (see Fig. 2). Thanks to the strong confinement, the atomic transition Doppler frequency shift is suppressed. Moreover, the trapping lifetime of the atoms in the lattice, usually longer than 1 s, allows long interrogation pulses. Furthermore, this spatially confining technique lets trap thousands of neutral atoms: the interrogation of this atom sample leads to a high resolution signal. Finally, the frequency shift caused by the interaction between the lattice laser and the atoms is cancelled at the first order by selecting appropriately the lattice wavelength. At this so-called *magic wavelength*, the two clock states are shifted by the same amount, keeping the clock transition unperturbed [14]. For Yb, the magic wavelength has been experimentally determined to be 759 nm [11, 15, 16].

Thousands of atoms in the lattice sites are thus ready to be interrogated by the clock laser (see Fig. 2). In the process, a single laser pulse of duration within 60 ms and 100 ms interacts with the atoms: if the laser is resonant, atoms will be excited along the clock transition, otherwise a computer controlled loop move the laser frequency closer to resonance, locking it to the atomic reference.

Since residual atoms perturbations are present, the systematic frequency shifts generated by these effects have to be evaluated and applied to the clock frequency as a correction term. While the physical parameter leading to the frequency shift is under control, the shift measurement is performed by comparing two Yb clock frequency measurements where the parameter under test has been changed. This is done by interleaving two frequency measurements in the two different operating regimes. For example, interleaving a clock cycle with a high number of atoms in the lattice with a clock cycle with a low number of atoms in the lattice allows to evaluate the frequency shift due to atomic collisions, which are proportional to the atomic density of the interrogated sample. The uncertainty of frequency shift measurements performed with this method is statistically limited by the measuring time, thus the physical effects concerned are not limiting issues. If the effect perturbing the atoms is not controlled, the frequency shift evaluation can be performed indirectly by measuring correlated quantities. The uncertainty of such evaluations would depend on the knowledge about the effect and the capacity to set-up the required specific experiment; as a result, these effects would probably contribute to the total uncertainty to a greater extent. The total uncertainty coming from residual frequency shift measurements represents the clock accuracy limit.

The ^{171}Yb optical lattice clock developed at INRIM has been recently characterized with a fractional uncertainty limit of 1.6×10^{-16} [11]. Main contributions to the uncertainty budget are the second order shifts due to the atoms interaction with the lattice laser and the shift generated by the black body radiation. Both effects are under investigation in order to be tackled properly. Concerning the latter for example, at the National Institute of Standards and Technology (NIST, USA) researchers decided to enclose atoms in a black body chamber into vacuum, in order to have a high degree of control of the black body temperature affecting atoms [17], while at RIKEN (Japan) researchers decided to interrogate atoms in a cryogenic environment [3], directly cancelling the black body radiation source.

3. Absolute frequency measurement of the ^{171}Yb clock transition

^{171}Yb clock transition, among other atomic transitions, has been recommended as a secondary representation of the second by the International Committee for Weights and Measures (CIPM) [18], which is in charge of the establishment of the international metrological conventions. This recommendation follows the availability of ^{171}Yb clock frequency measurements performed independently around the world [19 - 22]. At INRIM a direct absolute frequency measurement of the Yb clock transition against the INRIM Cs fountain clock as been performed [11]: the measurement value results to be 518 295 836 590 863.59(31) Hz, which is in agreement with the recommended value for ^{171}Yb [18]. This is the first direct measurement of the ^{171}Yb clock transition at this level of uncertainty. This measurement further confirm previous measurements and contribute to reaffirm the ytterbium transition as an interesting candidate for the redefinition of the SI second.

4. Conclusions

Optical clocks nowadays are outperforming Cs clocks by many order of magnitude both in term of accuracy and stability, raising the question about a redefinition of the SI second based on an optical atomic transition. Many different optical transitions, both in the visible and in the UV region, are being investigated. The ^{171}Yb transition has been already measured in many independent experiments with an high accuracy; at INRIM we demonstrated an ^{171}Yb optical lattice clock with a fractional uncertainty limit of 1.6×10^{-16} and we measured its frequency against the Cs Italian primary frequency standard with a fractional uncertainty of 5.9×10^{-16}. In addition, ytterbium has the potential to see further increased the control of its perturbations thanks to its atomic structure and properties. Thus, Yb clocks are competitive in the world scenario in the fullfilment of the requirements leading to the SI second redefinition.

References

[1] The international system of units (SI) Bureau International des Poids et Mesures, 2008
[2] Bauch A 2003 *Meas. Sci. Technol.* **14** 1159
[3] Ushijima I, Takamoto M, Das M, Ohkubo T and Katori H 2015 *Nature Photonics* **9** 185–189
[4] Nicholson T, Campbell S, Hutson R, Marti G, Bloom B, McNally R, Zhang W, Barrett M, Safronova M, Strouse G, Tew W and Ye J 2015 *Nature Communications* **6** 6896
[5] Levi F, Calonico D, Calosso C E, Godone A, Micalizio S and Costanzo G A 2014 *Metrologia* **51** 270
[6] Diddams S A, Jones D J, Ye J, Cundiff S T, Hall J L, Ranka J K, Windeler R S, Holzwarth R, Udem T and Hänsch T W 2000 *Phys. Rev. Lett.* **84**(22) 5102–5105
[7] Riehle F 2015 *Comptes Rendus Physique* **16** 506 – 515
[8] Clivati C, Costanzo G A, Frittelli M, Levi F, Mura A, Zucco M, Ambrosini R, Bortolotti C, Perini F, Roma M and Calonico D 2015 *IEEE Transactions on Ultrasonics, Ferroelectrics, and Frequency Control* **62** 1907–1912
[9] Chou C W, Hume D B, Rosenband T and Wineland D J 2010 *Science* **329** 1630–1633
[10] Blatt S, Ludlow A D, Campbell G K, Thomsen J W, Zelevinsky T, Boyd M M, Ye J, Baillard X, Fouché M, Le Targat R, Brusch A, Lemonde P, Takamoto M, Hong F L, Katori H and Flambaum V V 2008 *Phys. Rev. Lett.* **100**(14) 140801
[11] Pizzocaro M, Thoumany P, Rauf B, Bregolin F, Milani G, Clivati C, Costanzo G A, Levi F and Calonico D 2016 Absolute frequency measurement of the $^1S_0 - {}^3P_0$ transition of ^{171}Yb Accepted in *Metrologia*
[12] Drever R, Hall J, Kowalski F, Hough J, Ford G, Munley A and Ward H 1983 *Applied Physics B* **31** 97–105
[13] Metcalf H J and Van der Straten P 1999 *Laser cooling and trapping* (Springer Science & Business Media)
[14] Katori H, Takamoto M, Pal'chikov V G and Ovsiannikov V D 2003 *Phys. Rev. Lett.* **91**(17) 173005
[15] Barber Z W, Stalnaker J E, Lemke N D, Poli N, Oates C W, Fortier T M, Diddams S A, Hollberg L, Hoyt C W, Taichenachev A V and Yudin V I 2008 *Phys. Rev. Lett.* **100**(10) 103002
[16] Takamoto M, Ushijima I, Das M, Nemitz N, Ohkubo T, Yamanaka K, Ohmae N, Takano T, Akatsuka T, Yamaguchi A and Katori H 2015 *Comptes Rendus Physique* **16** 489 – 498
[17] Beloy K, Hinkley N, Phillips N B, Sherman J A, Schioppo M, Lehman J, Feldman A, Hanssen L M, Oates C W and Ludlow A D 2014 *Phys. Rev. Lett.* **113**(26) 260801
[18] Consultative commitee for time and frequency (CCTF): Report of the 19th meeting Tech. rep. Bureau International des Poids et Mesures, 2012
[19] Lemke N D, Ludlow A D, Barber Z W, Fortier T M, Diddams S A, Jiang Y, Jefferts S R, Heavner T P, Parker T E and Oates C W 2009 *Phys. Rev. Lett.* **103**(6) 063001
[20] Park C Y, Yu D H, Lee W K, Park S E, Kim E B, Lee S K, Cho J W, Yoon T H, Mun J, Park S J, Kwon T Y and Lee S B 2013 *Metrologia* **50** 119
[21] Akamatsu D, Yasuda M, Inaba H, Hosaka K, Tanabe T, Onae A and Hong F L 2014 *Opt. Express* **22** 7898–7905
[22] Nemitz N, Ohkubo T, Takamoto M, Ushijima I, Das M, Ohmae N and Katori H 2016 *Nature Photonics* **10** 258–261

IOP Conf. Series: Journal of Physics: Conf. Series **841** (2017) 012016 doi:10.1088/1742-6596/841/1/012016

Low energy quantum regimes of 1D dipolar Hubbard model with correlated hopping

S Fazzini[1], L Barbiero[2,3] and A Montorsi[1]

[1] Institute for condensed matter physics and complex systems, DISAT, Politecnico di Torino, I-10129, Italy

[2] CNR-IOM DEMOCRITOS Simulation Center and SISSA, Via Bonomea 265, I-34136 Trieste, Italy

[3] Dipartimento di Fisica e Astronomia "Galileo Galilei", Università di Padova, 35131 Padova, Italy

E-mail: serena.fazzini@polito.it

Abstract. We apply the bosonization technique to derive the phase diagram of a balanced unit density two-component dipolar Fermi gas in a one dimensional lattice geometry. The considered interaction processes are of the usual contact and dipolar long-range density-density type together with peculiar correlated hopping terms which can be generated dynamically. Rigorous bounds for the transition lines are obtained in the weak coupling regime. In addition to the standard bosonization description, we derive the low energy phase diagram taking place when part of the interaction is embodied non-perturbatively in the single component Hamiltonians. In this case the Luttinger liquid regime is shown to become unstable with respect to the opening of further gapped phases, among which insulating bond ordered wave and Haldane phases, the latter with degenerate edge modes.

1. Introduction

Experiments with cold atoms have disclosed a new way of investigating strongly correlated systems [1]. The possibility to cool down both fermionic and bosonic atomic gases to very low temperature and to trap them onto optical lattices, together with the ability to tune the interactions and the dimensionality with high accuracy, has allowed to simulate a great variety of interacting many-body lattice Hamiltonians [2]. Particular theoretical efforts have been devoted to identifying fermionic Hubbard-like systems [3 - 5] which ground state exhibits exotic [6] and topological phases [7], described by string-like order parameters [8]. Also particles with dipolar long range interaction like polar molecules and magnetic atoms [9] are currently available in laboratories. This has allowed for the experimental realization of a paradigmatic model in condensed matter physics namely the extended Bose-Hubbard model [10].

On the other hand, bosonization [11, 12] is a well-established analytical technique to investigate the low energy regime of one dimensional interacting fermionic systems. It consists in linearizing the spectrum around the Fermi points, passing to the continuum limit, and finally expressing the fermionic operators in terms of bosonic fields. In this way one has an efficient and general setting to study the low energy excitations induced by different types of interaction. Despite the fact that in most cases further numerical analyses are needed in order to get the full zero temperature phase diagram, bosonization is very useful to understand the nature of many quantum phases

of matter and evaluate the correlation functions which characterize them.
Within bosonization approach, usually the interaction is included in a perturbative way, starting from two non-interacting single component Luttinger liquids (LLs). In fact, as noticed in [11], in some cases part of the interaction can be included non-perturbatively already in the single component Hamiltonian, as long as it remains in a LL regime. This possibility has been exploited for instance in [13] to predict the presence of a Bond Ordered Wave (BOW) phase induced by dipolar interaction already within one-loop bosonization. Here we adopt it to derive the zero temperature phase diagram of a one dimensional Hubbard model in presence of correlated hopping – induced by a periodical modulation of the on-site interaction [14], as recently shown experimentally [15]– and long-range dipole-dipole interaction.
The paper is organized as follows. In Sec. 2 we introduce the model and rewrite the Hamiltonian in normal ordered form. In Sec. 3 we review its bosonization phase diagram, which coincides with that of two spin-charge decoupled sine- Gordon models, due to the peculiar r^{-3} nature of the power law decay of the dipolar interaction. The different gapped and partly gapped phases are characterized in terms of string and parity [16, 17] non-local orders. In Sec. 4 we then revisit the phase diagram by including part of both dipolar and correlated hopping interaction non perturbatively in the single component Hamiltonians. In this case, the phase diagram obtained from the study of Renormalization Group (RG) flow equations exhibit a richer structure. In Sec. 5 we give some conclusions.

2. The model

We consider a balanced unit density two-component dipolar Fermi mixture. Once these particles are trapped in a one dimensional lattice, an accurate description of the system is given by the following Hamiltonian [5]

$$H = -J \sum_{j,\sigma} Q_{j,j+1,\sigma} \left[1 - X(n_{j,\bar{\sigma}} - n_{j+1,\bar{\sigma}})^2\right] + U_0 \sum_{j} n_{j,\uparrow} n_{j,\downarrow} + V \sum_{j,r>0} \frac{n_j n_{j+r}}{r^3} \tag{1}$$

where $Q_{j,j+1,\sigma} = c^\dagger_{j,\sigma} c_{j+1,\sigma} + h.c.$; $\sigma =\uparrow, \downarrow$ is the index species and $\bar{\sigma}$ denotes its opposite, $c^\dagger_{j,\sigma}$ and $c_{j,\sigma}$ are the creation and annihilation operators, respectively, $n_{j,\sigma}$ counts the number of particles of species σ and $n_j = \sum_\sigma n_{j,\sigma}$. The coupling coefficients J, U_0, V, X, independently tunable in the experiments [1, 18, 19], describe the tunneling probability, on-site and dipolar interactions, and correlated hopping processes, respectively. Upon normal ordering of the operators, $Q_{j,j+1,\sigma} =: Q_{j,j+1,\sigma} : + \langle Q_{j,j+1,\sigma}\rangle$, $n_{j,\sigma} =: n_{j,\sigma} : + \langle n_{j,\sigma}\rangle$ (with $\langle Q_{j,j+1,\sigma}\rangle = 2/\pi$, $\langle n_{j,\sigma}\rangle = 1/2$), and omitting the constant terms, we get

$$\begin{aligned} H = &-\left(1-\frac{X}{2}\right)\sum_{j,\sigma} : Q_{j,j+1,\sigma} : + \sum_{j,\sigma}\sum_{r} V_{\parallel}(r) : n_{j,\sigma} :: n_{j+r,\sigma} : + \\ &+ \sum_{j,\sigma}\sum_{r} V_{\perp}(r) : n_{j,\sigma} :: n_{j+r,\bar{\sigma}} : + U_0 \sum_{j} : n_{j,\uparrow} :: n_{j,\downarrow} : \\ &- 2X \sum_{j,\sigma} : Q_{j,j+1,\sigma} :: n_{j,\bar{\sigma}} :: n_{j+1,\bar{\sigma}} : \end{aligned} \tag{2}$$

where we have set $J = 1$ and have defined $V_{\parallel}(r) = \frac{V}{r^3} - \frac{4X}{\pi}\delta_{r,1}$, $V_{\perp} = \frac{V}{r^3}$.

3. Weak coupling phase diagram

In the standard bosonization approach, one starts from the non-interacting Hamiltonian and consider the effect of interactions in a perturbative manner. The first step is to perform the continuum limit: $\sum_j \longrightarrow \frac{1}{a}\int dx$; $c_{j,\sigma} \longrightarrow \sqrt{a}\left[e^{ik_F x}\Psi_{R\sigma}(x) + e^{-ik_F x}\Psi_{L\sigma}(x)\right]$ (with $x = ja$,

$a \to 0$ being the lattice constant). Here $\Psi_{R\sigma}$ and $\Psi_{L\sigma}$ are the fermionic field operators for the right and left movers, respectively. As claimed before, we will consider the particular case in which the system is at half filling; hence $k_F = \pi/(2a)$. We finally write the fermionic fields $\Psi_{\chi\sigma}$ in terms of the bosonic ones ϕ_σ and θ_σ:

$$\Psi_{\chi\sigma}(x) = \frac{\eta_{\chi\sigma}}{\sqrt{2\pi\alpha}} e^{\mathrm{i}\sqrt{\pi}[\chi\phi_\sigma(x)+\theta_\sigma(x)]} \tag{3}$$

where the generic index χ can denote right (R) or left (L) movers, with the usual convention that it assumes a positive sign in the first case and a negative sign in the second one. Here $\alpha \sim a$ is an ultraviolet cutoff and $\eta_{\chi\sigma}$ are the Klein factors, which guarantee proper anti-commutation relations. With this notation, the kinetic operator $:Q_{j,j+1,\sigma}:$ and the density operator $:n_{j,\sigma}:$ appearing in Hamiltonian (2) have the following expressions in terms of the bosonic fields

$$:Q_{j,j+1,\sigma}:= -\frac{2}{\pi}(-1)^j \cos\left(2\sqrt{\pi}\phi_\sigma(x)\right) - a^2\left[(\nabla\phi_\sigma(x))^2 + (\nabla\theta_\sigma(x))^2\right] + \dots \tag{4}$$

$$:n_{j,\sigma}:= a\left[\frac{1}{\sqrt{\pi}}\nabla\phi_\sigma(x) - \frac{(-1)^j}{\pi a}\sin\left(2\sqrt{\pi}\phi_\sigma(x)\right)\right] + \dots \tag{5}$$

Here we have used dots to denote the higher order terms in expansion with respect to a, which will be neglected. Bosonization of two and three body-terms in the Hamiltonian (2) involves calculating the product of operators of the form (4) and (5). When the latters act on different fermionic species, the calculation is straightforward. When acting on the same species instead, the operator product expansion is needed. In deriving the bosonized expression for $:n_{j,\sigma}::n_{j+r,\sigma}:$, we make use of the fusion rule $\sin(2\sqrt{\pi}\phi_\sigma(x))\sin(2\sqrt{\pi}\phi_\sigma(x+R)) = \frac{a^2}{2R^2} - \frac{1}{2}\cos(4\sqrt{\pi}\phi_\sigma(x)) - \pi a^2(\nabla\phi_\sigma(x))^2 + \dots$, with $R = ra$. Thus we get

$$:n_{j,\sigma}::n_{j+r,\sigma}:\simeq a^2\left\{\frac{1-(-1)^r}{\pi}(\nabla\phi_\sigma(x))^2 - \frac{(-1)^r}{2\pi^2 a^2}\cos\left(4\sqrt{\pi}\phi_\sigma(x)\right) + \frac{(-1)^r}{2\pi^2 R^2} + (-1)^j\dots\right\}. \tag{6}$$

In the three-body term, also the oscillating part of (6) (with $r = 1$) contributes. To compute it, we apply the following operator product expansion, $\nabla\phi_\sigma(x)\sin(2\sqrt{\pi}\phi_\sigma(x+a)) = -\sin(2\sqrt{\pi}\phi_\sigma(x))\nabla\phi_\sigma(x+a) = \frac{1}{\sqrt{\pi}a}\cos(2\sqrt{\pi}\phi_\sigma(x)) + \dots$. One obtains $(-1)^j\frac{2}{\pi^2 a^2}\cos(2\sqrt{\pi}\phi_\sigma(x))$, from which the three body term contribution is evaluated. As customary, it is now convenient to introduce charge and spin field operators, defined as $\phi_c(x) = (\phi_\uparrow(x) + \phi_\downarrow(x))/\sqrt{2}$ and $\phi_s(x) = (\phi_\uparrow(x) - \phi_\downarrow(x))/\sqrt{2}$, respectively. Similar relations hold for the dual fields θ_ν ($\nu = c, s$) as functions of θ_σ ($\sigma = \uparrow, \downarrow$). In terms of these operators, the Hamiltonian can be separated into the sum of two independent Hamiltonians in the charge and spin sectors plus a coupling term:

$$H = H_c + H_s + H_{cs} \quad . \tag{7}$$

In each sector H_ν has the the form of a sine-Gordon model:

$$H_\nu = \frac{v_\nu}{2}\int dx\left[\left(\sqrt{K_\nu}\nabla\theta_\nu\right)^2 + \left(\frac{\nabla\phi_\nu}{\sqrt{K_\nu}}\right)^2\right] + \frac{m_\nu v_\nu}{2\pi a^2}\int dx\cos\left(\sqrt{8\pi}\phi_\nu\right), \quad \nu = c, s; \tag{8}$$

whereas the coupling Hamiltonian H_{cs} reads $H_{cs} = \frac{M_{cs}}{\pi a}\int dx\cos(\sqrt{8\pi}\phi_c)\cos(\sqrt{8\pi}\phi_s)$. The coefficients of the Luttinger and mass terms are defined as follows:

$$K_\nu = 1 + \frac{1}{4\pi}\left[\frac{16X}{\pi} - s_\nu U_0 - \frac{3\zeta(3)V}{2} - 4\zeta(3)V\delta_{\nu,c}\right] \quad ; \quad v_\nu = 2a\left[2 - \frac{X}{2} - \frac{X}{\pi^2} - K_\nu\right] \tag{9}$$

$$m_\nu = \frac{1}{2\pi}\left[\frac{16X}{\pi} - s_\nu\left(U_0 - \frac{3\zeta(3)V}{2}\right)\right] \quad ; \quad M_{cs} = \frac{1}{2\pi}\left[\frac{3\zeta(3)V}{2} - \frac{8X}{\pi}\right]. \tag{10}$$

Table 1. Correspondence between ground state quantum phases and nonlocal operators that manifest long range order (LRO) [4]. We have indicated the unpinned fields with the letter u.

	$\sqrt{2\pi}\Phi_c$	$\sqrt{2\pi}\Phi_s$	Δ_c	Δ_s	LRO
LL	u	u	0	0	none
LE	u	0	0	open	O_P^s
MI	0	u	open	0	O_P^c
HI	$\pi/2$	u	open	0	O_S^c
BOW	0	0	open	open	O_P^c, O_P^s
CDW	$\pi/2$	0	open	open	O_S^c, O_P^s

where $s_c = 1$, $s_s = -1$ and $\zeta(n)$ denotes the Riemann zeta function. The competition of the kinetic and mass terms in (8) can be discussed by analyzing the RG flow equations as in [4]. Apart from the gapless LL phase, two insulating charge gapped phases are associated to the pinning of solely the charge field to the two values $\phi_c = 0, \sqrt{\frac{\pi}{8}}$. In a uniform unit-density background, the latter values of ϕ_c describe respectively a Mott insulating (MI) regime with localized holon-doublon fluctuations and a Haldane insulator (HI) phase [20] with "dilute" hidden antiferromagnetic order of holons and doublons in analogy to the XXZ spin-1 chain behavior [6, 8]. When ϕ_c instead is unpinned, a metallic spin gapped phase is associated to the pinning of solely the spin field to the value $\phi_s = 0$. It actually describes the Luther Emery (LE) [21] liquid phase where spin-up and spin-down single particle quantum fluctuations take place in a uniform background of condensed holons and doublons. The $\phi_s = 0$ value also supports two insulating fully gapped regimes, occurring when the charge field is pinned as well. In this case for $\phi_c = 0$ the bond ordered wave (BOW) phase takes place. Whereas for $\phi_c = \sqrt{\frac{\pi}{8}}$ a charge density wave (CDW) phase with holon-doublon antiferromagnetic ordering is obtained. The five gapped or partly gapped phases can be characterized by the non-vanishing of the appropriate parity or string non-local order parameter, defined respectively as

$$O_P^\nu(j) = \langle \prod_{k=0}^{j-1} e^{i\pi S_{z,k}^\nu} \rangle \sim \langle \cos(\sqrt{2\pi}\phi_\nu) \rangle \quad , \quad O_S^\nu(j) = \langle \left(\prod_{k=0}^{j-1} e^{i\pi S_{z,k}^\nu} \right) S_{z,j}^\nu \rangle \sim \langle \sin(\sqrt{2\pi}\phi_\nu) \rangle \quad . \tag{11}$$

This is reported in Tab. 1, following the procedure outlined in [17, 4]. In particular, the HI phase turns out to have non-trivial topological properties, as the presence of degenerate edge modes [22]. Depending on the values of the coupling constants J, U_0, V, X all the above regimes can be realized. For instance, as reported in Fig. 1, already at $V = 0$ by varying U_0 and X the LL, MI and LE regimes are achieved.

4. Including interaction non-perturbatively

To gain further insight into the zero temperature phase diagram, we may regard Hamiltonian (2) as the sum of two single-species Hamiltonians, already containing part of the interaction non perturbatively, plus an inter-species part:

$$H = \sum_\sigma H_\sigma + H_{\uparrow\downarrow} \quad . \tag{12}$$

Here $\sum_\sigma H_\sigma$ is given by the first two terms in (2). Up to a multiplicative constant $(1 - \frac{X}{2})$, the single-species Hamiltonian H_σ is a long-range "t-V" model, which is known to have a gapless Luttinger liquid phase for small enough interaction strength. In this case the Luttinger parameter $K(V, X)$ can be evaluated numerically with high precision [23, 24], and analytically

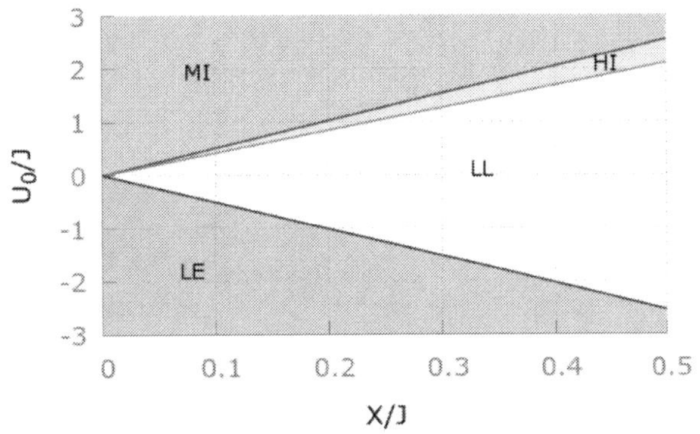

Figure 1. Phase diagram at $V = 0$ from bosonization analysis of Sec. 4. In case of standard bosonization of Sec. 3, the Haldane insulator phase HI would still be a LL.

both for vanishing X [25], and for vanishing V [11]. Explicitly:

$$K(V,0) = \left[1 + \frac{6\zeta(3)V}{\pi^2}\right]^{-1/2} \quad , \quad K(0,X) = \left[\frac{2}{\pi}\arccos\frac{2X}{\pi\left(1-\frac{X}{2}\right)}\right]^{-1} \quad . \tag{13}$$

Of course, K can also be obtained within a bosonization approximation [26]. In this case $K(X,V) \simeq 1 + \frac{1}{4\pi}\left[\frac{16X}{\pi} - \frac{7\zeta(3)V}{2}\right]$. At this point one can proceed to bosonization of the inter-species interaction $H_{\uparrow\downarrow}$. It can be added to the Luttinger liquid phase of the remaining part of the Hamiltonian, which Luttinger coefficients may be determined either analytically or numerically [23, 24]. In so doing one ends up with a Hamiltonian which is fully decoupled in the charge and spin fields. Explicitly:

$$H = H'_c + H'_s \tag{14}$$

with

$$H'_\nu = \frac{v'_\nu}{2}\int dx\left[K'_\nu\left(\partial_x\theta_\nu\right)^2 + \frac{1}{K'_\nu}\left(\partial_x\phi_\nu\right)^2\right] + \frac{m_\nu v_\nu}{2\pi a^2}\int dx\cos\left(\sqrt{8\pi}\phi_\nu\right) \tag{15}$$

and

$$K'_\nu = \sqrt{K}\left\{1 - \frac{X}{4\pi^2} + \frac{K^2}{8\pi}\left[\frac{2X}{\pi} - s_\nu\left(U_0 + 2\zeta(3)V\right)\right]\right\} \quad , \quad v'_\nu = v_\nu\frac{K_\nu}{K'_\nu} \tag{16}$$

where $s_c = 1$, $s_s = -1$. Now the study of RG flow equations can be done by inserting the non-perturbative dependence of K on V (X), while keeping to first order the remaining interaction in K, K'_ν and m_ν. The case $X = 0$ has already been treated in this way in [13]. At variance with what obtained with standard bosonization, the latter approach allows to identify a further BOW phase already within one-loop expansion, thanks to the decoupling of the opening of the spin gap from the Gaussian transition of the charge gap. As an application of our findings at generic X and V, here we report in Fig. 1 the phase diagram obtained at $V = 0$, where K is given by (13). In this case, the comparison with the results obtained within the standard weak coupling approach of Sec. 3 shows the appearance of a further non-trivial HI phase for $\frac{4\pi}{3}X \leq U_0 \leq \frac{16}{\pi}X$. From Tab. 1 it is seen that the latter corresponds to a non-vanishing string order parameter in the charge sector, which is known to amount to the presence of degenerate edge modes [22].

5. Conclusions

We have derived the bosonization phase diagram of a unit density balanced two-component Fermi gas with correlated hopping, on-site, and long-range dipolar interactions. The power law decay of the dipolar interaction with exponent greater than 1 allows to re-sum its contributions into an effective short range potential. Moreover, spin-charge coupling terms can be neglected having, in general, larger scaling dimension. The resulting bosonized Hamiltonian consists of two spin-charge separated sine Gordon models, which phase diagram can be discussed according to Tab. 1, depending on mass and Luttinger parameters. We derived the sine-Gordon models both within standard bosonization, in which case the single component Hamiltonians were the non-interacting up and down spin models; and by including non-perturbatively part of the interaction already at the level of the single component Hamiltonians. In the latter case further features emerge in the phase diagram, noticeably a non-trivial Haldane charge gapped phase also in absence of dipolar interaction. We expect that other 3- and 4-body processes [27] could be included non-perturbatively within the LL regime of the single component Hamiltonian, possibly inducing further orders in the ground state phase diagram. The present results should be compared with numerical findings. For instance in [5] by means of a density matrix renormalization group [28] analysis it was found that further exotic phases, not present in the classification given here (Tab. 1), appear. In such cases, one should go beyond one loop bosonization, including higher order harmonics which were neglected here. Finally it is relevant to underline that the previous quantum phases could be studied and probed [29] with the currently available experimental setups as proposed in [5].

References

[1] Bloch I, Dalibard J and Zwerger W 2008 *Rev. Mod. Phys.* **80** 885
[2] Dutta O *et al.* 2015 *Rep. Prog. Phys.* **78** 066001
[3] Chhajlany R W, Grzybowski P R, Stasiska J, Lewenstein M and Dutta O 2016 *Phys. Rev. Lett.* **116** 225303
[4] Barbiero L, Montorsi A and Roncaglia M 2013 *Phys. Rev.* B **88** 035109
[5] Fazzini S, Montorsi A, Roncaglia M and Barbiero L 2016 *Preprint* arXiv:1607.05682v2
[6] Haldane F D M 1983 *Phys. Rev. Lett.* **50** 1153
[7] Wen XG 1995 *Adv. in Phys.* **44** 405
[8] den Nijs M and Rommelse K 1989 *Phys. Rev.* B **40** 4709
[9] Lahaye T, Menotti C, Santos L, Lewenstein M and Pfau T 2009 *Rep. Prog. Phys.* **72** 126401
[10] Baier S, Mark M J, Petter D, Aikawa K, Chomaz L, Cai Z, Baranov M, Zoller P and Ferlaino F 2016 *Science* **352** 201-5
[11] Giamarchi T 2003 *Quantum Physics in One Dimension* (Oxford: Oxford University Press)
[12] Gogolin A O, Nersesyan A A and Tsvelik A M 1998 *Bosonization and Strongly Correlated Systems* (Cambridge: Cambridge University Press)
[13] Di Dio M, Barbiero L, Recati A and Dalmonte M 2014 *Phys. Rev.* A **90** 063608
[14] M. Di Liberto M, Creffield C E, Japaridze G I and Morais Smith C 2014 *Phys. Rev.* A **89** 013624
[15] Meinert F, Mark M J, Lauber K, Daley A J and Nägerl HC 2016 *Phys. Rev. Lett.* **116** 205301
[16] Berg E, Dalla Torre E G, Giamarchi T and Altman E 2008 *Phys. Rev.* B **77** 245119
[17] Montorsi A and Roncaglia M, 2012 *Phys. Rev. Lett.* **109** 236404
[18] Góral K, Santos L and Lewenstein M 2002 *Phys. Rev. Lett.* **88** 170406
[19] Bartolo N, Papoular D J, Barbiero L, Menotti C and Recati A 2013 *Phys. Rev.* A **88** 023603
[20] Dalla Torre E G, Berg E and Altman E 2006 *Phys. Rev. Lett.* **97** 260401
[21] Luther A and Emery V J 1974 *Phys. Rev. Lett.* **33** 589
[22] Montorsi A, Dolcini F, Iotti R and Rossi F 2016 *Preprint* arXiv:1610.05706v1
[23] Citro R, Orignac E, De Palo S and Chiofalo M L 2007 *Phys. Rev.* A **75** 051602(R)
[24] Otterbach J, Moos M, Muth D and Fleischhauer M 2013 *Phys. Rev. Lett.* **110** 156402
[25] Dalmonte M, Pupillo G and Zoller P 2010 *Phys. Rev. Lett.* **105** 140401
[26] Capponi S, Poilblanc D and Giamarchi T 2000 *Phys. Rev.* B **61** 13410
[27] Dolcini F and Montorsi A 2001 *Nucl. Phys.* B **592** 563 ; 2001 *Phys. Rev.* B **63** 121103; 2013 *Phys. Rev.* B **88** 115115 ; 2002 *Phys. Rev.* B **66**, 075112
[28] White S R 1992 *Phys. Rev. Lett.* **69** 2863
[29] M. Endres *et al.* 2011 *Science* **334** 200

IOP Conf. Series: Journal of Physics: Conf. Series **841** (2017) 012017 doi:10.1088/1742-6596/841/1/012017

Adaptive CFD schemes for aerospace propulsion

A Ferrero[1] and F Larocca[2]

[1] INRIA Bordeaux Sud-Ouest and University of Bordeaux, Talence, France
[2] Politecnico di Torino, Turin, Italy

E-mail: andrea.ferrero@inria.fr; francesco.larocca@polito.it

Abstract. The flow fields which can be observed inside several components of aerospace propulsion systems are characterised by the presence of very localised phenomena (boundary layers, shock waves,...) which can deeply influence the performances of the system. In order to accurately evaluate these effects by means of Computational Fluid Dynamics (CFD) simulations, it is necessary to locally refine the computational mesh. In this way the degrees of freedom related to the discretisation are focused in the most interesting regions and the computational cost of the simulation remains acceptable. In the present work, a discontinuous Galerkin (DG) discretisation is used to numerically solve the equations which describe the flow field. The local nature of the DG reconstruction makes it possible to efficiently exploit several adaptive schemes in which the size of the elements (h-adaptivity) and the order of reconstruction (p-adaptivity) are locally changed. After a review of the main adaptation criteria, some examples related to compressible flows in turbomachinery are presented. An hybrid hp-adaptive algorithm is also proposed and compared with a standard h-adaptive scheme in terms of computational efficiency.

1. Introduction

Several components of aerospace propulsion systems are characterised by flow fields in which very localised phenomena can be observed. Shock waves, boundary layers, shear layers, turbulent coherent structures are some examples. These phenomena show strong gradients in space and they usually have a spatial extension which is reduced with respect to the size of the propulsion system. However, the performances of the propulsion system are strongly influenced by the effects of these structures.

In order to accurately evaluate the flow field by means of Computational Fluid Dynamics (CFD) tools, it is necessary to refine sufficiently the computational mesh in the regions of interest. An efficient approach is based on automatic adaptive algorithms which can automatically refine the mesh only in the regions which require high resolution. This approach improves significantly the reliability of the simulation and there is a growing interest in the use of these techniques. Indeed, the design procedures adopted nowadays by the industries are often based on automatic optimisation algorithms which perform a systematic exploration of the design parameters. In this framework, it is fundamental to have tools which can automatically adapt the mesh to the several system configurations which are tested during the optimisation process. An overview of some adaptive techniques can be found in the results of the European Project ADIGMA (Adaptive Higher-Order Variational Methods for Aerodynamic Applications in Industry) [1].

Furthermore, the rapid development of High Performance Computing systems has recently introduced the possibility to perform large scale simulations based on massive parallelisation. While the classical industrial approach to fluid dynamics is based on Reynolds Averaged Navier-Stokes (RANS) simulations, Large Eddy Simulations (LES) are nowadays affordable [2]. The use of LES reduces the need of empirical turbulence models but requires a huge number of degrees of freedom. Furthermore, the fact that turbulent flow fields are unsteady makes it difficult to generate a good mesh a priori. For these reasons, several research efforts are nowadays devoted to the development of adaptive schemes.

In the present work, a Discontinuous Galerkin (DG) scheme is used to numerically solve the compressible Navier-Stokes equations. The main idea of the DG scheme is to describe the numerical solution inside each element as a linear combination of basis functions which do not have any continuity constraint at the interfaces between different elements. This feature makes the scheme extremely compact even when high-order reconstructions are employed.
Furthermore, the local nature of the DG reconstruction is particularly useful when working with adaptive schemes. For example, non-conforming meshes (also known as hanging-nodes meshes) can be produced by adaptive algorithms based on element splitting and they can be easily managed by a DG scheme. This is a classical example of h-adaptivity, in which the size of the elements is locally changed in order to adapt the grid resolution to the local requirements. However, the DG discretisation offers also an alternative path to the development of adaptive schemes. Since the reconstruction order (p) inside each element is independent from the reconstruction order in the neighbouring elements, it is possible to locally change the order and to adapt the resolution level without changing the mesh. This approach, known as p-adaptivity, allows to arbitrarly choose the number of degrees of freedom (and so the order p) inside each element. In this paper an example of hybrid hp-adaptive algorithm is proposed and compared with a classical h-adaptive algorithm.

2. Space and time discretisation

The results presented in this work are obtained by a research code in which the governing equations are numerically solved by means of a semidiscrete approach. In particular, the equations are discretised in space by means of a discontinuous Galerkin method. The resulting ordinary differential equations are integrated in time by means of a fully implicit backward Euler scheme. This choice is related to the fact that only steady examples are presented in this work. The resulting linear system is solved by means of the restarted GMRES method with the ILU0 preconditioner provided by the free version of the PARALUTION library [3].

2.1. DG space discretisation

Consider a partial differential equation which describes a general conservation law:

$$\frac{\partial u(\mathbf{x},t)}{\partial t} + \nabla \cdot \mathbf{F}(u, \nabla u) = q(u, \nabla u) \qquad \mathbf{x} \in \Omega \subset R^3, t \in R_0^+ \tag{1}$$

where u is the conservative variable, q is a source term and $\mathbf{F}$ is the flux vector (which can include both convective and diffusive terms). The solution u_h can be described by a linear combination of N_e basis functions:

$$u_h(\mathbf{x},t) = \tilde{\mathbf{u}} \cdot \mathbf{\Phi} = \sum_{i=1}^{N_e} \tilde{u}_i \Phi_i \tag{2}$$

The Galerkin approach consists in setting to zero the projection of the residuals of the discretised equation on the space spanned by the basis functions. The final variational formulation can be obtained by using integration by parts:

$$\int_{\Omega_e} \frac{\partial u_h}{\partial t} \Phi_i d\Omega + \int_{\partial\Omega_e} (\hat{\mathbf{F}} \cdot \mathbf{n}) \Phi_i ds - \int_{\Omega_e} (\mathbf{F}_h \cdot \nabla \Phi_i +) d\Omega - \int_{\Omega_e} q \Phi_i d\Omega = 0 \qquad 1 \leq i \leq N_e \qquad (3)$$

The volume integrals which appear in Eq. 3 are computed on the element Ω_e and the basis functions are defined only inside the element. For these reasons, there are not constraints related to the continuity of the solution between adjacent elements. The basis used in this work is a hierarchical modal basis obtained by applying the modified Gram-Schmidt orthonormalisation to a set of monomials defined in the physical space, according to Bassi et al. [4]. Local conservativity is guaranteed by using conservative numerical fluxes ($\hat{\mathbf{F}}$) across the interfaces. An upwind numerical flux based on the approach of [5] and [6] is used for the approximation of the convective fluxes. The diffusive fluxes are computed by means of a recovery based method [7]. The meshes are generated by the Gmsh tool [8] and then adapted in the CFD code.

3. Sensors for adaptive schemes

In order to adapt the discretisation according to the evolution of the solution it is necessary to use a sensor which can identifies the regions which need more resolution. In the framework of DG methods there are three main classes of sensors for adaptive schemes.

The first class is based on the definition of an adjoint equation which relates the error on a target function with the local properties of the discretisation. In this way, it is possible to perform a goal-oriented optimisation of the mesh in order to minimise the discretisation error on a quantity of interest [9]. This approach can be extremely efficient in steady RANS simulations. The extension to unsteady problems is challenging because of large memory requirements and the coupling of space-time adaptation [10].

The second approach is based on feature-based indicators which can identify some characteristic features of the flow field by performing a check on some phisical variables. Even if feature-based indicators can appear as ad-hoc tools, they can be quite effective and extremely cheap to evaluate [11]. This makes them interesting for unsteady problems.

Finally, it is possible to use the residuals of the discretisation to locally estimate the discretisation error on the conservative variables. This approach is not oriented to a particular target but aims at minimising all the sources of error in the field. This method is employed in this work. In particular the error indicator proposed by Leicht and Hartmann [12] is used:

$$\eta_e = h_e |\mathbf{R}|_{\Omega_e} + \sqrt{h_e} |\mathbf{r}|_{\partial\Omega_e} \qquad (4)$$

where h_e is the characteristic element size, $\mathbf{R}$ and $\mathbf{r}$ are the volume and interface residuals respectively which are defined in [12].

4. An hybrid hp-adaptive scheme

The local nature of the DG reconstruction offers the possibility to easily implement both h-adaptive and p-adaptive schemes. When a p-adaptive scheme is used, the reconstruction order is increased in the regions with the largest error sources: if the solution is sufficiently smooth, this approach leads to exponential convergence towards the exact solution.

However, the regularity of the solution is often reduced by the presence of singularities in the flow field (for example shocks) or in the geometry (for example sharp corners). In these circumstances, high-order reconstructions are not particularly useful because they cannot get the theoretical accuracy order. For this reason, it could be better to switch to an h-adaptive algorithm which refines the mesh close to the non-smooth features. The convergence rate obtained by the h-adaptive algorithm is not exponential but it is limited by the reconstruction order of the scheme and by the regularity of the solution. However, h-adaptivity is an effective way to keep under

control the error related to a singularity in the flow field.
The previous example suggests that an optimal apporach could be obtained by an hybrid algorithm which uses both h-adaptivity and p-adaptivity. In particular, the residual based error indicator described in the previous section is used to identify the elements which require more accuracy. Then it is necessary to decide if this element should be refined by splitting (h-adaptivity) or by increasing the reconstruction order (p-adaptivity). In order to make this choice it is possible to use an element wise smoothness sensor. Persson and Peraire [13] proposed a smoothness sensor for controlling artificial viscosity in shock capturing methods. Here, this sensor is applied to pressure in order to control the hybrid scheme:

$$S_{\Omega_e} = \frac{\int_{\Omega_e} (P - \bar{P})^2 d\Omega}{\int_{\Omega_e} P^2 d\Omega} \tag{5}$$

where P and $\bar{P}$ represent the pressure computed with the full DG expansion and with a truncated expansion in which the higher-order terms are neglected. The signal given by the sensor is related to the magnitude of the higher-order term in the expansion. These terms decay quickly on a smooth solution but they remain large in the presence of shock waves or other singularities. In the present work, the signal obtained by the smoothness indicator is compared with a threshold to decide whether the element should be refined by splitting or by p-refinement.

5. Transitional flow on the T106c cascade

In this Section the flow field in the T106c high-lift turbine cascade is studied. This test case shows several phenomena which are crucial for the design of low pressure gas turbines in turbofan engines. The working condition is described in [14] and is characterised by an inlet angle $\alpha_i = 32.7°$ and an isentropic exit Mach number $M_{2s} = 0.65$. The Reynolds number based on the axial chord and the isentropic exit conditions is $Re_{2s} = 80000$.
The flow field is described by the RANS equations coupled with a three equations turbulence model [15]. A second order accurate DG scheme is used for these simulations.
The chosen working condition is characterised by laminar separation on the suction side which is followed by transition to turbulence. In this test case it is fundamental to get enough resolution in the boundary layer and in the separated shear layer. The first requirement can be easily satisfied during the mesh generation process because it is sufficient to define the mesh element size as a function of the wall distance. However, the position of the shear layer is not know a priori and it is a result of the simulation. For this reason, it is useful to employ adaptive algorithms which can refine the mesh automatically during the simulation. In this test case an h-adaptive algorithm driven by the residual based error indicator of Section 3 is used. The original mesh and the adapted mesh after two refinement steps are shown in Fig. 1 and 2. At each refinement step a fixed fraction of the elements with the largest error indicator are splitted. The entropy field (normalised with respect to the constant volume specific heat) and some streamlines obtained on the final adapted mesh are reported in Fig. 3.

6. Transonic flow on a NACA0012 airfoil

Shock waves can deeply influence the performances of compressors and high-pressure turbines in turbojet engines. In this test case, the transonic flow field around the NACA0012 airfoil is considered according to the working condition described in [16]. The fluid is assumed inviscid. The free stream Mach number and angle of attack are $M_\infty = 0.8$ and $\alpha_\infty = 1.25°$. A second order accurate (p=1) scheme is used to study the flow field and the mesh is adapted by using an h-adaptive algorithm based on the residual indicator described in Section 3. In Fig. 4 and 5 the original mesh (1060 elements) and the adapted mesh (11806 elements) after five refinement steps are reported. The corresponding Mach fields are reported in Figs 6 and 7.

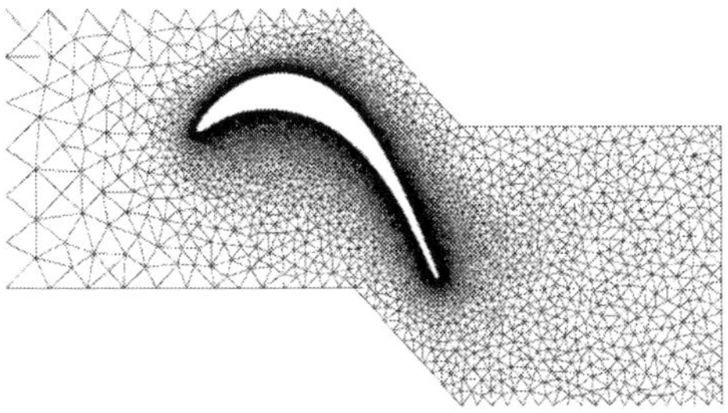

Figure 1. Initial mesh for the T106c test case.

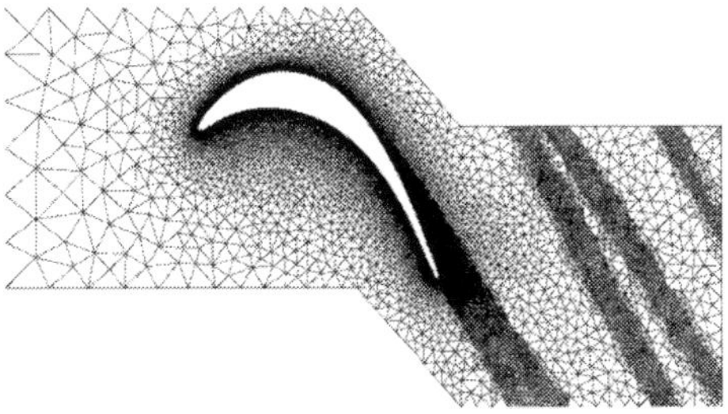

Figure 2. Adapted mesh for the T106c test case after two refinement steps.

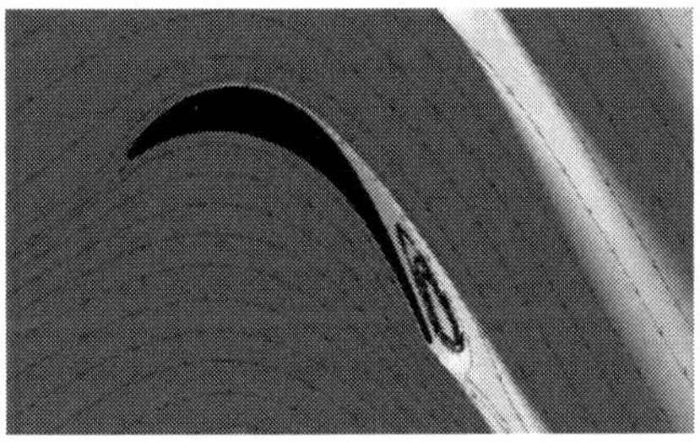

Figure 3. Normalised entropy field ans streamlines for the T106c cascade on the adapted mesh. The recirculation bubble and the shear layer can be clearly identified. The transition is localised in the shear layer at the edge of the separation bubble and it is physically related to a Kelvin-Helmholtz instability.

The same study is perfomed with a third order accurate DG scheme (p=2) and with the hybrid hp-adaptive algorithm described in Section 4. In Fig. 8 the error on the drag coefficient (computed with respect to a grid converged reference value) is reported as a function of the computational time for the different approaches. The plot shows that the implemented third order (p=2) scheme is not convenient with respect to the second order scheme (p=1) with the chosen adaptation strategy. This could be probably due to the presence of shock waves in the solution which limit the practical convergence order and reduce the benefits of the higher-order scheme. The hybrid scheme seems to be more efficient than the classical h-adaptive scheme for both reconstruction orders.

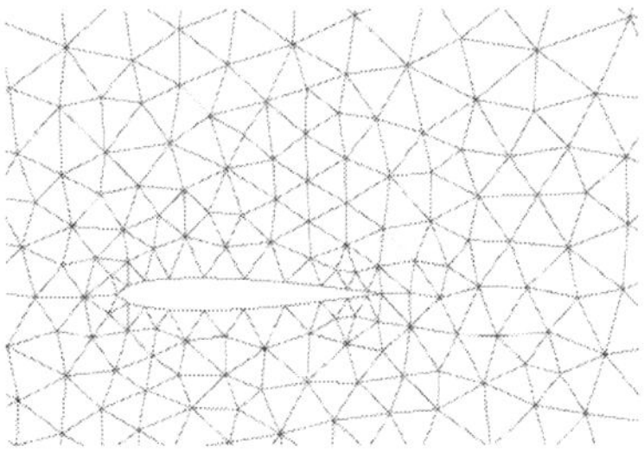

Figure 4. Initial mesh for the transonic NACA0012 airfoil.

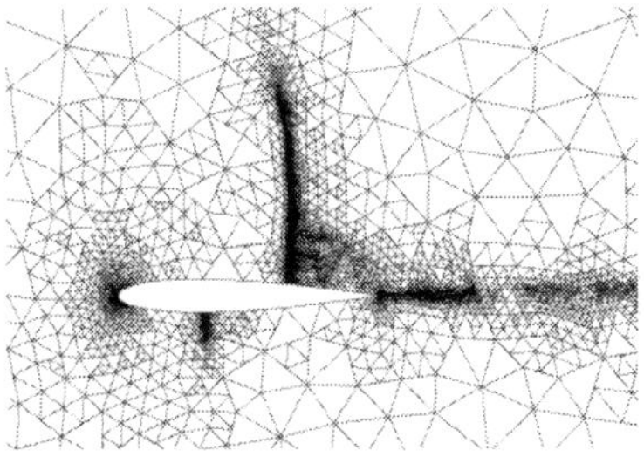

Figure 5. Adapted mesh for the transonic NACA0012 airfoil after 5 refinement steps.

7. Conclusions

The main adaptation strategies for discontinuous Galerkin discretisations are discussed and their importance in the simulation of turbulent compressible flows in propulsion systems is highlighted.

IOP Conf. Series: Journal of Physics: Conf. Series **841** (2017) 012017 doi:10.1088/1742-6596/841/1/012017

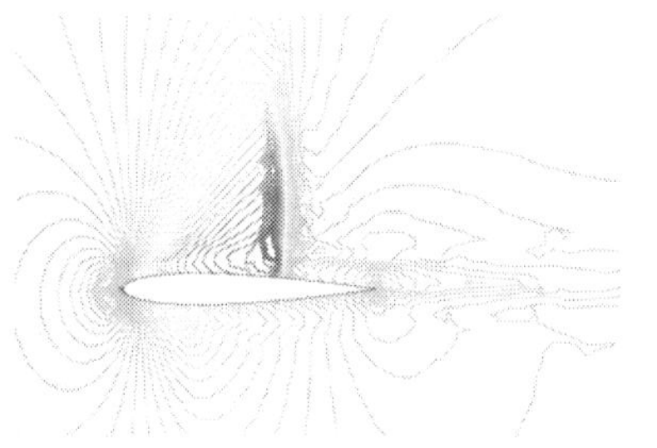

Figure 6. Mach field on the initial mesh (p=1).

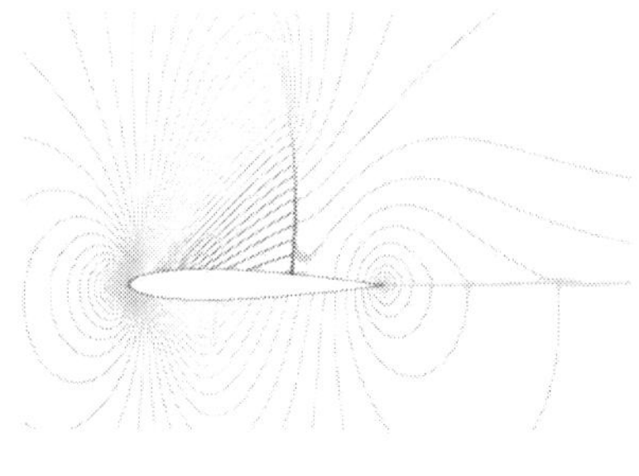

Figure 7. Mach field on the adapted mesh after 5 refinement steps (p=1).

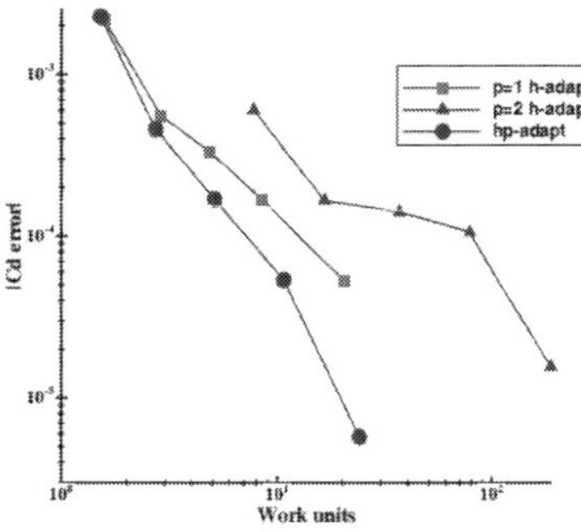

Figure 8. Error on the drag coefficient as a function of the normalised computational time. Comparison between an h-adaptive scheme (with both second ($p = 1$) and third order ($p = 2$) DG methods) and an hybrid hp-adaptive scheme ($1 \leq p \leq 2$).

Some applications of h-adaptive and hp-adaptive algorithms are presented. The proposed hybrid hp-adaptive algorithm is compared with an h-adaptive algorithm on a test case with shock waves. The hybrid algorithm performs better than the classical h-adaptive algorithm since it allows to get the same discretisation error with a lower computational cost. Future work will be devoted to dynamic load balancing to improve performances in adaptive parallel simulations.

References

[1] Kroll N 2009 *47th AIAA Aerospace Sciences Meeting* vol 176 (Orlando, Florida, USA)
[2] Hillewaert K 2013 *Development of the discontinuous Galerkin method for high-resolution, large scale CFD and acoustics in industrial geometries* vol 422 (Presses univ. de Louvain)
[3] Lukarski D and Trost N 2014 *PARALUTION project*
[4] Bassi F, Botti L, Colombo A, Di Pietro D A and Tesini P 2012 *J. Comput. Phys.* **231** 45–65
[5] Osher S and Solomon F 1982 *Math. comput.* **38** 339–374
[6] Pandolfi M 1984 *AIAA J.* **22** 602–610
[7] Ferrero A, Larocca F and Puppo G 2015 *Int. J. Numer. Meth. Fluids* **77** 63–91
[8] Geuzaine C and Remacle J F 2009 *Int. J. Numer. Meth. Eng.* **79** 1309–1331
[9] Hartmann R, Held J and Leicht T 2011 *J. Comput. Phys.* **230** 4268–4284
[10] Fidkowski K J and Darmofal D L 2011 *AIAA J.* **49** 673–694
[11] Ching E, Lv Y and Ihme M 2016 *54th AIAA Aerospace Sciences Meeting* (San Diego, California, USA)
[12] Leicht T and Hartmann R 2010 *J. Comput. Phys.* **229** 7344–7360
[13] Persson P O and Peraire J 2006 *AIAA paper* **2006-112**
[14] Hillewaert K, de Wiart C C and Arts T 2012 *Second International Workshop on High-Order CFD Methods* (Cologne, Germany)
[15] Pacciani R, Marconcini M, Fadai-Ghotbi A, Lardeau S and Leschziner M A 2011 *J. Turbomach.* **133** 031016
[16] May G and Woopen M 2012 *Second International Workshop on High-Order CFD Methods* (Cologne, Germany)

IOP Conf. Series: Journal of Physics: Conf. Series **841** (2017) 012018 doi:10.1088/1742-6596/841/1/012018

Effects of pre- and post-electrospinning plasma treatments on electrospun PCL nanofibers to improve cell interactions

M Asadian*[1], S Grande*[1], R Morent[1], A Nikiforov[1], H Declercq[2], N De Geyter[1]

[1] Faculty of Engineering & Architecture, Department of Applied Physics, Research Unit Plasma Technology (RUPT), Ghent University, Sint-Pietersnieuwstraat 41 B4, 9000 Ghent, Belgium

[2] Faculty of Medicine and Health Science, Department of Basic Medical Sciences, Tissue Engineering and Biomaterials Group, Ghent University, De Pintelaan 185 6B3, 9000 Ghent, Belgium

E-mail: Mahtab.Asadian@UGent.be, Silvia.Grande@UGent.be

Abstract. In this study, liquid plasma treatment was used to improve the morphology of Poly-ϵ-CaproLactone (PCL) NanoFibers (NFs), followed by performing a Dielectric Barrier Discharge (DBD) plasma surface modification to enhance the hydrophilicity of electrospun mats generated from plasma-modified PCL solutions. Cell interaction studies performed after 1 day and 7 days clearly revealed the highly increased cellular interactions on the double plasma-treated nanofibers compared to the pristine ones due to the combination of (1) a better NF morphology and (2) an increased surface hydrophilicity.

1. Introduction

The application of electrospun nanofibers (NFs) in tissue engineering is increasing owing to their morphological similarity to the ExtraCellular Matrix (ECM) as most of the ECM proteins have a fibrous structure with diameters in the nanometer or sub-micrometer range [1]. Nevertheless, a great challenge in electrospinning is the preparation of a suitable polymer solution because the system requires a balance of forces controlled by rheological behavior, conductivity and surface tension. In many cases, researchers have used mixed solvents or even additives to make solutions compatible to the electrospinning process [2]. However, within this work, a non-thermal Atmospheric Pressure Plasma Jet (APPJ) directly generated in a polymer solution will be explored trying to improve the electrospinnability and final NF quality. The applied APPJ set-up allows a close, intense contact between the plasma and the polymer solution which can lead to a very efficient effect on the quality of the resultant fibers [3 - 5]. As PCL is a good polymeric candidate for tissue engineering applications due to its numerous advantages such as good mechanical properties (highly elastic), chemical stability, excellent biocompatibility, research will be focused on PCL in this study. Nonetheless, due to its highly hydrophobic structure, PCL shows very poor cellular interactions. To cope with this problem, a Dielectric Barrier Discharge (DBD) plasma surface treatment will also be performed on selected nanofibrous samples after

IOP Conf. Series: Journal of Physics: Conf. Series **841** (2017) 012018 doi:10.1088/1742-6596/841/1/012018

the electrospinning process [6, 7]. For this purpose, PCL-NFs will be exposed to an argon plasma in an effort to improve their hydrophilicity as this is known to positively affect cellular interactions [8]. Plasma surface modification is a very attractive technique for this purpose as it does not significantly change the scaffold morphology nor its mechanical properties [9, 10]. In this study, the PCL solution viscosity and conductivity will be monitored before and after APPJ treatment, while Scanning Electron microscopy (SEM) will be used to evaluate the morphology of the resultant NFs. Furthermore, PCL-NFs will also be seeded with HFFs (Human foreskin fibroblasts) to perform cell adhesion tests after 1 and 7 days. For this purpose, 4 different samples will be examined: (1) untreated scaffolds, (2) scaffolds plasma-treated before the electrospinning process, (3) scaffolds plasma-treated after the electrospinning process and (4) scaffolds plasma-treated before and after the electrospinning step. Cell interactions on the different samples will be assessed by fluorescence microscopy and SEM.

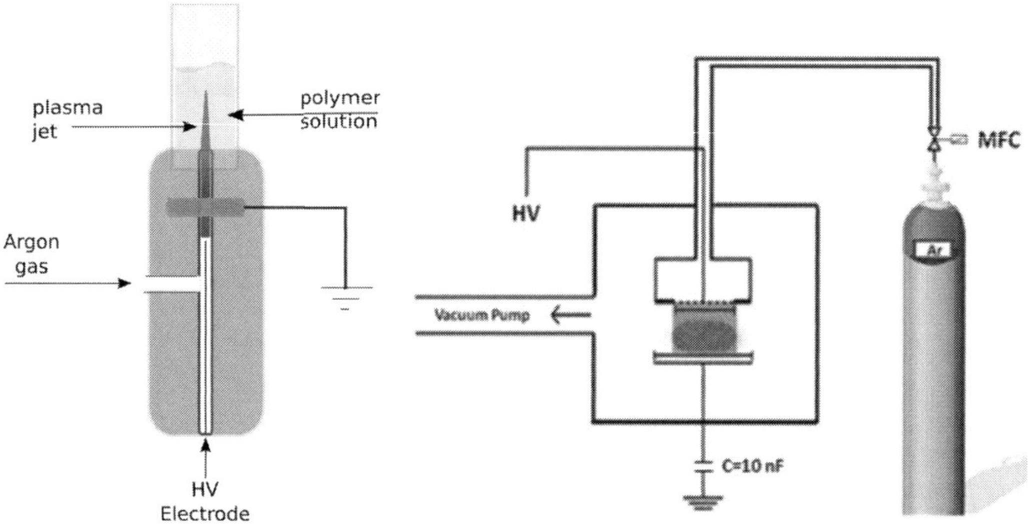

Figure 1. A schematic of the APPJ (left) to treat the PCL solution and the DBD (right) to improve the hydrophilicity of the PCL nanofibers.

2. Experimental

PCL with a MW of 80000 g/mol, chloroform and N,N-Dimethylformamide were purchased from Sigma-Aldrich and used without further purification. A solution of 5% of PCL in CHCl3+DMF (ratio of 9:1 (w/w)) was prepared at room temperature by stirring for 2 hours. Half of the obtained polymer solution was electrospun without any plasma pre-treatment step, while the other half of the polymer solutions was treated with the atmospheric pressure plasma jet (APPJ) schematically presented in Fig. 1 for 5 minutes making use of an argon flow rate of 0.9 slm (standard liters per minute) and an applied voltage of 7.5 kV. Before and after the APPJ modification step, the solution was characterized by measuring its conductivity and viscosity making use of a Mettler Toledo FiveEasy conductivity meter equipped with an InLab 720 probe and a Brookfield DV2T extra viscometer, respectively. After performing the APPJ modification step, untreated and plasma-modified PCL solutions were immediately electrospun making use of a Nanospinner 24 electrospinning machine manufactured by Inovenso (Turkey). This machine consists of a copper tip which is fed with the PCL polymer solution at a rate of 3.5 ml/h. A high voltage of 25 kV is applied to this tip, while a grounded rotating drum (rotation speed: 100 rpm) placed at 20 cm from the tip acts as collector plate. Round cover glasses with a diameter of 12 mm were fixed to the rotating drum and used as collecting substrates.

In a next step, the obtained PCL nanofibers were exposed to a DBD plasma schematically presented in Fig. 1(right) to increase their hydrophilicity. The DBD treatments were performed in argon at a discharge power of 1.5 W at medium pressure (50 kPa). A treatment time of 30 seconds was selected since this treatment duration was found to result in the highest possible hydrophilicity combined with negligible changes in NF morphology. The morphology of the PCL-NFs before and after APPJ liquid treatment and DBD plasma surface treatment were characterized with a scanning electron microscope (SEM; JSM-6010 PLUS/LV; JEOL). SEM images are acquired with an accelerating voltage of 7 kV, after coating the samples with a thin layer of gold making use of a sputter coater (JFC-1300 autofine coater, JEOL). The average diameter of nanofibers is calculated over 20 fibers using ImageJ (National Institutes of Health, USA) analysis software.

In a final step, HFF cells were cultured on untreated, pre-electrospinning plasma treated, post-electrospinning plasma treated and pre- and post-electrospinning plasma treated NFs after exposure to UV light for 30 minutes as a sterilization step. For this purpose, a density of 40.000 cells/500 μl of culture medium per scaffold was used. The cell seeded scaffolds were incubated at 37°C under 5% CO_2 atmosphere for 1 and 7 days and Tissue Culture PolyStyrene (TCPS) was used as positive control. Live/dead cell staining was applied to evaluate cell viability: the cell seeded scaffolds were incubated in 2 μl (1 mg/ml) of calcein-AM and 2 μl (1mg/ml) propidium iodide in 1 ml phosphate buffered saline (PBS), after which they were imaged with a fluorescence microscope (Olympus; IX 81) with appropriate filters [8]. The morphology of the HFF cells on the scaffolds after 1 and 7 days was also evaluated by SEM after performing a cell fixation and dehydration step. The preparation procedure included cell washing with PBS followed by fixation with 2.5% glutaraldehyde in 0.1 M cacodylate buffer for 1 hour. The cells were subsequently dehydrated by immersion through increasing ethanol series (50%, 70%, 85%, 95% and 100%) for 10 min. Afterwards, samples were immersed in hexamethyldisilazane (HDMS) for 10 min.

3. Results and discussion

3.1. APPJ treatment of the PCL polymer solution: solution properties and PCL-NF morphology

The conductivity and viscosity of the PCL polymer solution before and after APPJ treatment can be found in Table 1: from reported values it is evident that plasma exposure can strongly increase these parameters. SEM images of electrospun PCL-NFs generated from untreated and plasma treated PCL solutions are also displayed in Fig. 2. This figure clearly shows that the nanofibrous scaffold obtained from the untreated polymer solution presents a high density of beads (on average 2000 beads per mm^2) and an average fiber diameter of approximately 300 nm (high standard deviation due to the presence of beads). After 5 minutes of treatment with the plasma jet, the beads however completely disappear resulting in a network of elongated, uniform fibers with an average fiber diameter of approximately 500 nm.

Table 1. Polymer solution characteristics before and after APPJ treatment

	Untreated	**APPJ for 5 min**
Conductivity (μS)/cm	0.25 ± 0.05	5.1 ± 0.1
Viscosity (cP)	85 ± 5	320 ± 5
Fibers diameter (nm)	300 ± 150	500 ± 50

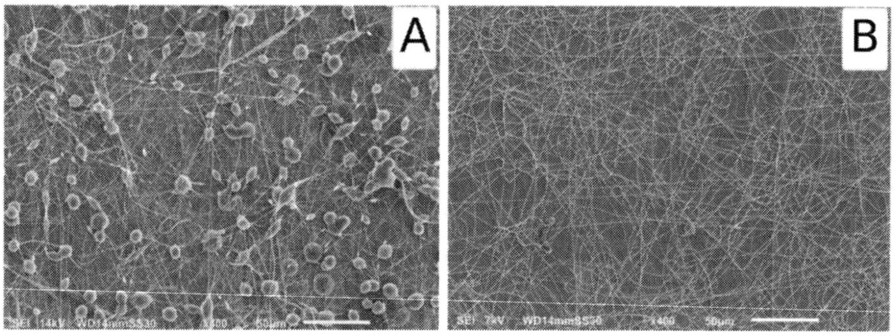

Figure 2. SEM images of nanofibers generated from A) an untreated PCL solution and B) a PCL solution treated with an APPJ for 5 minutes.

As mentioned before, plasma treatment is a very useful method to modify the chemical properties of a polymer surface. Plasma incorporated polar chemical groups are known to increase the surface energy of polymers and thus increase the surface wettability. The untreated nanofibers have a contact angle of approximately 127° which decrease to values of 55°, 64° and 95° for the nanofibers plasma treated with combined plasma treatments, DBD treated and APPJ treated respectively.

3.2. Live/dead immune microscopy and cell morphology using SEM

The cellular interactions on differently prepared PCL scaffolds have also been investigated 1 day and 7 days after cell seeding and the images obtained by SEM and fluorescent microscopy are presented in Figs. 3 and 4 respectively. These images clearly demonstrate the differences in cellular interactions between untreated, APPJ treated, DBD treated and combined plasma treatments on PCL-NFs. On untreated samples, after 1 day of cell seeding, mostly round and thus poorly attached cells can be observed. In addition, after 7 days of cell seeding, most of the present cells still show a round morphology while also quite a few cells appear to be dead (red color in Fig. 4). Cellular interactions on untreated samples are thus very poor. On samples which have been treated with the DBD, one can see that after 1 day and 7 days of cell seeding, more live cells are present on the samples and the cells adhere better to the sample surface, which can be evidenced from their spread morphology. These better cellular interactions compared to the untreated sample is most likely due to the enhanced hydrophilicity, which is known to positively affect cell-material interactions. When looking at the sample which has been treated with the APPJ alone, one can see that also in this case, cells nicely adhere to the sample as they present an elongated morphology, which is most likely due to the presence of uniform fibers instead of beads. However, the best cell-material interactions can be observed on the combined

APPJ and DBD plasma-treated samples: in this case, a high number of nicely elongated cells can be observed on the surface, even after only 1 day after cell seeding. After 7 days, a layer of cells is almost completely covering the nanofibers (as can be nicely observed in Fig. 3), which was not at all the case for the other investigated samples. These enhanced cellular interactions compared to the APPJ treated samples can again be related to the enhanced hydrophilicity upon DBD treatment. One can therefore conclude that a double plasma treatment is able to strongly enhance the cellular interactions on PCL nanofibers.

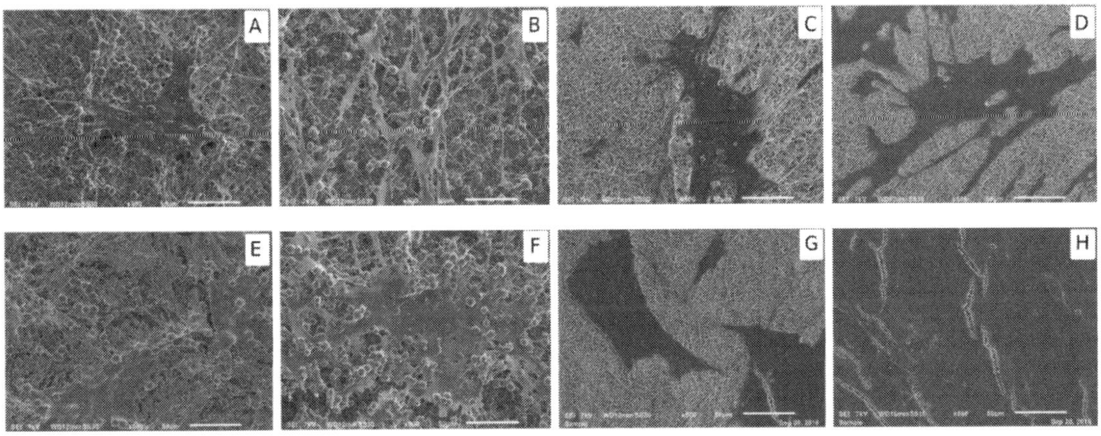

Figure 3. SEM images. First row: after 1 day, A) untreated samples, B) fibers with a DBD plasma treatment, C) APPJ plasma treatment, D) APPJ+DBD plasma treatment. Second row: after 7 days of seeding, E) untreated, F) DBD treatment, G) APPJ treatment, H) APPJ+DBD plasma treatment.

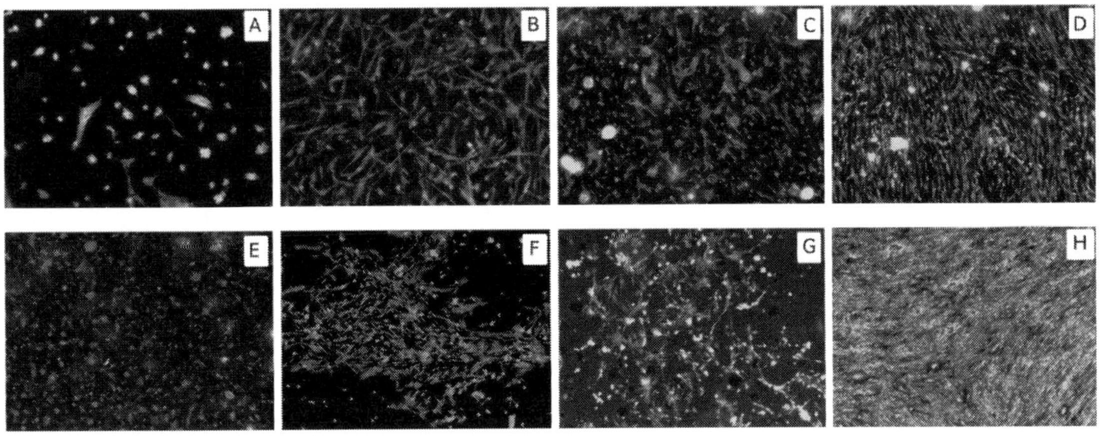

Figure 4. Fluorescence microscopy images. First row: after 1 day, A) untreated samples, B) fibers with a DBD plasma treatment, C) APPJ plasma treatment, D) APPJ+DBD plasma treatment. Second row: after 7 days of seeding, E) untreated, F) DBD treatment, G) APPJ treatment, H) APPJ+DBD plasma treatment.

4. Conclusion

After electrospinning an untreated PCL solution, the generated NFs presented a large amount of beads. It was however found that in the liquid plasma-treated fiber mesh, beads completely disappeared leading to more uniform fibers in diameter and in morphology. However, the resultant, elongated NFs still showed to have a strong hydrophobic behavior which is unwanted in the foreseen biomedical applications. DBD plasma treatment in argon of the NFs is therefore also performed to significantly improve their hydrophilicity. Cell studies revealed that HFFs cultured on PCL-NFs present a considerably higher cell adhesion and proliferation on combined APPJ and DBD plasma-treated NFs compared to pristine NFs. These results suggest that double plasma-modified NFs may have a large potential for the efficient replacement of damaged tissues in the future.

Acknowledgments

The research leading to these results has received funding from the European Research Council under the European Unions Seventh Framework Program (FP/2007-2013): ERC Grant Agreement number 335929 (PLASMATS).

References

[1] Teo W E, Ramakrishna S 2006, *Nanotechnology*, **17** (14), R89
[2] MacDiarmid A G *et al.* 2001, *Synthetic Metals*, **119**, 27-30
[3] Shi Q *et al.* 2011, *Journal of Polymer Science Part B: Polymer Physics*, **49** (2), 115-122
[4] Hsu C M, Shivkumar S 2004, *Macromolecular Materials and Engineering*, **289** (4), 334-340
[5] Colombo V, Fabiani D, Focarete M L, Gherardi M, Gualandi C, Laurita R and Zaccaria M 2014, *Plasma Processes and Polymers*, **11** (3), 247-255
[6] Desmet T, Morent R, De Geyter N, Leys C, Schacht E, Dubruel P 2009, *Biomacromolecules*, **10** (9), 2351-2378
[7] Jacobs T, Morent R, De Geyter N, Dubruel P, Leys C 2012, *Plasma Chemistry and Plasma Processing*, **32** (5), 1039-1073
[8] Yildirim E, Ayan H, Vasilets V, Fridman A, Guceri S, Sun W 2008, *Plasma Processes and Polymers*, **5** (1), 58-66
[9] Morent R, De Geyter N, Verschuren J, De Clerck K, Kiekens P, Leys C 2008, *Surface and Coatings Technology*, **202** (14), 3427-3449
[10] Morent R, De Geyter N, Leys C, Gengembre L, Payen E 2007, *Textile Research Journal*, **77** (7), 471-488

IOP Conf. Series: Journal of Physics: Conf. Series **841** (2017) 012019 doi:10.1088/1742-6596/841/1/012019

Thermodynamics of a qubit undergoing dephasing

S Marcantoni[1,2]

[1] Dipartimento di Fisica, Università di Trieste, I-34151 Trieste, Italy

[2] Istituto Nazionale di Fisica Nucleare, Sezione di Trieste, I-34151 Trieste, Italy

E-mail: stefano.marcantoni@ts.infn.it

Abstract. The thermodynamics of a qubit undergoing dephasing due to the coupling with the external environment is discussed. First of all, we assume the dynamics of the system to be described by a master equation in Lindblad form. In this framework, we review a standard formulation of the first and second law of thermodynamics that has been known in literature for a long time. After that, we explicitly model the environment with a set of quantum harmonic oscillators choosing the interaction such that the global dynamics of system and bath is analytically solvable and the Lindblad master equation is recovered in the weak-coupling limit. In this generalized setting, we can show that the correlations between system and bath play a fundamental role in the heat exchange. Moreover, the internal entropy production of the qubit is proven to be positive for arbitrary coupling strength.

1. Introduction

Standard thermodynamics is a phenomenological theory describing transformations of macroscopic systems in equilibrium according to few simple laws [1]. In particular, the first law of thermodynamics deals with the conservation of total energy that can be exchanged in the form of heat or work. The second law of thermodynamics instead regards the entropy, a state function that can never decrease in isolated systems, thus providing an arrow of time.

In the last years many attempts have been made to study the fate of thermodynamics when the system of interest can no longer be considered macroscopic and is out of equilibrium. This research is motivated by the high degree of control reached in many experimental setups such as ultracold atoms [2], optomechanical systems [3] and trapped ions [4], that allows to test the laws of thermodynamics in a completely new scenario. At the nanoscale quantum effects are expected to become important and the formulation of a consistent theory of quantum thermodynamics has recently attracted lot of attention [5]. From a fundamental point of view, one can address the issue of equilibration and thermalization in closed systems [6]. Moreover, it is interesting to study how the phenomenological laws of thermodynamics emerge from the underlying microscopic dynamics and in what sense they can be generalized beyond the usual thermodynamic limit [7].

In the following, we discuss two different approaches to quantum thermodynamics by means of an explicit example, namely a qubit undergoing dephasing due to the interaction with the external environment. In Section 2 a master equation in Lindblad form is used to describe the time-evolution of the qubit and the thermodynamic formalism developed in [8] for a driven open quantum system is reviewed. In this case the environment is taken implicitly into account. In Section 3 an explicit model for the interaction between the qubit and a bath of harmonic oscillators is introduced, such that the unitary dynamics of the global system is analytically solvable [9] and the Lindblad master equation for the qubit alone is recovered in the weak-coupling limit. In this more general framework we can highlight the important role of bipartite correlations in the heat balance following the definitions given in [10]. Moreover, we can show that the internal entropy production for the qubit is positive without resorting to the weak-coupling approximation.

2. Lindblad master equation

Consider a qubit evolving in time according to the following master equation in Lindblad form [11, 12]

$$\partial_t \varrho(t) = -i\Big[\frac{\omega_0}{2}\sigma_z, \varrho(t)\Big] + \frac{\gamma}{2}\big(\sigma_z \varrho(t)\sigma_z - \varrho(t)\big), \tag{1}$$

whose solution is easily found to be

$$\varrho(t) = \varrho_{00}|0\rangle\langle 0| + \varrho_{11}|1\rangle\langle 1| + \mathrm{e}^{-\gamma t}(\varrho_{10}\mathrm{e}^{i\omega_0 t}|1\rangle\langle 0| + \varrho_{01}\mathrm{e}^{-i\omega_0 t}|0\rangle\langle 1|), \tag{2}$$

where $\sigma_z|\ell\rangle = (-)^\ell|\ell\rangle$, with $\ell \in \{0,1\}$. This is a model for dephasing, since the populations remain constant while coherence decays exponentially in time. The thermodynamics of this quantum system can be studied using the standard formulation presented in [8]. In particular, the first law of thermodynamics for a driven open quantum system with Hamiltonian $H(t)$ reads

$$\partial_t \mathbb{U}(t) = \partial_t \mathbb{Q}(t) + \partial_t \mathbb{W}(t), \tag{3}$$

where the internal energy $\mathbb{U}(t)$, the heat flux $\partial_t\mathbb{Q}(t)$ and the work power $\partial_t\mathbb{W}(t)$ are defined respectively

$$\mathbb{U}(t) := \mathrm{Tr}\left[\varrho(t)\, H(t)\right], \tag{4}$$
$$\partial_t\mathbb{Q}(t) := \mathrm{Tr}\left[\partial_t\varrho(t)\, H(t)\right], \tag{5}$$
$$\partial_t\mathbb{W}(t) := \mathrm{Tr}\left[\varrho(t)\, \partial_t H(t)\right]. \tag{6}$$

This separation between work and heat contributions to the energy variation is quite reasonable. Indeed, the work power (Eq. (6)) is vanishing as expected in absence of an external field modelled by a time-dependent Hamiltonian, while the heat flux (Eq. (5)) is zero for the unitary dynamics generated by $H(t)$, namely if the system is closed.

For the qubit described in equation (2), the preservation of populations implies that the internal energy, namely the mean value of $\sigma_z\omega_0/2$, is constant in time. Moreover, the work power is zero, since there is no time-dependence in the Hamiltonian, and as a consequence the heat exchange with the environment is also vanishing. Up to now, it seems that dephasing happens without energy transfer; however, we will come back to this point later on, when we explicitly consider the presence of the bath.

For an open quantum system interacting with a thermal bath at inverse temperature β, the internal entropy production σ can be defined as the difference between the total variation of the entropy and the entropy flux due to the heat exchange [13]

$$\sigma(t) := \partial_t\mathbb{S}(t) - \beta\,\partial_t\mathbb{Q}(t), \tag{7}$$

where $\mathbb{S}$ is the von Neumann entropy

$$\mathbb{S}(t) := -\mathrm{Tr}\left[\varrho(t)\log\varrho(t)\right]. \tag{8}$$

The second law of thermodynamics states that $\sigma \geq 0$. In our model there is no heat exchange, so that the internal entropy production equals the variation of the von Neumann entropy and the second law reads $\partial_t\mathbb{S}(t) \geq 0$. Such a property is easily verified using the eigenvalues $(1 \pm r(t))/2$ of the density matrix $\varrho(t)$:

$$\partial_t\mathbb{S}(t) = -\frac{1}{2}\log\left(\frac{1+r(t)}{1-r(t)}\right)\partial_t r(t) \geq 0, \tag{9}$$

where

$$r(t) = \sqrt{1 - 4\left(\varrho_{00}\varrho_{11} - \mathrm{e}^{-2\gamma t}|\varrho_{01}|^2\right)}, \qquad \partial_t r(t) = -\frac{4\gamma\mathrm{e}^{-2\gamma t}|\varrho_{01}|^2}{r(t)}. \tag{10}$$

3. Explicit model for the environment

Instead of considering the thermodynamics of the qubit alone, a complete thermodynamic description of the qubit and the environment together could be given, in order to include possible effects due to the correlations between them. Indeed, as shown in [10], in a generic bipartite quantum system the correlations between the subsystems play a fundamental role and a consistent formulation of the first and second law of thermodynamics explicitly accounting for that can be given.

In the following, we study the exactly solvable model of a qubit in interaction with a thermal bosonic bath presented in [9]. In the weak-coupling limit the dynamics of the qubit is given by equation (1), but this more general approach highlights some interesting thermodynamic features. For instance, we show that due to correlations a non zero heat flux in the bath can happen without changing the internal energy of the qubit. Moreover, the internal entropy production of the qubit is found to be positive for an arbitrary coupling strength λ, namely without performing the standard Markov approximations.

3.1. The model

Consider a total Hamiltonian given by $H_{\rm tot} = H_S + H_B + H_{\rm int}$ with

$$H_S = \frac{\omega_0}{2}\sigma_z, \quad H_B = \sum_{k=1}^{\infty} \omega_k \mathrm{a}_k^\dagger \mathrm{a}_k\,, \quad H_{\rm int} = \lambda\sigma_z \otimes \sum_{k=1}^{\infty}\left(f_k^* \mathrm{a}_k + f_k \mathrm{a}_k^\dagger\right),$$

where a_k is the bosonic annihilation operator of mode k, satisfying the canonical commutation relations $[\mathrm{a}_k, \mathrm{a}_l^\dagger] = \delta_{kl}$, and the complex parameters f_k are such that $\sum_{k=1}^{\infty}|f_k|^2 < \infty$. We assume that the initial state of the total system can be written as $\varrho_{SB}(0) = \varrho_S(0) \otimes \varrho_B^\beta$, where $\varrho_S(0)$ is the initial state of the qubit and ϱ_B^β is the Gibbs state of the thermal bath at inverse temperature β,

$$\varrho_S(0) = \sum_{\ell,\ell'=0}^{1} \varrho_{\ell\ell'}|\ell\rangle\langle\ell'|\,, \qquad \sigma_z|\ell\rangle = (-)^\ell|\ell\rangle, \qquad \varrho_B^\beta = \frac{\mathrm{e}^{-\beta\sum_k \omega_k \mathrm{a}_k^\dagger \mathrm{a}_k}}{\mathrm{Tr}\left[\mathrm{e}^{-\beta\sum_k \omega_k \mathrm{a}_k^\dagger \mathrm{a}_k}\right]}. \tag{11}$$

The dynamics of the total system can be analytically solved (see [9] or [10] for all the details) and it turns out that the density matrix $\varrho_{SB}(t)$ can be written as follows

$$\varrho_{SB}(t) = \sum_{\ell,\ell'=0}^{1} \varrho_{\ell\ell'}\mathrm{e}^{-i\omega_0\zeta_{\ell\ell'}t/2}|\ell\rangle\langle\ell'| \otimes D_\ell(g_t)\,\varrho_B^\beta\,D_{\ell'}^\dagger(g_t), \tag{12}$$

where $\zeta_{\ell\ell'} = (-)^\ell - (-)^{\ell'}$ and $D_\ell(g_t)$ is the displacement operator

$$D_\ell(g_t) = \mathrm{e}^{(-)^\ell\lambda\sum_k[g_k(t)\mathrm{a}_k^\dagger - g_k^*(t)\mathrm{a}_k]}, \quad g_k^*(t) = f_k^*(\mathrm{e}^{-i\omega_k t} - 1)/\omega_k. \tag{13}$$

By partial tracing one can obtain the reduced density matrices of the two subsystems

$$\varrho_S(t) = \varrho_{00}|0\rangle\langle 0| + \varrho_{11}|1\rangle\langle 1| + \mathrm{e}^{-8\lambda^2\Gamma(t)}\left(\varrho_{10}\mathrm{e}^{i\omega_0 t}|1\rangle\langle 0| + \varrho_{01}\mathrm{e}^{-i\omega_0 t}|0\rangle\langle 1|\right), \tag{14}$$

$$\varrho_B(t) = \sum_{\ell=0}^{1}\varrho_{\ell\ell}D_\ell(g_t)\,\varrho_B^\beta\,D_\ell^\dagger(g_t), \tag{15}$$

where $\mathrm{Tr}\left[D_\ell(g_t)\,\varrho_B^\beta\,D_{\ell'}^\dagger(g_t)\right] = \mathrm{e}^{-8\lambda^2\Gamma(t)}$ for $\ell \neq \ell'$, with

$$\Gamma(t) = \sum_k \frac{|f_k|^2}{\omega_k^2}\coth(\beta\omega_k/2)\sin^2(\omega_k t/2). \tag{16}$$

The bath at any time is described by a convex combination of displaced thermal states, while the dynamics of the qubit is similar to that one in equation (2) but with a time-dependent damping $\Gamma(t)$.

It is possible to recover equation (2) from equation (14) in the so called weak-coupling limit. First of all, we should substitute the discrete sum in equation (16) with the following integral

$$\Gamma(t) = \int_0^\infty d\omega \frac{1}{\omega} \coth(\beta\omega/2) \sin^2(\omega t/2)\, e^{-\epsilon\omega}, \quad \epsilon \geq 0 \tag{17}$$

where a regularized Ohmic spectral density given by $f_k \simeq \sqrt{\omega}\, e^{-\omega\epsilon/2}$ has been used as in [9]. Then, a new time scale τ is defined such that $t = \tau/\lambda^2$ and in the limit $\lambda \to 0$ one finds $\lambda^2\Gamma(\tau/\lambda^2) \simeq \pi\tau/(2\beta)$. The solution of the Lindblad master equation (1) is immediately found identifying the constant damping rate $\gamma = 4\pi/\beta$.

3.2. Heat balance

Using the model above, we can exploit an interesting feature of bipartite quantum systems related to the heat balance. The total energy of the composite system $\mathbb{U}_{\text{tot}}$ defined as follows

$$\mathbb{U}_{\text{tot}} := \text{Tr}\left[H_{\text{tot}}\varrho_{SB}(t)\right] \tag{18}$$

is conserved, so that its time derivative is vanishing $\partial_t \mathbb{U}_{\text{tot}} = 0$. Explicitly, one can write

$$\begin{aligned}\partial_t \mathbb{U}_{\text{tot}} =& \text{Tr}\left[H_{\text{tot}}\, \partial_t\varrho_S(t) \otimes \varrho_B(t)\right] + \text{Tr}\left[H_{\text{tot}}\, \varrho_S(t) \otimes \partial_t\varrho_B(t)\right] + \text{Tr}\left[H_{\text{tot}}\, \partial_t\chi(t)\right] = \\ =& \text{Tr}\left[\partial_t\varrho_S(t)\, H'_S(t)\right] + \text{Tr}\left[\partial_t\varrho_B(t)\, H'_B(t)\right] + \text{Tr}[H_{\text{int}}\, \partial_t\chi(t)] = 0,\end{aligned} \tag{19}$$

where the operator $\chi(t) := \varrho_{SB}(t) - \varrho_S(t) \otimes \varrho_B(t)$ has been introduced to describe the amount of correlations between S and B, and a modified Hamiltonian has been defined for each subsystem

$$H'_{S,B}(t) := H_{S,B} + \text{Tr}_{B,S}\left[\varrho_{B,S}(t)H_{\text{int}}\right]. \tag{20}$$

The first two terms in the second line of equation (19) can be associated to heat exchanged by S and B respectively, indeed they are formally similar to equation (5). However, the modified Hamiltonians appear instead of the free Hamiltonians H_S and H_B; this is expected since the interaction Hamiltonian should contribute to the internal energy of each subsystem. Given the second line of equation (19) and defining a binding energy as

$$\mathbb{U}_\chi(t) := \text{Tr}[\chi(t)H_{\text{int}}], \tag{21}$$

the heat balance can be stated as follows

$$\partial_t \mathbb{Q}_S(t) + \partial_t \mathbb{Q}_B(t) = -\partial_t \mathbb{U}_\chi(t). \tag{22}$$

The binding energy is interpreted as an amount of energy stored in the correlations between S and B that can be exchanged with both subsystems in the form of *heat* [10].

For the model of interest, denoting the qubit polarization at time $t = 0$ by $\langle\sigma_z\rangle$ and using equations (14) and (15), the modified qubit Hamiltonian takes the form

$$H'_S(t) = \left(\frac{\omega_0}{2} - 4\lambda^2\langle\sigma_z\rangle\,\Delta(t)\right)\sigma_z, \tag{23}$$

where $\Delta(t)$ corresponds to

$$\Delta(t) = -\frac{\text{Tr}\left[\varrho_B(t)\sum_k\left(f_k^* a_k + f_k a_k^\dagger\right)\right]}{4\lambda\langle\sigma_z\rangle} = \sum_k \frac{|f_k|^2}{\omega_k}\sin^2(\omega_k t/2), \tag{24}$$

while $H'_B(t)$ reads

$$H'_B(t) = \sum_k \omega_k \mathrm{a}_k^\dagger \mathrm{a}_k + \lambda \langle \sigma_z \rangle \sum_k \left(f_k^* \mathrm{a}_k + f_k \mathrm{a}_k^\dagger \right). \tag{25}$$

As a consequence, the heat fluxes for both S and B can be explicitly computed and turn out to be

$$\partial_t \mathbb{Q}_S(t) = 0, \tag{26}$$

$$\partial_t \mathbb{Q}_B(t) = 4\lambda^2 \left(1 - \langle \sigma_z \rangle^2\right) \partial_t \Delta(t). \tag{27}$$

Notice that the heat exchanged by the qubit is vanishing even in this more general situation, because the modified Hamiltonian is proportional to σ_z and $\mathrm{Tr}\,[\partial_t \varrho_S(t) \sigma_z] = 0$ since populations are preserved (see Eq. (14)). Nevertheless, the heat transfer for the bath of harmonic oscillators does not vanish as expected, indeed the correlations built in during the time-evolution effectively act as a third subsystem exchanging energy with B in the form of heat according to equation (22).

3.3. Entropy production

Another interesting comparison between the present model and the weak-coupling limit discussed in Section 2 regards the internal entropy production for the qubit. Again, the internal entropy production equals the time derivative of the von Neumann entropy because $\partial_t \mathbb{Q}_S(t) = 0$. Using equation (14), the entropy variation of the qubit $\partial_t \mathbb{S}_S$ can be calculated from the eigenvalues $(1 \pm r_S(t))/2$ of $\varrho_S(t)$ with

$$r_S(t) = \sqrt{1 - 4\left(\varrho_{00}\varrho_{11} - \mathrm{e}^{-16\lambda^2 \Gamma(t)} |\varrho_{01}|^2\right)}. \tag{28}$$

Explicitly, one finds

$$\partial_t \mathbb{S}_S(t) = -\frac{1}{2} \log\left(\frac{1 + r_S(t)}{1 - r_S(t)}\right) \partial_t r_S(t) = \lambda^2 \frac{16 |\varrho_{01}|^2 \,\mathrm{e}^{-16\lambda^2 \Gamma(t)}}{r_S(t)} \log\left(\frac{1 + r_S(t)}{1 - r_S(t)}\right) \partial_t \Gamma(t); \tag{29}$$

therefore, the sign of $\partial_t \mathbb{S}_S(t)$ corresponds to the sign of $\partial_t \Gamma(t)$. The behaviour of the latter can be studied considering the integral expression in equation (17). By means of the substitution $\omega t = \tilde{\omega}$ and taking the derivative inside the integral it turns out that

$$\partial_t \Gamma(t) = \int_0^\infty \mathrm{d}\tilde{\omega} \frac{1}{\tilde{\omega}} \sin^2(\tilde{\omega}/2)\, \partial_t \left[\coth(\beta\tilde{\omega}/2t)\, \mathrm{e}^{-\epsilon\tilde{\omega}/t}\right] \geq 0. \tag{30}$$

The second law of thermodynamics in the form $\partial_t \mathbb{S}_S(t) \geq 0$ is hence satisfied without invoking the weak-coupling limit.

4. Conclusions

We compared two different approaches to study the thermodynamics of a quantum system. In particular, a qubit undergoing dephasing has been firstly analyzed in the Markovian regime where the time evolution is well described by a Lindblad master equation and the paradigm of [8] can be used. After that, an explicit model for the environment has been introduced such that the global dynamics of system and bath is analytically solvable and the previous qubit dynamics is recovered in the weak-coupling limit. In this more general framework, the important role of correlations in the heat balance has been highlighted using the definitions given in [10]. Moreover, the second law of thermodynamics for the qubit has been proven to hold independently of the coupling strength.

IOP Conf. Series: Journal of Physics: Conf. Series **841** (2017) 012019 doi:10.1088/1742-6596/841/1/012019

References

[1] Callen H B 1985 *Thermodynamics and an Introduction to Thermostatistics* (New York: John Wiley & Sons).
[2] Bloch I, Dalibard J and Zwerger W 2008 *Rev. Mod. Phys.* **80**, 885.
[3] Aspelmeyer M, Kippenberg T J and Marquardt F 2014 *Rev. Mod. Phys.* **86** 1391.
[4] Leibfried D, Blatt R, Monroe C and Wineland D 2003 *Rev. Mod. Phys.* **75** 281.
[5] Vinjanampathy S and Anders J 2016 *Contemp. Phys.* **57** 1-35.
[6] Gogolin C and Eisert J 2016 *Rep. Prog. Phys.* **79** 056001.
[7] Gemmer J, Michel M and Mahler G 2009 *Quantum Thermodynamics* (Berlin: Springer).
[8] Alicki R 1979 *J. Phys. A: Math. Gen.* **12** 103.
[9] Palma G M, Suominen K A and Ekert A 1996 *Proc. R. Soc. Lond.* A **452** 567.
[10] Alipour S, Benatti F, Bakhshinezhad F, Afsary M, Marcantoni S and Rezakhani A T 2016 *Sci. Rep.* **6** 35568.
[11] Lindblad G 1976 *Commun. Math. Phys.* **48** 119.
[12] Gorini V, Kossakowski A and Sudarshan E C G 1976 *J. Math. Phys.* **17** 821.
[13] Spohn H and Lebowitz J L 1979 *Adv. Chem. Phys.* **38** 109

IOP Conf. Series: Journal of Physics: Conf. Series **841** (2017) 012020 doi:10.1088/1742-6596/841/1/012020

Investigation of ferrofluid nanostructure by laser light scattering: medical applications

E K Nepomnyashchaya[1], E N Velichko[1], I V Pleshakov[1,2], E T Aksenov[1], E A Savchenko[1]

[1]Department of Quantum Electronics, Peter the Great Saint Petersburg Polytechnic University, Saint Petersburg 195251, Russia

[2]Laboratory of Quantum Electronics, Ioffe Institute, Saint-Petersburg, 194021, Russia

E-mail: elina.nep@gmail.com

Abstract. Investigation of ferrofluids nanostructure by the laser light scattering technique is presented. Experimental studies involved measurements of the intensity of the laser radiation scattered by ferrofluid particles in interaction with albumin and under the influence of magnetic field. The effects of the magnitude and duration of the applied magnetic field on the formation of aggregates of magnetic nanoparticles and also the influence of magnetic fluids of different concentrations on blood proteins are considered. The findings may be useful for medical applications.

1. Introduction

Magnetic fluids (also known as ferrofluids) represent a special category of smart nanomaterials in colloidal suspensions composed of single-domain magnetic nanoparticles [1]. Owing to their unique properties, magnetic fluids have found a widespread use in medicine. Their ability to be going to the point where the magnetic field intensity is the highest and to follow the field in their motion allow to direct magnetic nanoparticles with the blood flow to any point of the body [2]. Drug Coated magnetic particles can act as a drug delivery system. Their ability to effectively absorb electromagnetic radiation at specified frequencies and convert the energy into heat allows their use in magnetic fluid hyperthermia for the treatment of localized cancerous tumors [3]. By fixing antigens on a magnetic particle it is possible to effectively detect cancer cells, bacteria, and viruses via flow cytometry. Recent studies have shown that the introduction of magnetic nanoparticles with cancer antigens into the blood flow allows one to detect single tumor cells and even destroy them with the help of single pulses at the absorption frequency of a ferromagnetic nanoparticle. By using this technique referred to as theranostics not only the tumor but also all possible metastases can be destroyed and the patient can be completely cured [4, 5].

Despite the advantages offered by magnetic nanoparticles for cancer therapy, their use in clinical practice is still at the proof-of-concept stage. In order to reveal the consequences of the introduction of magnetic fluids into a body, experiments with the participation of animals and model objects are being carried out. In this paper, specific features of changes in the structure of magnetic nanoparticles in a liquid exposed to magnetic fields of different powers and durations are considered. In addition, the interaction of magnetic nanoparticles with human blood proteins is studied in order to reveal a possible influence of nanoparticles on the properties of proteins [6]. It is known that as soon as extraneous particles are injected into the blood, they become a target of all safety systems of an organism. The basic mechanism is the spontaneous adsorption of plasma proteins (opsonization process). Thus, the introduction of a large number of nanoparticles into a body can lead to a malfunction of these proteins.

2. Materials and method

2.1 Laser correlation spectroscopy

To study the nanostructure of magnetic fluids and their conglomerates, we used the laser correlation spectroscopic technique. A typical scheme of the device is shown in Fig. 1. In this setup, a coherent linearly polarized light beam from a laser with a diameter of 1.2 mm was transmitted through a converging lens and focused on the sample. A He-Ne laser with wavelength λ = 632.8 nm, 5 mW power, high stability and narrow spectral line was chosen for laser correlation measurements. The scattered light at angle θ = 15° passed through an iris diaphragm and was detected by a photomultiplier. The signal from the photomultiplier arrived to a computer for the correlation analysis.

We modified the scheme by adding coils to generate a magnetic field in the sample. The applied field was varied in the range 0 – 300 Oe with accuracy ± 0.5 Oe. The field distribution between the magnet poles was measured by a Hall sensor. It was found that the field gradient at the sample location was less than 1 Oe/cm. More information about measurement of light scattering on magnetic nanoparticle could be found in [7].

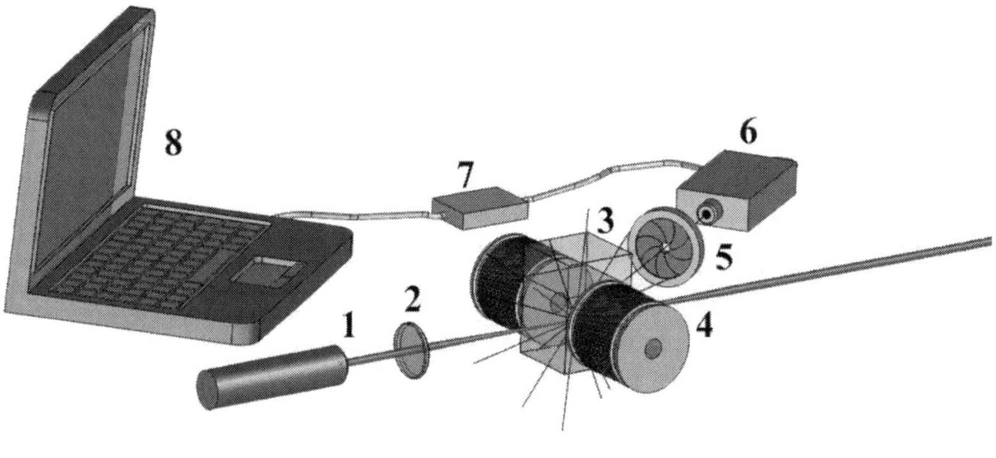

Figure 1. Scheme of a laser correlation spectrometer. 1 – laser radiation source; 2 – converging lens; 3 – sample; 4 – coils; 5 – iris diaphragm; 6 – photomultiplier; 7 – AD converter; 8 – computer

Laser correlation spectroscopy allows one to estimate sizes of nanoparticles in a fluid from measurements of scattered radiation parameters. The experiments involve recording of the time dependence of the scattered light intensity and the calculation of its autocorrelation function. According to the dynamic scattering theory, the autocorrelation function of the light scattered by particles in a solution will carry information on the diffusion coefficient (D) of these particles. By using the Stokes-Einstein relation

$$D = k_b T / 6\pi\eta R, \quad (1)$$

we can calculated the radii of nanoparticles or of their agglomerates. Here, η is the viscosity of the medium, k_b is the Boltzmann constant, T is the temperature, and R is the particle radius. Conclusions on the agglomeration and de-agglomeration of the particles and their degree of coupling can be drawn by measuring the sizes of scatters.

To derive information about sizes of nanoparticles the distribution of scattered light intensity on D coefficient I(D) is counted. Special algorithm based on Tikhonov regularization method [8] is used for calculation of intensity distribution I(D) from autocorrelation function. Then R is recounted from D by relation (1) and final dependence I(R) is calculated.

To evaluate the accuracy of experimental studies the test objects — water solutions of latex microspheres with specified sizes of 30 nm, 45 nm, and 75 nm were investigated. A series of 20 measurements were conducted for polydisperse solutions of microspheres of three sizes. The results of statistical analysis proved high accuracy of our method of analysis of particle sizes in polydisperse solutions — the percentage error was 6 % (at a confidence level of 95 %).

2.2 *Object of study*

The objects used in our studies were nanoparticles of magnetite Fe_3O_4 in water. The samples of ferrofluids were produced in our laboratory by the method described in [9]. It was shown in it that FF nanoparticles follow a log-normal distribution with a scaling parameter of 0.17. Aqueous dispersions are typically stable for long periods. However, at physiological pH nanoparticles form aggregates in the form of flocculates. Therefore, the oleic acid was used as a surfactant that prevented particle aggregation while storage. The average radius of the particles was about 4 nm ÷ 6 nm and typical concentrations were between 0.02 vol. % to 0.2 vol. %.

In order to study the interaction of magnetic nanoparticles with proteins of the blood plasma, albumin was used. Typical albumin size is 3.6 nm and the concentration is 100 mg/ml. This protein was chosen because of its high importance for the human organism. It performs a variety of functions, including the transport of substances in the body. Distortions in the structure of this protein can result in irreversible changes and cause a serious harm to the human health. Our task was to reveal changes in the protein structure which can occur at its mixing with a solution of magnetic fluid.

3. Results and discussion

It is known that it is preferable to use nanoparticles with radii of 5 – 50 nm for medical purposes because nanoparticles of these sizes are slowly removed by the reticuloendothelial system, which gives them sufficient time to accumulate in the tumor or other target region of the body [2]. In addition, large particles can damage small capillaries of both the extremities and the brain. Therefore, the nanoparticles should have a high colloidal stability in biological media because agglomeration processes are highly undesirable. We carried out measurements of the dependence of sizes of magnetic particle agglomerates on the applied magnetic field. Experimental results are shown in Fig. 2. R is the radii of particles in nanometers, H is the magnetic field value in Oe (given value), I(R) is the relative intensity of light scattered on each size of particles (counted from autocorrelation function).

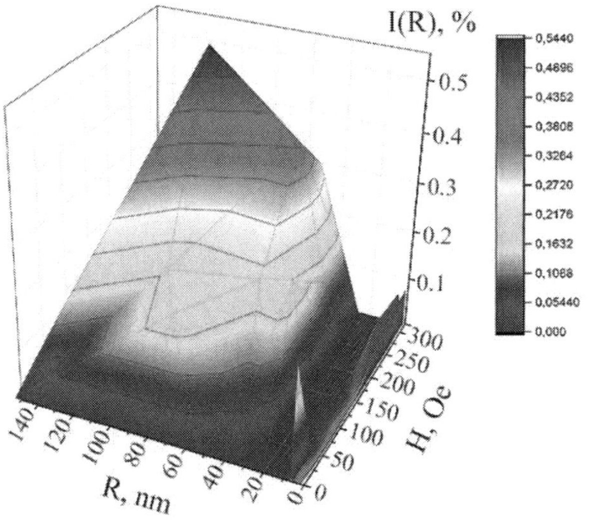

Figure 2. Dependence of sizes of magnetic nanoparticle agglomerates on applied magnetic field

It can be seen that under the fields lower than150 Oe the sizes of the majority of agglomerates do not exceed 40 nm. A further increase in the field gives rise to the formation of agglomerates with a radius of 60 – 140 nm, which exceeds the limit allowable in medical applications. Thus, it can be concluded that the maximum field should not exceed 150 Oe.

Then dynamic characteristics of the agglomerate formation were measured. A magnetic field of 150 Oe was applied to the sample for a long time (up to 40 minutes). The result is shown in Fig. 3, t stands for the field implementation time in minutes.

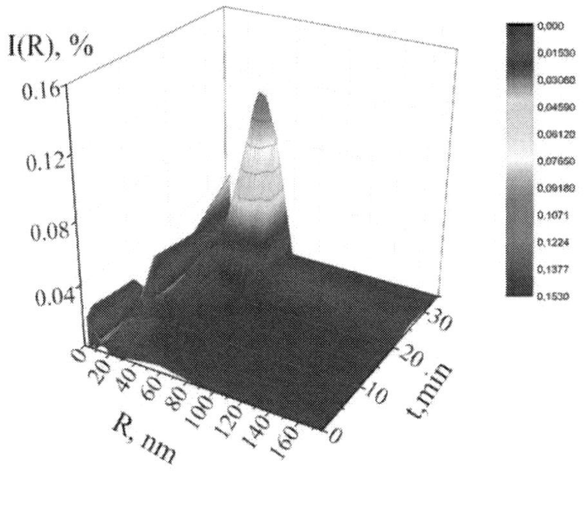

Figure 3. Dependence of sizes of magnetic nanoparticle agglomerates on magnetic field application period

After 20 minutes of exposure to the magnetic field, agglomerates of size larger than 40 nm began to appear in solution. After this, the number of small agglomerates also substantially increased. This may indicate that it is necessary to limit the exposure of a body to a magnetic field by a time period of twenty minutes. According to some studies [3], in order to destroy cancer cells by the method of Magnetic Fluid Hyperthermia, the field application during 30-40 minutes is sufficient for the local tumor heating to 41-54°C and cell destruction. The duration of exposure can be decreased by using a repeated procedure within 48 hours. This reduces the danger of colloidal stability disturbance and increases the safety of the procedure.

To investigate the effect of magnetic nanoparticles on albumin, we measured agglomerate sizes in a solution as functions of the magnetic fluid concentration. In the experiments, magnetic nanoparticles in water at a concentration of 0.02 vol. % were used. The solution of magnetic nanoparticles (from 0 to 1 μl) was added to 1 ml of albumin. The result is shown in Fig. 4. C is the volume of added magnetic fluid.

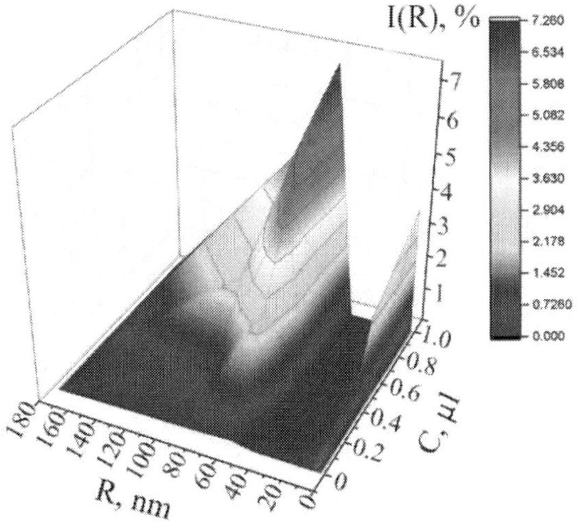

Figure 4. Dependence of sizes of magnetic nanoparticle agglomerates with albumin on concentration of magnetic particles

The size of albumin at a zero concentration of magnetic nanoparticles was about 4 nm. Addition of 0.1 μl of the ferrofluid gave rise to the formation of agglomerates with a radius of about 70 nm, which indicates that a part of magnetic nanoparticles was deposited on the surfaces of albumin molecules. The concentration of such agglomerates increased with increasing concentration of the ferrofluid and reached a dangerous limit after 0.4 μl of ferrofluid was added. Thus, it can be seen that magnetic nanoparticles exert a considerable influence on blood proteins even if they are coated with oleic acid. However, at low concentrations this effect is not so appreciable. When a specially designed protein layer is used as a surfactant instead of oleic acid, the influence of magnetic particles on proteins can be further reduced. In this case their use becomes relatively safe.

4. Conclusion

Our studies have revealed that the formation of aggregates of magnetic particles of size 40 nm occurs in a magnetic fluid under an applied magnetic field higher than 150 Oe and with an exposure time of 20 minutes or longer. These values can serve as thresholds for the magnetic field and the duration of exposure in biomedical research, in which the sizes of conglomerates of magnetic particles should not exceed 40 nm. It has also been found that magnetic nanoparticles coated with oleic acid exert a considerable influence on blood proteins (albumin). Extensive studies are needed to find the parameters of ferrofluids and the conditions of their use appropriate for their safe and efficient application in human therapy.

References

[1] Scherer C and Figueiredo A 2005 *Brazilian Journal of Physics* **35** 718–727
[2] Lunnoo T and Puangmali T 2015 *Nanoscale Research Letters* **10** 426
[3] Latorre M and Rinaldi C 2009 *PRHSJ* **28** 227–238
[4] Duguet E, Vasseur S, Mornet S and Devoisselle J-M 2006 *Nanomedicine* **1** 157–168
[5] Răcuciu M 2016 *Researchgate* (2015)26.03
[6] Markides H, Rotherham M and El Haj A J 2012 *Journal of Nanomaterials* 1–11
[7] Nepomnyashchaya E K, Prokofiev A V, Velichko E N, Pleshakov I V and Kuzmin Yu I 2016 *Journal of Magnetism and Magnetic Materials* (2016)10.002
[8] Nepomnyashchaya E, Velichko E and Aksenov E 2016 *IOP Journal of Physics: Conference Series* **769** 012025
[9] Agruzov P M, Pleshakov I V, Bibik E E, Stepanov S I and Shamrai A V 2015 *EPL* **111** 57003

IOP Conf. Series: Journal of Physics: Conf. Series **841** (2017) 012021 doi:10.1088/1742-6596/841/1/012021

Theory of photon-driven correlated electrons in one dimension

M Puviani[1] and F Manghi[1,2]

[1] Dipartimento di Scienze Fisiche, Informatiche e Matematiche, Università degli Studi di Modena e Reggio Emilia, Via Campi 213/A, I-41125 Modena, Italy
[2] CNR-Institute of NanoSciences S3, Via Campi 213/A, I-41125 Modena, Italy

E-mail: matteo.puviani@unimore.it

Abstract. In this paper we present a general theoretical framework to study interacting electrons under the influence of an external time-periodic driving, such as a homogeneous laser field. This is performed through a true many-body calculation and the use of Floquet theory. In particular, we consider a linear atomic chain using the Hubbard model to describe the short-ranged Coulomb interactions between electrons, plus Cluster Perturbation Theory to embed the many-body exact solution for a finite system into both an extended and an infinite lattice. Due to the presence of the external time-periodic perturbation, the electronic problem can be mapped into the study of photon-dressed quasiparticles thanks to Floquet theorem, keeping into account of all the virtual processes (absorption and emission of photons by electrons) with the laser field. This leads to an extension of the many-body static theories to out-of-equilibrium systems. This theoretical approach allowed us to show how the electronic properties of the system can be controlled and tuned varying the laser parameters. Above all, an inverse insulator-to-metal transition can be obtained for the one dimensional infinite lattice, and edge localized states appear as a finite size effect in an extended truncated chain.

1. Introduction

The possibility of driving materials out of equilibrium with the application of a homogeneuos laser field has been recently widely studied for simple non-interacting systems but with meaningful results [1, 2]. In addition, these problems have become of greater interest thanks to the experimental advances in time-and angle-resolved photoemission spectroscopy (Tr-ARPES) [3] and the possibility of using cold atoms in optical lattices in presence of periodic drivings. In these conditions, the electrons of the material interact with the monochromatic electromagnetic field reaching a nonequilibrium steady state to form "photon-dressed quasiparticles". These are a combination of electrons from the material and photons from the driving field: therefore they can be treated as free as a whole, replacing the concept of bare electrons interacting with an external field. This scheme allows us to study the problem using the Floquet theory, mapping the original out-of equilibrium time-periodic problem into an effective steady time-independent one [4,5]. On the other hand, correlated static systems have always been fascinating due to their wide range of properties and applications, as well as for the possibility to induce and control phase transitions.

Therefore, the study of correlated systems with a true many-body approach in out-of-equilbrium conditions is at the same time challenging and interesting for the possibility to control the new degrees of freedom due to the driving to obtain new properties and reach conditions otherwise not accessible in static conditions. Indeed, the presence of correlation between electrons is affected by the photon-dressing induced by the external time-periodic field, emerging in a competing effect between the local e-e repulsion and the absorption or emission of photons which leads to a renormalization of the kinetic energy of the electrons.

In this paper we considered the prototype system of a one dimensional atomic chain at half filling, described in static conditions by the Hubbard Hamiltonian and predicted to be insulating. Then, we included Floquet theory (section 2) to describe the exact interaction with an external laser field using an effectively time-independent scheme: to do this, the problem must be reformulated using the Green's function formalism. First, we calculate the one-particle Floquet-Green's function in the interacting framework for a finite size atomic cell (section 3). After that, this exact solution is embedded into the extended or infinite lattice with a Cluster Perturbation Theory (CPT) method [6–10], here applied to non-equilibrium conditions, in order to get the spectral functions and densities of states which describe the system (section 4). Eventually, our results are focused on the quasienergy band dispersion of the infinite chain and the comparison with the truncated chain, with highlight to finite-size effects (section 5).

2. The Floquet-Hubbard model

The problem of a lattice with a time-periodic external perturbation is characterized by an electronic time-periodic Hamiltonian $\hat{H}(t) = \hat{H}(t+T)$ ($T = \frac{2\pi}{\Omega}$ being the period of the external perturbation, Ω its frequency). Therefore, the time-dependent Schrödinger equation reads

$$\hat{H}(t)\Psi(\mathbf{x},t) = i\ \partial_t\Psi(\mathbf{x},t)\ , \tag{1}$$

where $\Psi(\mathbf{x},t) = \Psi(x^{(1)},x^{(2)},...,x^{(N)},t)$ is the many-body wave function for N particles. According to Floquet theorem, which can be interpreted as Bloch theorem in time domain, the solution can be written as a factorization of a quasiparticle wave function periodic in time and a complex exponential playing the role of the time-evolution of the quasiparticle:

$$\Psi(\mathbf{x},t) = e^{-iEt}\ \Phi(\mathbf{x},t)\ . \tag{2}$$

We have named E the quasiparticle energy and $\Phi(\mathbf{x},t)$ the quasiparticle wave function, which can be rewritten according to Fourier expansion as

$$\Phi(\mathbf{x},t) = \sum_{n=-\infty}^{+\infty} e^{-in\Omega t}\ \phi_n(\mathbf{x})\ . \tag{3}$$

Substituting in (1) the form of the solution provided in (2) we get, after some algebra, the following equation:

$$\left(\hat{H}(t) - i\ \partial_t\right)\Phi(\mathbf{x},t) = E\ \Phi(\mathbf{x},t)\ . \tag{4}$$

Defining the Floquet operator as $\hat{H}_F \equiv \hat{H}(t) - i\partial_t$, playing the role of an effective Hamiltonian, equation (4) is mapped into a time-independent eigenvalue problem as follows:

$$\hat{H}_F\ \Phi_\alpha(\mathbf{x},t) = E_\alpha\ \Phi_\alpha(\mathbf{x},t)\ , \tag{5}$$

where α are the labels for Floquet eigenstates and their corresponding quasi-energies. Moreover, it's important to notice that, according to Floquet theorem and the expansion in (3), if E_α is a solution, then $E_\alpha + n\Omega$ (with n any integer number) is a solution, too. This means that

IOP Conf. Series: Journal of Physics: Conf. Series **841** (2017) 012021 doi:10.1088/1742-6596/841/1/012021

there are infinite replicas of Floquet quasienergies (or, equivalently, that they're defined up to multiples of the photon energy, Ω), which physically correspond to the exchange (either emission or absorption) of photons of the electrons with the applied electromagnetic field. Eventually, it's noticeable that while the expectation values of the Hamiltonian calculated for Floquet states are time-dependent (they are not conserved), namely $\langle\Psi_\alpha(t)|\hat{H}(t)|\Psi_\alpha(t)\rangle = \varepsilon_\alpha(t)$; Floquet quasienergies are not, and they are therefore conserved.

Now, in order to study a one-dimensional insulating chain as a true many-body problem, we can use the Hubbard Hamiltonian with one-orbital per atom, nearest-neighbor hopping and on-site Coulomb interaction. In static conditions it is written in the second quantized form as

$$\hat{H} = -\sum_{\langle i,j\rangle}\sum_{\sigma=\uparrow,\downarrow} J_{i,j}\ \hat{c}^\dagger_{i,\sigma}\ \hat{c}_{j,\sigma} + U\sum_i \hat{n}_{i,\uparrow}\ \hat{n}_{i,\downarrow}\ , \tag{6}$$

J describing the unperturbed hopping (non-local), while U is the repulsive Hubbard term (local). We can now describe the external driving by the vector potential in one dimension $A(t) = A_0\sin(\Omega t)$: according to Peierls' substitution [11, 12] it can be included in the hopping term of the Hubbard Hamiltonian provided that $J_{i,j} \to J\ e^{iA(t)\cdot(x_i-x_j)}$. It can also be rewritten in an expanded form as follows:

$$J\ e^{iA(t)\cdot(x_i-x_j)} = \sum_{p=-\infty}^{+\infty} \mathrm{J}_p(A_0(x_i-x_j))\ e^{ip\Omega t}\ , \tag{7}$$

where $\mathrm{J}_p(z)$ is the Bessel function of first kind and order p, with argument z. Moreover, for the driven system we can expand the time-dependent creation and annihilation operators similarly to (3), as follows:

$$\hat{c}^\dagger_{i,\sigma}(t) = \sum_{n=-\infty}^{+\infty} e^{-in\Omega t}\ \hat{c}^\dagger_{i,\sigma,n}\ , \quad \hat{c}_{i,\sigma}(t) = \sum_{n=-\infty}^{+\infty} e^{in\Omega t}\ \hat{c}_{i,\sigma,n}\ . \tag{8}$$

The time-independent operators appearing in the expansion act both on electrons (creating or annihilating one with spin σ on site i) and on the photons of the driving field, absorbing or emitting n of them (attention must be paid since the sign of n is physically meaningful). With the forementioned Peierls' substitution and the time-dependent creation/anihilation operators defined in (8), starting from (6), we can write the Floquet-Hubbard effective Hamiltonian as

$$\hat{H}_{FH} = -J\sum_{\langle i,j\rangle}\sum_{\sigma=\uparrow,\downarrow}\sum_{n,m=-\infty}^{+\infty} e^{iA(t)\cdot(x_i-x_j)}\ \hat{c}^\dagger_{i,n,\sigma}\ \hat{c}_{j,m,\sigma}\ e^{i(n-m)\Omega t} + U\sum_i \hat{n}_{i,\uparrow}\ \hat{n}_{i,\downarrow} - i\partial_t\ . \tag{9}$$

Here we have put $J_{i,j} = J$, with the summation performed over i, j neighboring sites. Its solution can be written in the form provided in (3), where the spatial many-body wave function $\phi_n(\mathbf{x})$ can be espanded in a complete set of localized non-interacting N-particle states. However, this can be done via exact diagonalization only for a finite and small number of electrons.

3. Cluster Floquet-Green's Function

It is convenient for our theoretical development to use the Green's function formalism, here extended to correlated driven systems treated with Floquet aproach. In particular, we can write the Floquet-Green's function in the Lehmann representation as follows:

$$G_{ij,nm}(\omega) = \lim_{\eta\to0^+}\sum_{n',n'',m',m''}\sum_\xi \left(\frac{\langle\Phi^N_{0,m'}|\hat{c}^\dagger_{j,m}|\Phi^{N-1}_{\xi,m''}\rangle\langle\Phi^{N-1}_{\xi,n'}|\hat{c}_{i,n}|\Phi^N_{0,n''}\rangle}{\omega - E^N_0 + E^{N-1}_\xi - i\eta}\ \delta_{m''+m,m'}\ \delta_{n''-n,n'} \right.$$
$$\left. + \frac{\langle\Phi^N_{0,m'}|\hat{c}_{i,n}|\Phi^{N+1}_{\xi,m''}\rangle\langle\Phi^{N+1}_{\xi,n'}|\hat{c}^\dagger_{j,m}|\Phi^N_{0,n''}\rangle}{\omega + E^N_0 - E^{N+1}_\xi + i\eta}\ \delta_{m''-n,m'}\ \delta_{n''+m,n'} \right)\ . \tag{10}$$

In order to compute it, we need to know all the interacting $(N+1)$- and $(N-1)$-particles eigenstates and eigenvalues, plus the ground state of N particles and its energy. Actually, this last step requires us to define an interacting many-body ground state for a system in non-equilibrium conditions, whose energy is given up to multiples of the photon energy of the driving. Indeed, even though the result is independent on which replica of the ground state level is chosen, still the states of a Floquet mode are mixed with other replicas of different modes. This means that excited states of a given mode can have lower energies of the ground state of a higher mode (i.e.: if $E_{1,n} - E_{0,n} < \Omega$, then $E_{1,n-1} = E_1 + (n-1)\Omega < E_{0,n} = E_0 + n\Omega$). In order to solve this intrinsic ambiguity and make a meaningful physical choice, we have adopted an adiabatic criterion: once the ground state for the static condition (represented by a null vector potential $A_0 = 0$) is found (the problem can be solved in this limit even without Floquet approach), we switch the perturbation on incresing the vector potential and we calculate the energy levels for the different values of A_0. The ground state energy at each value of the external field is then obtained continuously and smoothly following the ground state level through the increasing of the driving, from $A_0 = 0$ up to the desired value.

4. CPT and Lattice Floquet-Green's Function

Once we have obtained the Floquet-Green's function to describe a small cluster of atoms in the presence of correlation and a time-periodic driving, the extension to a larger or infinite system can be performed through Cluster Perturbation Theory. In an operatorial form the following Dyson-like expression holds:

$$\hat{G}^{\mathrm{lat}}(\omega) = \hat{G}^{\mathrm{cl}}(\omega) + \hat{G}^{\mathrm{cl}}(\omega)\, \hat{T}\, \hat{G}^{\mathrm{lat}}(\omega) \,. \tag{11}$$

$\hat{G}^{\mathrm{cl}}(\omega)$ and $\hat{G}^{\mathrm{lat}}(\omega)$ are the cluster and the lattice Floquet-Green's functions respectively, and $\hat{T}$ represents the interaction between clusters. The latter contains solely the hopping term, which is the factor allowing for tunneling of electrons between the atomic sites of the lattice, and is the only non-local term in the Floquet-Hubbard Hamiltonian connecting neighboring clusters. Translated into the Floquet-Hubbard formalism, this becomes

$$G^{\mathrm{lat}}_{ij,nm}(\omega) = G^{\mathrm{cl}}_{ij,nm}(\omega) + \sum_{\langle i',j' \rangle} \sum_{n',m'} G^{\mathrm{cl}}_{ii',nn'}(\omega)\, T_{i'j',n'm'}(k,\omega)\, G^{\mathrm{lat}}_{j'j,m'm}(\omega) \,, \tag{12}$$

where the summation is performed over neighboring sites. $T_{i'j',n'm'}(k,\omega) = \sum_l \mathsf{J}_{n'-m'}(A_0(x_{i'} - x_{j'}))\, e^{ikR_l}$ is the effective time-independent hopping as appearing in expression (7) with the periodicity of the cluster. The lattice Floquet-Green's function can be computed numerically by equation (12): it is valid both for an infinite lattice and for an extended finite one. In the former case periodicity is included in the hopping term of interactions between clusters, while in the latter one a Floquet-Green's function for M-sites (M large) can be computed starting from a few-sites function explicitly including the interactions (which are here a finite number) between neighboring clusters (in this case the k-dependence in it vanishes). For an infinite lattice we can compute the spectral functions of the photon-dressed quasiparticles as

$$D(k,\omega) = -\frac{1}{\pi} \sum_b Im\{G^{\mathrm{lat}}_b(k,\omega)\} \,, \tag{13}$$

where the summation is intended over all the bands, and with

$$G^{\mathrm{lat}}_b(k,\omega) = \sum_{ij,nm} e^{-ik(x_i - x_j)}\, c^b_{i,n}(k)^*\, c^b_{j,m}(k)\, G^{\mathrm{lat}}_{ij,nm}(k,\omega) \,. \tag{14}$$

The coefficients $c^b_{i,n}(k)$ and $c^b_{j,m}(k)$ are obtained from the solution of the Floquet problem without interactions (single-particle). On the other hand, for a finite chain the local and the total density of states are given respectively by

$$n_i(\omega) = -\frac{1}{\pi}\sum_b \sum_{n,m} Im\{(c^b_{i,n})^* \; c^b_{i,m} \; G^{\text{lat}}_{ii,nm}(\omega)\} \; , \quad n(\omega) = \sum_i n_i(\omega) \; . \tag{15}$$

5. Results

Using this theoretical approach, we firstly computed the exact diagonalization for a cluster formed by two sites at half occupation, then we calculated the infinite lattice Floquet-Green's function and the corresponding spectral function. We performed a systematic calculation of the Floquet quasienergy bands for different values of the correlation ($0 < U < 4$) and the external field parameters ($0 < A_0 < 5$ and $1 < \Omega < 10$), analyzing their features: in particular, as pointed out in [13], an insulator-to-metal transition can be driven by the external field at low frequencies ($\Omega \lesssim J$) and values of A_0 non-trivially depending on these parameters. However,

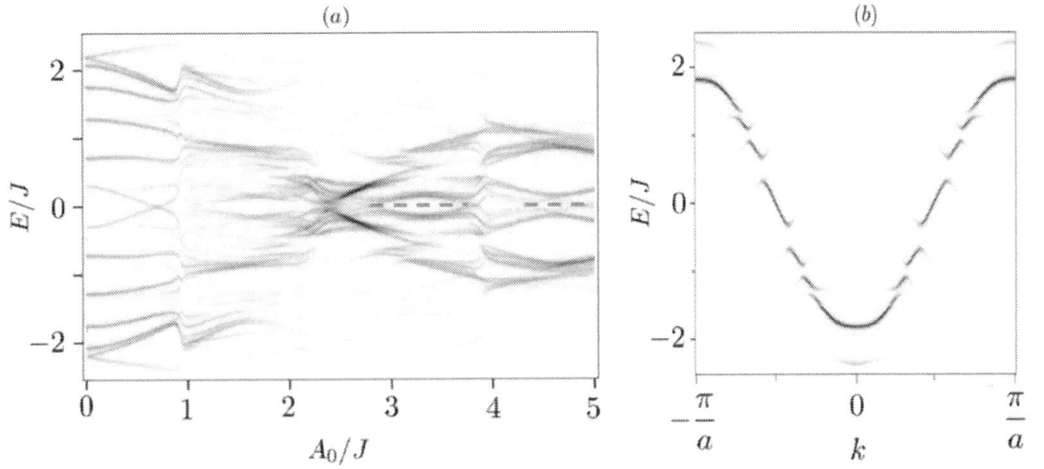

Figure 1. (a) Zero-mode Floquet quasienergy band dispersion for an infinte 1D chain calculated for $U/J = 2.0$ and $\Omega = 1.0\ J$ in 11 k-points in the FBZ, as a function of the field intensity A_0/J. Dashed blue lines are drawn in correspondence of the zero-energy states of the finite chain lying in the bulk gap (see Fig. 2(a)). (b) Zero-mode quasienergy band dispersion for $U/J = 2.0$, $\Omega = 1.0\ J$ and $A_0/J = 0.7$ within the FBZ.

this is not due to the Floquet quasienergies replicas, but to virtual processes, since it shows up also in the zero Floquet mode where the average number of photons exhanged is zero. In Fig. 1(a) is represented the quasienergy band dispersion for some k-points for different A_0 values in the case of $\Omega = 1.0\ J$ and $U/J = 2.0$: the closure of the gap in the Floquet zero-mode around the Fermi level in the quasienergy band dispersion occurs at $A_0/J \approx 0.7$, $A_0/J \approx 2.78$, $A_0/J \approx 3.84$ and higher values. In Fig. 1(b), instead, the band dispersion within the FBZ is reported for the specific value of $A_0/J = 0.7$, for which the gap is closed and the chain has a metal behaviour. With the same parameters of the driving field we calculated the density of states for an extended linear chain of 30 atoms, with the repetition of a 2-sites cluster: the results are shown in Fig. 2(a) as a function of the vector potential A_0. It can be seen that for this finite system zero-energy levels appear in correspondence of the gap of the inifite lattice band dispersion (reported as dashed blue lines in Fig. 1(a)). Moreover, analyzing for these parameters the local density of states for each site of the chain as represented in Fig. 2(b), it is clear that the these zero-energy states appearing in the finite chain are localized at the edges and exponentially decay in the inside.

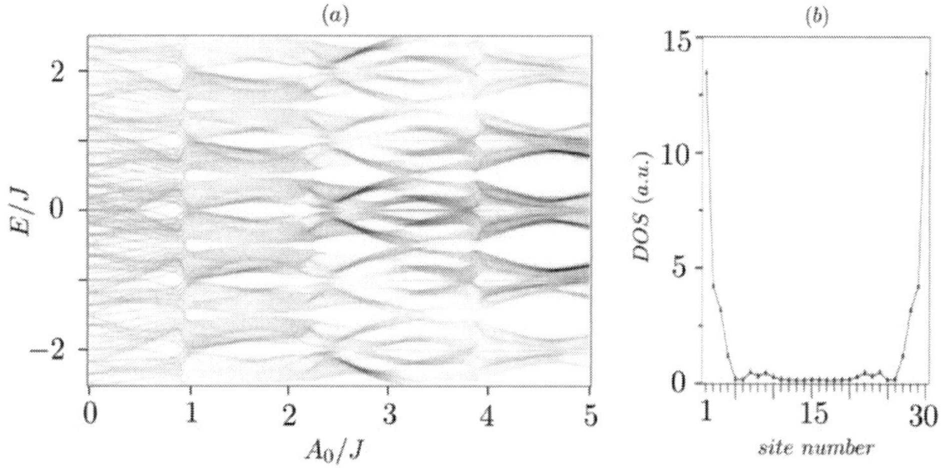

Figure 2. (a) Density of states of an open 1D chain with 30 atoms, calculated for $U/J = 2.0$ and $\Omega = 1.0\ J$. In panel (b) the local densities of states for the zero-energy level are reported as a function of the sites positions, for $U/J = 2.0$, $\Omega = 1.0\ J$ and $A_0/J = 3.0$.

6. Conclusion

In conclusion, we have developed a theoretical method to calculate Floquet quasienergies for extended correlated systems driven out of equilibrium by a periodic perturbation. In order to treat the many-body problem we used the exact diagonalization of the Hubbard model including Floquet theory for time-periodic systems in it. Moreover, we combined CPT to the Floquet-Green's function formalism in order to describe extended or infinite lattices. In this paper, we applied the theory to a one dimensional chain, comparing the infinite lattice to the truncated chain in order to characterize the finite size effects for this system. In fact, in the periodic lattice insulating and metallic phases can be driven by the external field, while gapless edge states appear in the bulk insulating phase. Since this procedure is fairly general, it can be extended to 2- or 3-dimensional lattices, while the Hubbard model could be also applied to superconductors describing attractive $e - e$ interactions.

References

[1] Gómez-León a and Platero G 2012 *Physical Review B* **86** 115318
[2] Delplace P, Gómez-León Á and Platero G 2013 *Physical Review B* **88** 245422
[3] Wang Y H, Steinberg H, Jarillo-Herrero P and Gedik N 2013 *Science* **342** 453–7
[4] Shirley J H 1963 *Interaction of quantum system with a strongly oscillating field* Ph.D. thesis
[5] Aoki H, Tsuji N, Eckstein M, Kollar M, Oka T and Werner P 2014 *Reviews of Modern Physics* **86** 779–837
[6] Manghi F 2014 *Journal of Physics: Condensed Matter* **26** 015602
[7] Grandi F, Manghi F, Corradini O and Bertoni C M 2015 *Phys. Rev. B* **91**(11) 115112
[8] Grandi F, Manghi F, Corradini O, Bertoni C M and Bonini A 2015 *New Journal of Physics* **17** 023004
[9] Potthoff M, Aichhorn M and Dahnken C 2003 *Phys. Rev. Lett.* **91**(20) 206402
[10] Sénéchal D, Perez D and Pioro-Ladrière M 2000 *Phys. Rev. Lett.* **84**(3) 522–5
[11] Hofstadter D R 1976 *Phys. Rev. B* **14**(6) 2239–49
[12] Jiménez-García K, LeBlanc L J, Williams R A, Beeler M C, Perry A R and Spielman I B 2012 *Phys. Rev. Lett.* **108**(22) 225303
[13] Puviani M and Manghi F 2016 *Phys. Rev. B* **94**(16) 161111

IOP Conf. Series: Journal of Physics: Conf. Series **841** (2017) 012022 doi:10.1088/1742-6596/841/1/012022

ICRF Traveling Wave launcher for fusion devices

R Ragona[1,2]
[1] Laboratory for Plasma Physics, LPP-ERM/KMS, Bruxelles, BE
[2] Department of Applied Physics, Ghent University, Ghent, BE

E-mail: riccardo.ragona@rma.ac.be

Abstract. Ion Cyclotron Resonance Heating and Current Drive is a method that has the ability to heat directly the ions in the Deuterium-Tritrium fuel to the high temperature needed for the fusion reaction to works. The capability of efficiently couple the Radio Frequency power to the plasma plays a big role in the overall performance of a fusion device. A Traveling Wave Antenna in a resonant ring configuration is a good candidate for an Ion Cyclotron Resonance Heating and Current Drive system. It has the capability to increase the coupled power with respect to present designs and to have a highly selective power spectrum that can be peaked around the maximally absorbed wave. It is also insensitive to the loading variations due to fluctuation of the plasma edge increasing the reliability and the efficiency of the system. It works as a low power density launcher due to the possible large number of current carrying elements.

1. Introduction

One of the most deeply investigated methods to confine the plasma is the use of magnetic field. In particular, in the last decades, one configuration among others has received more attention: the *tokamak*. This donut shaped machine uses a combination of toroidal and poloidal magnetic fields to confine the plasma. The poloidal component of the confinement field is produced by inducing a current in the plasma itself, by means of a transformer action, via a central solenoid. This is a peculiar feature of the tokamak configuration. Inducing a current has incidentally the advantage of heating up the plasma due to its resistivity η. This method is called *ohmic heating*. Unfortunately the power P_{ohmic} is related to the temperature by $P_{\text{ohimic}} \propto \eta J^2 \propto T^{-3/2}$, being J the induced current density and $\eta \propto T^{-3/2}$ [1]. The maximum level of achievable temperature with this method in a tokamak reactor is on the order of $2-3$ keV. To reach the higher values of temperature needed for the D-T reaction to minimise the requirements on pressure and confinement time, auxiliary heating systems are required.

There are two main typologies: injection of neutral particle beams and injection of Radio Frequency (RF) waves [2]. Neutral particles are unaffected by the magnetic field, they can be injected inside the plasma where they can be ionised by collisions with the background plasma. Being now confined, they can transfer their energy to the plasma via Coulomb collisions. In this case the major problem is to have neutral particles with energy high enough to reach the core of the plasma. For fusion reactors, beams energy in the order of 1 MeV are required to have good penetration and this gives rise to serious technological challenges that are now faced up by the research community [3]. As far as the second option is concerned, the injection of RF waves has a simple idea behind: high-frequency waves are launched from an external source into the plasma where they can be absorbed, via a collision-less damping mechanism, when the source

IOP Conf. Series: Journal of Physics: Conf. Series **841** (2017) 012022 doi:10.1088/1742-6596/841/1/012022

frequency matches one of the natural resonant frequency of the plasma. There are many natural resonant frequencies in a plasma like the cyclotron frequency (and harmonics) of electrons and ions. Depending on the chosen frequency, it is possible to have Electron Cyclotron Resonant Heating (ECRH) or Ion Cyclotron Resonant Heating (ICRH). The latter is the one of interest for this paper.

2. ICRF Heating

Particles gyrate around the magnetic field lines with a gyro (Larmor) radius $r_L = v_\perp/\omega_c$, where $v_\perp$ is the component of the velocity perpendicular to the field line and $\omega_c = qB_0/m$ is the cyclotron angular frequency of a particle of charge q and mass m, being B_0 the background magnetic field. For electrons and negative ions the angular frequency has a negative value. The idea behind heating a plasma with RF waves exploits the resonant wave-particle interactions when the injected angular frequency is $\omega = N\omega_{ci} + k_\parallel v_{th}$ (with $N = 0, 1, 2, \ldots$ the cyclotron harmonic number), where v_{th} is the thermal velocity of the particles and $k_\parallel$ is the component of the wave vector $\mathbf{k}$ parallel to $\mathbf{B_0}$. In a tokamak configuration, the magnetic field is inhomogeneous with $B(R) \propto 1/R$ with R the major radius of the torus. This has an effect on the cyclotron frequencies giving the possibility to control the location of a cyclotron resonance by varying the value of the magnetic field.

To study how waves propagate inside a inhomogeneous magnetised plasma and how power is absorbed by resonant species one has to derive the dispersion relation $\mathcal{D}(\omega, \mathbf{k}) = 0$ or $\omega = \omega(\mathbf{k})$ taking into account the plasma properties [4]. A simplified version of the dispersion relation can be obtained with the so called "cold plasma approximation" where all thermal effects are removed. Taking a reference system in which the z-axis is aligned with the background magnetic field and the x-axis is aligned with the direction of the inhomogeneity of the plasma and considering, for simplicity, a slab geometry as an unfolded torus geometry then $k_x \equiv k_\perp$ and $k_z \equiv k_\parallel$, where for simplicity but without loss of generality $k_y = 0$, begin the parallel and perpendicular direction referred to the magnetic field lines (see the following section for a schematic view). A wave is launched from the outside of the plasma with a ω fixed by the RF source and a $k_\parallel$ imposed by the launcher configuration ($k_\parallel$ is a real number). The dispersion relation is used to compute the value of $k_\perp = k_\perp(\omega, k_\parallel, x)$. As long as $k_\perp$ remains real, the waves propagates inside the plasma. Two interesting phenomena occurs when $k_\perp$ changes sign crossing zero (cutoff) or passing through infinity (resonance). In the first case, at the cutoff point $k_\perp^2 = 0$ and after that the wave become evanescent with no further propagation but instead with an exponential decay of the amplitude. The resonance occurs when $k_\perp^2 = \infty$ with a slowing down of the energy flow opening, in principle, the possibility of energy absorption when some dissipation is introduced. A full kinetic description of the plasma reveals a different picture in which the wave resonance is modified in a mode conversion mechanism [5] where an incoming plasma wave is transformed into another different plasma wave. With carefully chosen plasma parameters there can be a strong wave-particle interaction with strong absorption, being the wave-particle resonance and not the wave resonance the mechanism responsible for the absorption of power.

The discussion will be focused only on one of the two roots of the dispersion relation, namely the one called *Fast Magnetosonic Wave* (FMW) [6] due to its lower value of k, thus higher phase velocity, and due to fact that the direction of propagation of wave is across magnetic field lines thus perpendicular to the background magnetic field B_0. This is the one of main interest in the Ion Cyclotron Range of Frequencies (ICRF). An important aspect to describe wave-particle interactions is the wave polarisation, defined as the E_+/E_- where E_+ is the component rotating in the direction of the gyro motion of the ions while E_- with the electrons. The absorption will then takes place where the polarisation of the wave field is favourable to interact with the resonant particles. In that location, the particles see, in their reference frame, an accelerating electric field. Unfortunately it is not possible to impose the polarisation of the field from the outside; the Electric field polarisation depends mainly on the plasma composition. In case of more than one ion species, with a different concentration, the polarisation will have the form $|E_+/E_-| \simeq |(\omega_{cm} - \omega_{cM})/(\omega_{cm} + \omega_{cM})|$. As example with minority (few %) of Hydrogen in D, $|E_+/E_-| \simeq 1/3$ recalling that $\omega_{cH} = 2\omega_{cD}$. The point here is that only part of the power carried by the wave is actually used to transfer energy to the particles. The ideal condition to heat ions is then to have the E-field made all of E_+ at the foreseen absorption location [7,8,9].

A systematic study of the wave propagation and absorption will lead to power deposition profiles. Solving the wave equation in realistic geometry taking into account kinetic effects requires a substantial computational effort and several codes are available [10,11,12]. The power deposition profiles are used to find the more promising heating scheme for a given scenario. This means finding, among other variables, the angular frequency ω and $k_\parallel$ that are then used in the ICRF system [7,8,9,13].

3. ICRF system

The most general schematic of an ICRH&CD system is shown in Fig. 1. The system is composed by a power generation unit connected to the antenna (launcher) with a transmission line. The matching unit has the role of provide impedance matching between the RF generator output impedance and the antenna input impedance to ensure maximum power transfer. The launcher is responsible of coupling the power from the generator to the plasma. A simple sketch of a ICRF launcher is represented in Fig. 2, on the right. It is made usually by a short array of metal straps aligned in the poloidal direction and facing the plasma edge. One end of the strap is connected to the metallic backplate and the other end is connected to input transmission lines. When an oscillating electric current density is driven on the straps, an oscillating electromagnetic field is generated. In Fig. 2 the two components E_y and B_z are represented at an arbitrary distance in front of the antenna. This electromagnetic (EM) field is the excitation source for the plasma FMW. The wave behaviour in the region of the antenna is very different from the one described above for the core plasma. Fig. 2, on the left, illustrates the problem. The strap inside the antenna box faces the inhomogeneous plasma with a given density profile that influences the dispersion relation. The region around the scrape-off layer is very important for the antenna coupling. On one side there is vacuum or very low density residual plasma, on the other side a bulk plasma where the FMW can propagate. The EM fields that excite the FMW are evanescent from the strap in the vacuum region up to a cutoff density where $k_\perp^2$ become positive and the wave starts to propagate. In Fig. 2 the amplitude of E_y and B_z component are depicted, with the density profile, showing an exponentially decaying behaviour followed by the characteristic

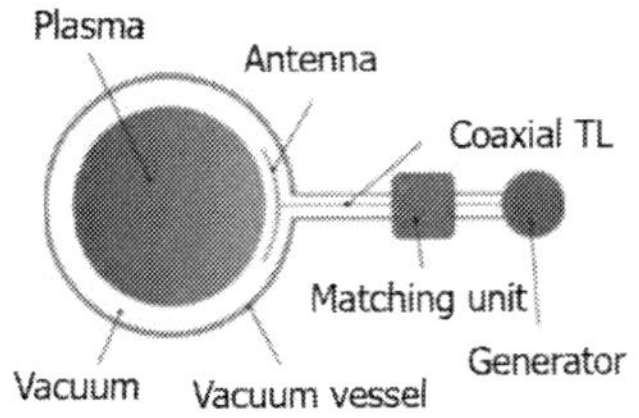

Figure 1. General schematic of an ICRH&CD system with the components highlighted.

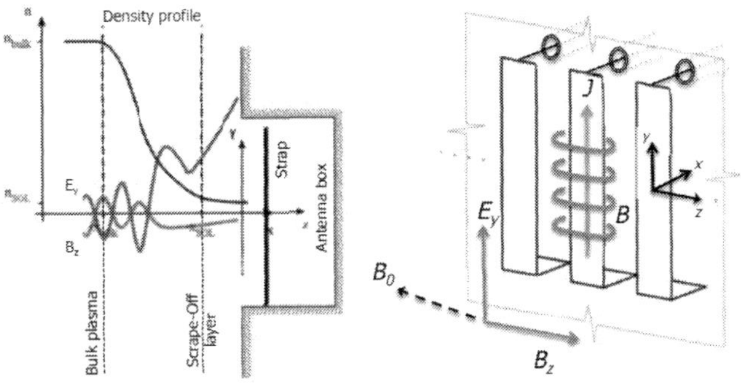

Figure 2. Schematic view (left) of the coupling mechanism and sketch (right) of a launcher.

IOP Conf. Series: Journal of Physics: Conf. Series **841** (2017) 012022 doi:10.1088/1742-6596/841/1/012022

oscillating propagative mode. The farther away the cutoff point is, towards the core, the higher are the fields at the antenna in order to couple the same amount of power. The limit is the maximum possible electric field on the antenna before breakdowns that is around 45 kV. An issue specific of this kind of antennae is the mutual coupling between radiating straps. As shown in Fig. 2 right, the magnetic field generated by the current on one strap will link the adjacent straps inducing a current on them. This gives rise to the problem of mutual coupling when the straps are operated at arbitrary current phasing. A way to compensate this problem is to use septa in between straps to isolate one from the others. The drawback is that the septa increase the inductance of the straps reducing the amount of coupled power. Sometimes also external decouplers are added to balance the power flow. The arbitrary phasing of the current is needed to obtain a radiation spectrum as close as possible to the optimal one that comes from the absorption analysis. A change in the position of the cutoff density always implies a modification in the amount of power coupled and continuously happens in reactor plasmas due to the fluctuation of the edge in front of the antenna. The effect is that the generator will see a rapidly changing loading. This is taken into account while designing a ICRF system and the capability of being not sensitive to those changes is called *load resilience*.

The power coupled to the plasma can be computed via the Poynting theorem as

$$P_{\text{rad}} = \frac{1}{4\pi^2\omega\mu_0}\Re\iint |E_y|^2\frac{1}{\xi_0^*}\,\mathrm{d}k_z\,\mathrm{d}k_y$$

where $1/\xi_0^*$ is the surface conductance at the aperture of the box and $\Re$ is the real part of the expression. An example of surface conductance computed for a typical reactor like plasma at a certain frequency is given in Fig. 3. It is a function of k_y and k_z rapidly decreasing for large $|k_z|$ and asymmetric in k_y due to the plasma anisotropy. The coupling spectrum of an array of eight straps is shown in Fig. 4, red solid line. The dashed orange line is the cut of the surface conductance of picture 3 that shows the shaping action of the surface conductance on the power spectrum of the array represented by the dashed blue line. The position of the maxima are given by the ratio of the current phase difference $\Delta\phi$ and the strap inter distance d plus an integer multiple of $2\pi/d$, $k_{z\max} = (\Delta\phi + p2\pi)/d$ where $p = 0$ is the main maximum and $p = \pm1, \pm2, \ldots$ are the secondary maxima (side lobes) [14]. Due to the effect of the surface conductance, a smaller d will push the side lobes outside reducing the power in those component of the spectrum. Increasing the number of the strap will increase the directivity of the antenna

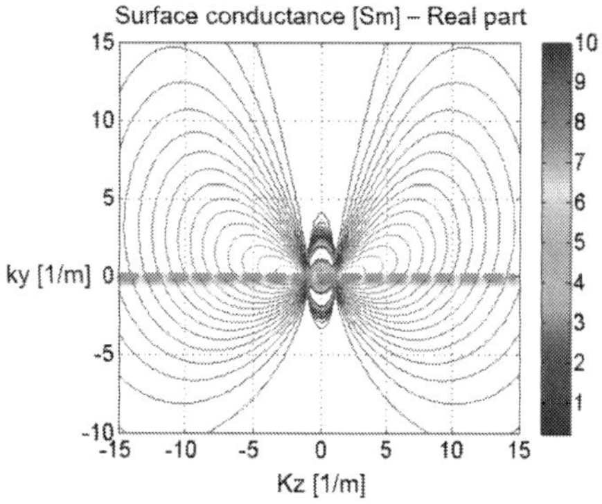

Figure 3. Surface conductance for a typical reactor like plasma. The dashed line is the cut $k_y = 0$ used in figure 4.

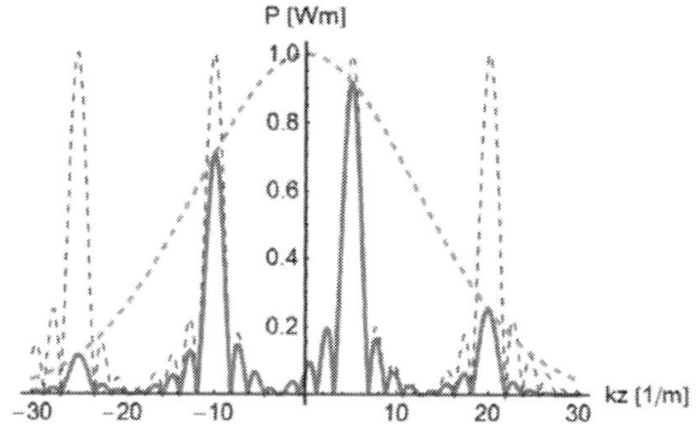

Figure 4. Example of coupling spectrum for eight straps. The dashed orange line il the cut for $k_y = 0$ of the surface conductance of figure 3.

decreasing the width of the peaks. A change of phase will move the position of the main maximum and consequently the secondary ones. A combination of d, $\Delta\phi$ and number of straps will give the optimum spectrum and a Traveling Wave Antenna is a convenient way to achieve that [14,15].

4. Traveling Wave Antenna

A Traveling Wave Antenna is a structure where the current that generates the radiation fields is in the form of a traveling wave. A sketch of a TWA is shown in Fig. 5. This particular configuration is called *combline* and it is vastly used in the RF and microwaves filter domain. An array of straps is tuned by capacitors to resonate at a certain frequency. If a signal is injected at the input port it travels along the structure up to the output port. The arrows on the straps depict the amplitude and phase difference, for a given time step, of the current density on each strap. The mutual coupling between elements is essential for this type of antennae differently from what for the conventional ones. The absence of septa between straps leads to an increased coupling capability [15]. The electrical response of the structure for two cases with different loading is shown in Fig. 6. The S_{11} (reflection) and S_{21} (transmission) coefficient of the structure are shown in red and blue respectively. The band-pass behaviour is similar to the one of a filter and is clearly visible around 50 MHz. Those results were obtained by full-wave simulations and confirmed by measurements on dedicated mock-ups. What is interesting is that the structure is not sensitive to changes of loading resistance [14]. This can be seen on the figure noting that, in the band-pass, the reflection coefficient does not change appreciably for the two different cases of loading resistance represented by the solid and dashed lines. The transmission coefficient presents instead a considerable effect. It can be shown that only the phase between the input and the output is changing. The behaviour of a TWA is then similar to a lossy transmission line. This fact is exploited in the resonant ring configuration where a power coupler is connected to a TWA, a generator and a dummy load in such a way to form a ring in which the power is recirculated while the generator is always seeing a matched load [14,15]. A sketch of this configuration is shown in Fig. 7 together with the contour plot of the B_z field component in the plane xz. The field is computed in TWA box and in the plasma in front of it (see also Fig. 2 for the boundary conditions). A coupling code was modified to self consistently compute the fields from the current distribution on the TWA structure. The resonant condition is ensured by a proper tuning of the line stretcher. The power flow in the resonant ring is shown by means of orange arrows. The system is load resilient, characteristic that is required for the use in a fusion reactor.

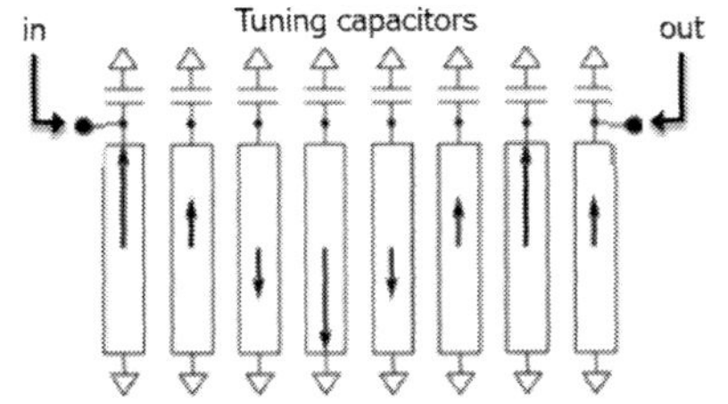

Figure 5. TWA schematic with connections and tuning capacitors.

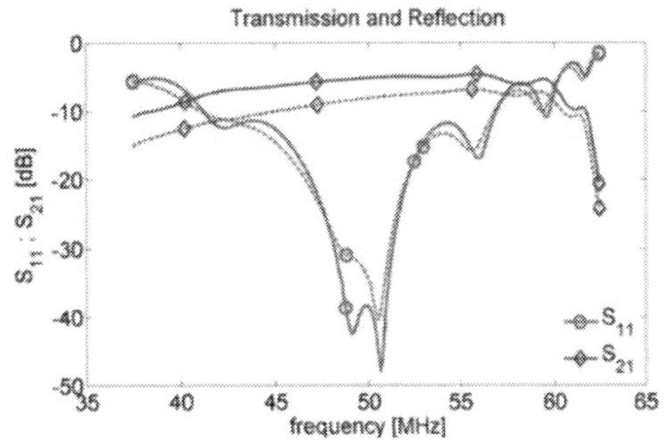

Figure 6. Electrical response of the TWA section.

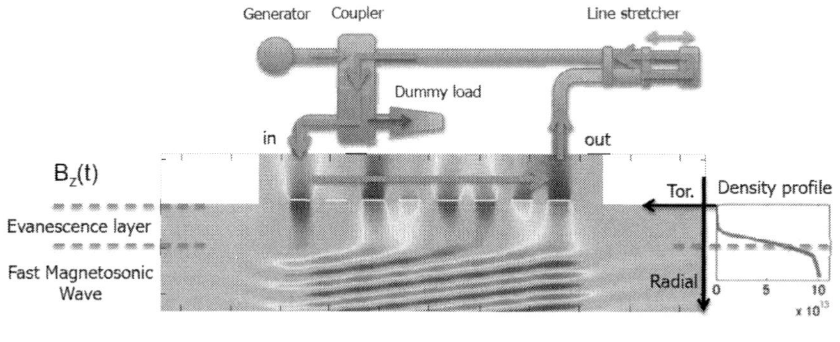

Figure 7. Resonant ring configuration and contour plot of the B_z in the TWA and the plasma in front of it.

5. Conclusions

A general overview of the ICRF wave propagation and system configurations was used to highlight the difficulties in the design of a launcher capable of providing a suitable coupling spectrum. The required current distribution is the key point to obtain the desired performance. The Travelling Wave Antenna is a good candidate for a launcher because it is capable of providing a current distribution without the necessity of feeding all the straps. If operated inside a resonant ring, the system shows load resilience and perfect matching of the generator. With this type of launcher, the absence of septa leads to higher coupling with respect to the conventional one. There are still open questions like how to properly design the tuning capacitor and what is the mutual effect of multiple neighbouring sections. The high value of auxiliary heating power likely to be injected in a fusion reactor requires ICRF systems capable of dealing with that power level without exceeding voltage standoffs and ensuring an high efficiency. The TWA in a resonant ring seems a good candidate showing a higher power coupling capability, an intrinsic load resilience and, due to the higher number of straps, more selective spectra. The limit for the number of straps depends on the coupling itself and on the feeding scheme. Further studies are ongoing to understand which configuration is the best in order to meet the requirement for a fusion reactor.

References

[1] Freidberg J P 2007 *Plasma Physics and Fusion Energy.* (Cambridge University Press)
[2] Dolan T J 2013 *Magnetic Fusion Technology.* (Springer)
[3] Sonato P et al 2016 submitted to the *26th IAEA Fusion Energy Conference* Kyoto, Japan
[4] Stix T H 1992 *Waves in Plasmas.* (AIP-Press)
[5] Perkins F W 1977 *Nucl. Fusion* **17** 6
[6] Bingren S H I 1998 *China Nuclear Science and Technology Report* S5
[7] Gallart D et al 2015 *AIP Conf. Proc.* **1689**
[8] Kazakov Y O, Kiptily V G, Sharapov S E, Van Eester D and JET EFDA Contributors 2015 *Nucl. Fusion* **55** 032001
[9] Kazakov Y O, Ongena J, Van Eester D, Bilato R, Dumont R, Lerche E, Mantsinen M and Messiaen A 2015 *Phys. Plasmas* **22** 082511
[10] Brambilla M 1999 *Plasma Phys. Controlled Fusion* **41** 1
[11] Brambilla M and Bilato R 2006 *Nucl. Fusion* **46** 7
[12] Van Eester D and Koch R 1998 *Plasma Phys. Controlled Fusion* **40** 1949-1975
[13] Lerche E, Van Eester D, Messiaen A, Franke T and EFDA-PPPT contributors 2014 *AIP Conf. Proc.* **1580** 338
[14] Ragona R, Messiaen A 2016 *Nucl. Fusion* **56** 076009
[15] Messiaen A, Ragona R 2016 *43rd EPS Conf.*

IOP Conf. Series: Journal of Physics: Conf. Series **841** (2017) 012023 doi:10.1088/1742-6596/841/1/012023

Investigating the low-energy K^- interactions in nuclear matter with AMADEUS

R Del Grande[1,2], M Bazzi[1], G Belloti[3], C Berucci[4], A M Bragadireanu[5], D Bosnar[6], A D Butt[3,7], M Cargnelli[4], C Curceanu[1], L Fabbietti[8,9], C Fiorini[3,7], F Ghio[10,11], C Guaraldo[1], R S Hayano[12], M Iliescu[1], M Iwasaki[13], P Levi Sandri[1], J Marton[4], M Miliucci[1,2], S Okada[13], D Pietreanu[5], K Piscicchia[1,14], A Romero Vidal[15], A Scordo[1], H Shi[1], D L Sirghi[1,5], F Sirghi[1,5], I Tucakovic[16], O Vazquez Doce[8,9], E Widmann[4], S Wycech[17], J Zmeskal[4], G Mandaglio[18,19], M Martini[1,20] and P Moskal[21]

[1] INFN Laboratori Nazionali di Frascati, Frascati, Rome, Italy
[2] Università degli Studi di Roma Tor Vergata, Rome, Italy
[3] Politecnico di Milano, Dip. di Elettronica, Informazione e Bioingegneria, Milano, Italy
[4] Stefan-Meyer-Institut für Subatomare Physik, Wien, Austria
[5] Horia Hulubei National Institute of Physics and Nuclear Engineering, Magurele, Romania
[6] University of Zagreb, Zagreb, Croatia
[7] INFN Sezione di Milano, Milano, Italy
[8] Excellence Cluster Origin and Structure of the Universe, Garching, Germany
[9] Physik Department E12, Technische Universität München, Garching, Germany
[10] INFN Sezione di Roma I, Rome, Italy
[11] Istituto Superiore di Sanità, Rome, Italy
[12] The University of Tokyo, Tokyo, Japan
[13] RIKEN, The Institute of Physics and Chemical Research, Saitama, Japan
[14] Museo Storico della Fisica e Centro Studi e Ricerche Enrico Fermi, Rome, Italy
[15] University of Santiago de Compostela, Santiago de Compostela, Spain
[16] Ruđer Bošković Institute, Zagreb, Croatia
[17] National Centre for Nuclear Research, Warsaw, Poland
[18] Dipartimento ChiBioFarAm dell'Università di Messina, Messina, Italy
[19] INFN Gruppo collegato di Messina, Messina, Italy
[20] Dipartimento di Scienze e Tecnologie applicate, Università 'Guglielmo Marconi', Rome, Italy
[21] Institute of Physics, Jagiellonian University, Kracow, Poland

E-mail: raffaele.delgrande@lnf.infn.it

Abstract. The AMADEUS collaboration studies low-energy K^- interactions with light nuclei in order to clarify some aspects related to the behaviour of hadrons containing strangeness in nuclear medium. One of the main topics is the quest about the possible formation of Kaonic Nuclear Clusters (KNC), which depends on the strength of the anti-kaons interaction with nucleons. In kaonic absorption experiments, the search for KNC is strictly connected with the measurement of the yields of the so-called multi-nucleon absorption processes. In this paper, the study of Σ^0p correlated production from K^- absorption in ^{12}C, using the KLOE 2004-2005 data set, is reported. The yield of the two nucleon absorption (2NA), when the produced Σ^0 and p particles are free from any final state interaction process, was measured for the first time. The contribution of a ppK$^-$ bound state was also tested. The best fit is obtained for a ppK$^-$

state with a binding energy of 45 MeV and a width of 30 MeV, but the statistical significance is at the level of 1σ only.

1. Introduction

The AMADEUS (Anti-kaonic Matter At DAΦNE: An Experiment with Unraveling Spectroscopy) [1] collaboration investigates low-energy negative kaons induced reactions in light nuclear targets in order to provide experimental constraints to the non-perturbative QCD in the strangeness sector.

The study of $\bar{K}N$ interaction has a special interest to understand the interplay between the spontaneous and explicit chiral symmetry breaking (induced by the large strange quark mass) and its role in the origin of hadron masses [2]. The existence of the $\Lambda(1405)$ state only few MeV below the $\bar{K}N$ threshold is related to the attractive antikaon-nucleon interaction, thus chiral perturbation theory is not applicable. Effective Lagrangian techniques as well as phenomenological potential models were developed to deal with this problem [3–10], leading to contrasting predictions for the $\Lambda(1405)$ parameters and the related possibility for $\bar{K}$ multi-nucleon bound states. The $\Lambda(1405)$ is experimentally found to exhibit different masses and widths, depending on the production channel as well as the chosen $(\Sigma\pi)^0$ observed decay. A detailed discussion regarding the theoretical and experimental determination of the $\Lambda(1405)$ properties can be found in [11]. Calculations of the Binding Energy (BE) and the Width (Γ) of the ppK^- state lead to different results, depending on the theoretical approach [12–19]. A strong experimental effort was devoted to the search for a ppK^- state, but a clear evidence is still missing [20–26].

The broad program of the AMADEUS collaboration is devoted to a complete experimental investigation of the $\bar{K}$ behaviour in light nuclear matter, ranging from the measurement of hyperon resonances properties [27–29] to the search for $\bar{K}$-multi-N bound states. As will be outlined in the following sections, the extraction of bound state signal, in kaon induced reactions, strongly depends on an accurate description of K^- multi-nucleon absorption processes, overlapping over a broad range of the allowed phase space [30, 31].

This contribution reports on the investigation of K^- absorptions in ^{12}C nuclei, searching for Σ^0 proton correlated production. The aim of the analysis is to disentangle the different K^- multi-nucleon absorption processes in order to extract the corresponding yields. A systematic search for signal of ppK^- bound state was performed. $\Sigma^0 p$ is the golden decay channel of an eventual ppK^- state, being free from the $\Sigma N \rightarrow \Lambda N'$ conversion processes which strongly affect the competing Λp decay. In Sec. 2, we will introduce the features of the DAΦNE accelerator and the KLOE detector. In Sec. 3 the event selection procedure is described; results and conclusions are discussed in Secs. 4 and 5, respectively.

2. DAΦNE and KLOE

AMADEUS takes advantage of the DAΦNE (Double Annular Φ-factory for Nice Experiments) collider [32], which is a double rings e^+ e^- collider, designed to work at the Center of Mass (CM) energy of the ϕ meson, which decays into K^+ K^- pairs with a Branching Ratio of $BR(\phi \rightarrow K^+ K^-) = 0.489 \pm 0.005$. Kaons are produced back-to-back with an almost monochromatic momentum of $p_K \sim 127$ MeV/c.

The KLOE (K LOng Experiment) [33] detector consists in a large cylindrical Drift Chamber (DC) [34] and a fine sampling lead-scintillating fibres calorimeter [35], all immersed in an axial magnetic field of 0.52 T, provided by a superconducting solenoid. The KLOE 4π geometry allows to cover almost the whole solid angle around the DAΦNE interaction point, therefore the acceptance is of ~ 98 %. The DC has an inner radius of 0.25 m, an outer radius of 2 m

IOP Conf. Series: Journal of Physics: Conf. Series **841** (2017) 012023 doi:10.1088/1742-6596/841/1/012023

and a length of 3.3 m. The DC entrance wall composition is 750 μm of carbon fibre and 150 μm of aluminium foil. Dedicated GEANT Monte Carlo (MC) simulations [36] of the KLOE apparatus were performed to estimate the percentages of K^- absorptions in the materials of the DC entrance wall (the K^- absorption physics were treated by the GEISHA package [37]). Out of the total number of kaons interacting in the DC entrance wall, about 81% results to be absorbed in the carbon fibre component and the residual 19% in the aluminium foil. The KLOE DC is filled with a mixture of helium and isobutane (90% in volume ^{4}He and 10% in volume C_4H_{10}). The chamber is characterised by excellent position and momentum resolutions. Tracks are reconstructed with a resolution in the transverse $R-\phi$ plane of $\sigma_{R\phi} \sim 150\,\mu$m and a resolution along the z-axis of $\sigma_z \sim$ 2mm. The transverse momentum resolution for low momentum tracks ((50 $< p <$ 300) MeV/c) is $\frac{\sigma_{pT}}{pT} \sim 0.4\%$. The KLOE calorimeter is composed of a cylindrical barrel and two end-caps. The volume ratio (lead/fibres/glue=42:48:10) is optimised for a high light yield and a high efficiency for photons in the range (20-300) MeV/c. The position of the cluster along the fibres can be obtained with a resolution $\sigma_{\parallel} \sim 1.4\,\mathrm{cm}/\sqrt{E/(1\mathrm{GeV})}$. The resolution in the orthogonal direction is $\sigma_{\perp} \sim 1.3$ cm. The energy and time resolutions for photon clusters are given by $\frac{\sigma_E}{E_\gamma} = \frac{0.057}{\sqrt{E_\gamma/(1\mathrm{GeV})}}$ and $\sigma_t = \frac{57\,\mathrm{ps}}{\sqrt{E_\gamma/(1\mathrm{GeV})}} \oplus 100$ ps.

The analysis presented below refers to a total integrated luminosity of $\mathcal{L} = 1.74$ pb^{-1}, from the 2004/2005 KLOE data taking campaign.

3. Events selection

The Σ^0p correlated events selection starts from the reconstruction of a $\Lambda(1116)$ particle through its charged decay $\Lambda \rightarrow$ p + π^- that occurs with a Branching Ratio (BR) of BR=63.9 $\pm$ 0.5 %, where the error contain both the statistic and systematic uncertainties. The proton and negative pion tracks are identified using both the dE/dx information from the DC wires and the measured energy release in the calorimeter. Details about p - π^- particle identification and vertex reconstruction can be found in [27, 38]. The measured pπ^- invariant mass is $m_{\mathrm{p}\pi^-} = 1115.723 \pm 0.002$ stat MeV/c^2 with a resolution of σ = 0.5 MeV/c^2. The systematic error is under evaluation. The hadronic interaction vertex is reconstructed through a backward extrapolation of the Λ path and an extra-proton track (with the same selection requirements optimized for the proton from Λ decay). The resolution achieved for the radial coordinate of the Λp vertex ($\rho_{\Lambda\mathrm{p}}$) is 1.2 mm. Σ^0 candidates are identified through their decay into $\Lambda\gamma$ pairs. After the Λp pair, the photon selection is carried out via its identification in the calorimeter [39]. In order to select K^- captures in ^{12}C the cut $\rho_{\Lambda\mathrm{p}}$=(25 $\pm$ 1.2) cm was applied, optimised by means of MC simulations and a study of the Λ decay path.

4. Results

A simultaneous fit of the measured invariant mass $m_{\Sigma^0\mathrm{p}}$, angular correlation $\cos(\theta_{\Sigma^0\mathrm{p}})$, Σ^0 momentum P_{Σ^0} and proton momentum $\mathrm{P_p}$ was performed to the following simulated processes:

1. K^- + "pp" $\rightarrow \Sigma^0$ + p (2NA),
2. K^- + "ppn" $\rightarrow \Sigma^0$ + p + n (3NA),
3. K^- + "ppnn" $\rightarrow \Sigma^0$ + p + n + n (4NA),

where 2NA, 3NA and 4NA refers to the number of nucleons involved in the multi-nucleon K^- absorptions. The interacting nucleon are bound in the Carbon nucleus. Final State Interactions (FSI) of the Σ^0 or p particles, produced in 2NA processes, with the residual nucleus, were also considered.

Two background sources are considered whose shapes were obtained from experimental data samples:

4. $K^- + A \rightarrow \Sigma^0(\pi) + p + R$,
5. $K^- + A \rightarrow \Lambda + p + \pi^0 + R$,

where R represents the residual nucleus. Process number 4 corresponds to the uncorrelated production of a Σ^0 and of a proton, in this case the proton originates from the fragmentation of the residual nucleus. Process number 5 describes Σ^0 misidentification when a $\Lambda\pi^0$ pair is produced. We refer to [39] for more details. The fit is shown in Fig. 1 and the obtained results are summarised in Table 1, where the yields of each contribution are normalised to the total number of stopped negative kaons. The best fit delivers a reduced χ^2 of 0.85.

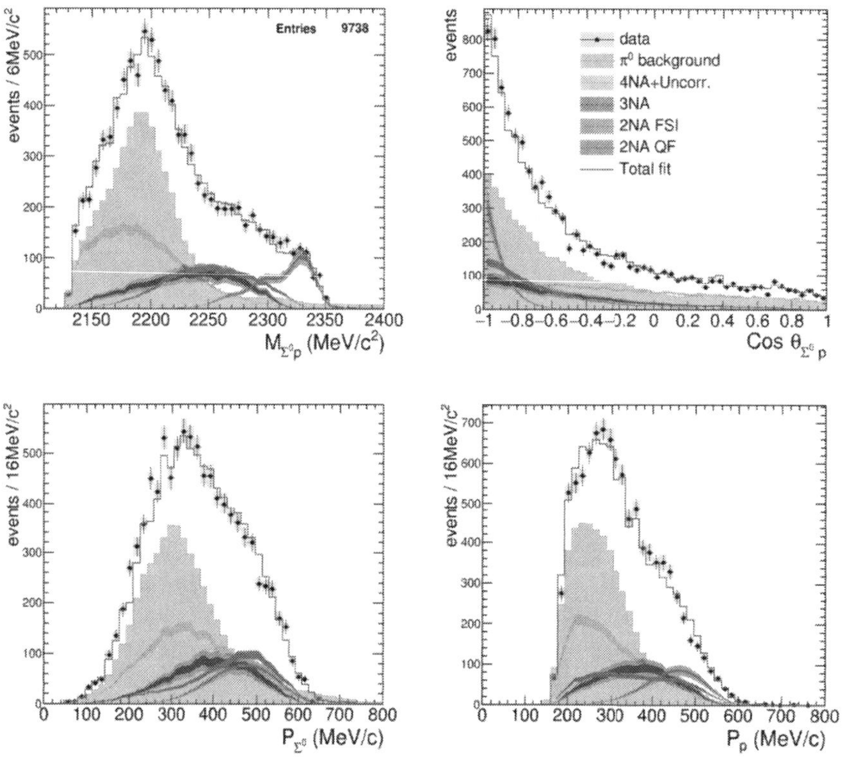

Figure 1. Fit of the Σ^0p invariant mass, $\cos(\theta_{\Sigma^0 p})$, Σ^0 and proton momenta is shown. Data points are represented by black circles, the systematic errors by boxes, coloured histograms correspond to the simulated processes (light-coloured bands show the statistical errors and the darker bands represent the symmetrised systematic errors). The gray line is the total fit.

The 2NA production probability in ^{12}C, when the produced Σ^0 and p particles are free form any FSI process (2NA-QF), was measured for the first time, with good precision. A second fit was performed to include the contribution of a ppK$^-$ bound state, decaying into Σ^0p. The ppK$^-$ mass was sampled according to a Breit-Wigner distribution. The event kinematic is obtained by imposing total energy and momentum conservation for the ppK$^-$-R system. Binding Energy (BE) and Width (Γ) were varied on a grid of values, $\Delta(\mathrm{BE}) = (15 \div 75)\,\mathrm{MeV/c^2}$ and $\Delta(\Gamma) = (30 \div 70)\,\mathrm{MeV/c^2}$ in steps of 15 and 20 MeV/c^2, respectively. The best fit (reduced $\chi^2 = 0.807$) is obtained for a ppK$^-$ candidate with a binding energy of 45 MeV/c^2 and a width of 30 MeV/c^2. Out of the four simultaneously fitted distributions, the Σ^0p invariant mass and proton momentum distributions are displayed in Fig. 2 a) and b), respectively. The ppK$^-$

Table 1. Yields of the various Σ^0p production processes normalised to the number of stopped K^- in the DC wall. The statistical and systematic errors are also shown.

Process	yield / $K^-_{stop} \times 10^{-2}$	$\sigma_{stat} \times 10^{-2}$	$\sigma_{syst} \times 10^{-2}$
2NA-QF	0.127	±0.019	$^{+0.004}_{-0.008}$
2NA-FSI	0.272	±0.028	$^{+0.022}_{-0.023}$
Tot 2NA	0.399	±0.033	$^{+0.023}_{-0.032}$
3NA	0.274	±0.069	$^{+0.044}_{-0.021}$
Tot 3 body	0.546	±0.074	$^{+0.048}_{-0.033}$
4NA + bkg.	0.773	±0.053	$^{+0.025}_{-0.076}$

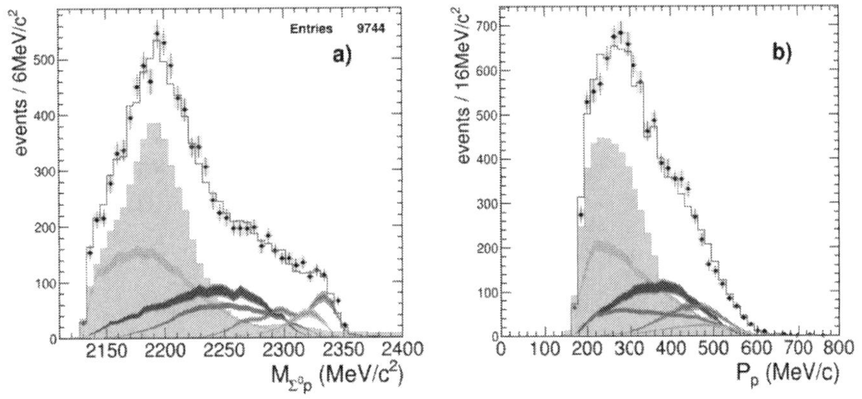

Figure 2. Experimental distributions of the Σ^0p invariant mass and proton momentum together with the results of the global fit including the ppK^-. The contributing distributions are labelled as in Fig. 1. The green curve represent the ppK^- signal.

bound state contribution is shown in green. The resulting yield normalised to the number of stopped K^- is pp$K^-/K^-_{stop} = (0.044 \pm 0.009 \ \ \text{stat}^{+0.004}_{-0.005} \ \text{syst}) \times 10^{-2}$. An F-test was performed in order to estimate the significance of the measured bound state component with respect to a statistical fluctuation, i.e. the null hypothesis. The ppK^- signal was found to be significant at the level of 1σ only. Although the measured spectra are compatible with the hypothesis of a bound state contribution, the significance of the result is not sufficient to claim the discovery of this state.

5. Conclusions

The main goals of the analysis described in this work consisted in the extraction of the relative yields of the K^- multi-nucleon absorption processes and the search for ppK^- state signal. A sample of K^- captures in ^{12}C was selected and the Σ^0p final state was chosen because free from conversion (one of the main background sources of the Λp channel). We isolated the contribution of the free 2NA (2NA-QF) processes, without FSI, from the fit of the KLOE 2004-2005 data, using simulations to reproduce the kinematic shapes for the multi-nucleon absorption processes and the experimental data samples for the backgrounds. The yield of the 2NA process, when the produced Σ^0 and p particles are free form any FSI, was measured for the first time. A search for a ppK^- bound state contribution was carried out with a scan in binding energies and widths ($\Delta(\mathrm{BE}) = (15 \div 75)\,\mathrm{MeV/c^2}$ and $\Delta(\Gamma) = (30 \div 70)\,\mathrm{MeV/c^2}$ in steps of 15 and 20 $\mathrm{MeV/c^2}$, respectively). The best value of the reduced χ^2 is achieved for a binding energy of 45 $\mathrm{MeV/c^2}$ and a width of 30 $\mathrm{MeV/c^2}$. The corresponding yield is $\mathrm{ppK^-/K^-_{stop}} = (0.044 \pm 0.009\ \mathrm{stat}^{+0.004}_{-0.005}\mathrm{syst})\times 10^{-2}$. In order to extract the significance of the $\mathrm{ppK^-}$ signal an F-test was performed. A significance of 1σ only was obtained for the measured $\mathrm{ppK^-}$ yield, which is not sufficient to claim the observation of this state.

Acknowledgments

We thank all the KLOE Collaboration and the DAΦNE staff for the fruitful collaboration.

Part of this work was supported by the European Community-Research Infrastructure Integrating Activity "Study of Strongly Interacting Matter" (HadronPhysics2, Grant Agreement No. 227431, and HadronPhysics3 (HP3) Contract No. 283286) under the EU Seventh Framework Programme.

References

[1] Curceanu C *et al* 2015 *Acta Phys. Polon.* B **46 no.1** 203
[2] Weise W 2015 *Hyperfine Interact.* **233 no.1-3** 131
[3] Oller J A and Meißner U-G 2001 *Phys. Lett.* B **500** 263
[4] Jido D, Hosaka A, Nacher J C, Oset E and Ramos A 2002 *Phys. Rev.* C **66** 025203
[5] Jido D, Oller J A, Oset E, Ramos A and Meißner U-G 2003 *Nucl. Phys.* A **725** 181
[6] Ikeda Y *et al* 2012 *Nucl. Phys.* A **881** 98
[7] Jido D, Oset E and Sekihara T 2011 *Eur. Phys. J.* A **47** 42
[8] Jido D, Oset E and Sekihara T 2013 *Eur. Phys. J.* A **49** 95
[9] Esmaili J *et al* 2010 *Phys. Lett.* B **686** 23
[10] Esmaili J *et al* 2011 *Phys. Rev.* C **83** 055207
[11] Hyodo T and Jido D 2012 *Prog. Part. Nucl. Phys.* **67** 55
[12] Yamazaki T *et al* 2007 *Phys. Rev.* C **76** 045201
[13] Doté A *et al* 2009 *Phys. Rev.* C **79** 014003
[14] Wycech S *et al* 2009 *Phys. Rev.* C **79** 014001
[15] Barnea N *et al* 2012 *Phys. Lett.* B **712** 132
[16] Shevchenko N V *et al* 2007 *Phys. Rev. Lett.* **98** 082301
[17] Ikeda Y *et al* 2009 *Phys. Rev.* C **79** 035201
[18] Oset E *et al* 2012 *Nucl. Phys.* A **881** 127
[19] Sekihara T *et al* 2016 arXiv:1607.02058
[20] Agakishiev G *et al (HADES Coll.)* 2015 *Phys. Lett.* B **742** 242
[21] Agnello M *et al (FINUDA Coll.)* 2005 *Phys. Rev. Lett.* **94** 212303
[22] Yamazaki T *et al* 2010 *Phys. Rev. Lett.* **104** 132502
[23] Ichikawa Y *et al* 2015 *Prog. Theor. Exp. Phys.* **021D01**
[24] Tokiyasu A O *et al* 2014 *Phys. Lett.* B **728** 616
[25] Fabbietti L *et al* 2013 *Nucl. Phys.* A **914** 60
[26] Hashimoto T *et al* 2015 *Prog. Theor. Exp. Phys.* 061D01
[27] Piscicchia K *et al* 2013 *PoS Bormio2013* 034
[28] Scordo A *et al* 2014 *PoS Bormio2014* 039
[29] Piscicchia K, Wycech S and Curceanu C 2016 *Nucl. Phys.* A **954** 75
[30] Katz P A *et al* 1970 *Phys. Rev.* D **1** 1267
[31] Vander Velde-Wilquet C *et al* 1977 *Nuovo Cimento* A **39** 538
[32] Gallo A *et al* 2006 *Conf. Proc.* **C060626** 604
[33] Bossi F *et al (KLOE Coll.)* 2008 *Riv. Nuovo Cim.* **31** 531
[34] Adinolfi M *et al (KLOE Coll.)* 2002 *Nucl. Inst. Meth.* A **488** 51
[35] Adinolfi M *et al (KLOE Coll.)* 2002 *Nucl. Inst. Meth.* A **482** 368
[36] Allison J *et al* 2016 *Nucl. Inst. Meth.* A **835** 186
[37] Fesefeldt H *et al* 1990 *Nucl. Inst. Meth.* A **292** 279
[38] Curceanu C *et al* 2015 *Acta Phys. Polon.* B **46 no.1** 203
[39] Vazquez Doce O *et al* 2016 *Phys. Lett.* B **758** 134

IOP Conf. Series: Journal of Physics: Conf. Series **841** (2017) 012024 doi:10.1088/1742-6596/841/1/012024

Dimuon production in Pb-Pb collisions at 20-160 AGeV at the CERN SPS: Mapping the QCD phase diagram in the transition region with a new experiment

E Incani[1,2], **G Usai**[1,2]
[1] Dipartimento di Fisica, Università degli Studi di Cagliari, 09042 Monserrato, Italy
[2] INFN - Sezione di Cagliari, Italy
E-mail: elisa.incani@ca.infn.it, gianluca.usai@ca.infn.it

Abstract.
In this paper new ideas to experimentally investigate the issues of chiral symmetry restoration and the first order phase transition in the region of moderate-large baryon density of the phase diagram of strongly interacting matter are presented. The experimental strategy to address these points is to use a new fixed-target experiment at the CERN SPS (Super Proton Synchrotron) dedicated to the measurement of the production of muon pairs with unprecedented precision. Dileptons offer the possibility to measure temperature to obtain a caloric curve and to probe chiral symmetry restoration by studying for the first time the mass modifications in a simultaneous measurement of the vector meson ρ and its axial vector partner a_1.

1. Introduction

The theory of strong interactions, Quantum ChromoDynamics (QCD), predicts a phase transition at high temperature and/or baryon density between hadronic matter (where quarks and gluons are confined inside hadrons), and a deconfined state of matter, the so-called Quark-Gluon Plasma (QGP). The possible phases of strongly interacting matter can be displayed in a phase diagram as a function of its temperature and the baryon density (Fig. 1) [1].

At present, the QCD phase diagram was studied mainly in the region of low baryon density. In this regime the phase transition between hadronic matter and the QGP is a cross-over around a critical temperature of 155 MeV [1]. On the other hand, for moderate temperatures and high baryon densities, a first order transition is expected with the coexistence of a mixed-phase.

In vacuum chiral symmetry of the QCD is spontaneously broken. Most of the baryonic mass in the Universe arises from this breacking rather than from the Higgs coupling [2]. Corresponding to the phase transition to the deconfined phase, it is expected that chiral symmetry is restored.

The doublet of the vector meson and its axial vector partner, split in vacuum due to chiral symmetry breaking, should become degenerate at chiral restoration. Fig. 2 displays qualitatively two possibilities for the ρ and a_1, known as dropping mass and melting resonance scenarios.

The basic aspects of the phase diagram structure not yet confirmed experimentally are:

- how chiral restoration affects the hadron spectrum;

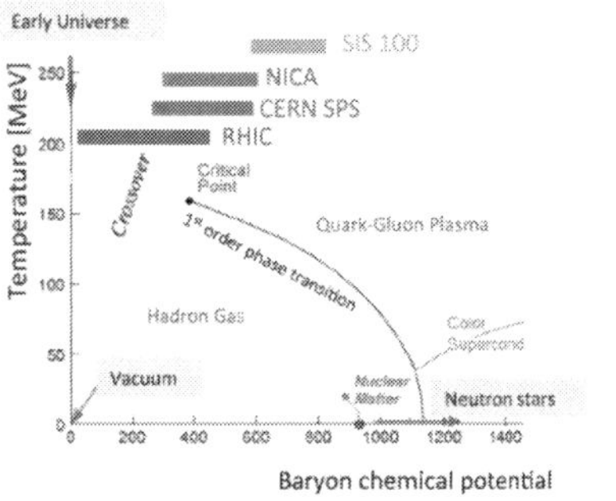

Figure 1. The phase diagram of strongly interacting matter and the regions covered by different facilities in forthcoming years.

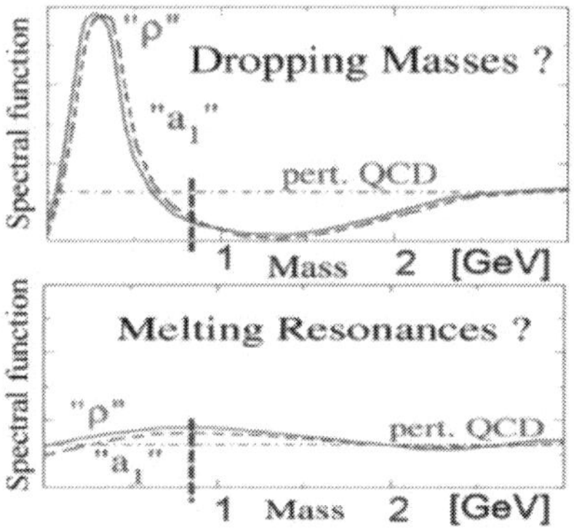

Figure 2. Possible scenarios for vector ρ and axial-vector a_1 spectral functions when chiral symmetry is restored.

- the existence of the critical point and the first order transition.

In order to shed light on these fundamental phenomena, it is proposed a new experimental apparatus (here denoted NA60+) to measure the production of muon pairs with unprecedented precision in nucleus-nucleus collisions at the CERN SPS (Pb-Pb collisions over the energy range from $\sqrt{s_{NN}} \sim 6$ GeV up to 17 GeV in the centre-of-mass system, $\sim (20 \div 160)$ GeV/nucleon in the laboratory system) [1].

The presence of dileptons provides a probe of the expanding system, in fact they are produced at all stages of the evolution and also retain the primary informations of the system, because of the absence of any final state interaction.

In the low mass region (M<1 GeV) the dilepton mass spectrum is dominated by the ρ meson. Its spectral function has been precisely measured in Indium-Indium collisions at 160 GeV/nucleon by the NA60 experiment at the CERN SPS [3]. The intermediate mass region (1.1 GeV<M<3 GeV) is populated by the radiation emitted in the QGP via quark-antiquark annihilation and by the radiation emitted at a later stage from the hadronic medium. In particular, between 1.1 GeV<M<1.5 GeV the dominant hadronic process in the confined medium is $a_1\pi^0 \to \mu^+\mu^-$. Contrary to ρ meson, in high energy nuclear collisions no measurements exist for the axial-vector a_1. NA60+ proposes to perform a precision measurement of the a_1 at a collision energy close to the onset of deconfinement, where the QGP yield is very small or negligible.

For M>1.5 GeV the mass spectrum can be described using a simple formula: $dN/dM \propto M^{3/2}exp(-M/T)$, where T represents the average temperature of the emitting sources [4]. The temperature measurement was first pioneered by NA60 at the full SPS energy of 160 GeV/nucleon, the fit of the spectrum giving $T = (205 \pm 12)$ MeV [5, 6]. This is above critical temperature, thus showing that the QGP is already produced at this collision energy.

NA60+ proposes to measure for the first time the caloric curve with accurate temperature measurements as a function of energy density with a beam energy scan ($\sqrt{s_{NN}} = (6.2 \div 17.3)$ GeV), to search for a possible flattening of temperature where the phase transition occurs in the phase diagram. The energy density will be derived from the charged particle multiplicity density through Bjorken scaling or more advanced models [7].

In addition, the study of J/ψ suppression and open-charm production are also expected to be sensitive to the onset of deconfinement and might be studied with the proposed apparatus [1].

Various facilities can in principle investigate the high baryon densities region of the QCD phase diagram: CERN SPS, RHIC, NICA and FAIR (Fig. 1):

- The CERN SPS with its new injection scheme can deliver intense ion beams leading to interaction rates exceeding 1 MHz with a good energy coverage from 4.5 GeV to 17.3 GeV in centre-of-mass system. At present the NA61 experiment is running at the CERN SPS and it measures the production of hadronic and has a physics program complementary to the one proposed for NA60+.
- RHIC (energy coverage from $\sqrt{s_{NN}} \sim (7.5 \div 200)$ GeV), and NICA (reaching $\sqrt{s_{NN}}$ =11 GeV) provide good energy coverage but the interaction rates are 2-3 orders of magnitude smaller than the SPS, so that high-precision measurements are not possible.
- FAIR SIS/100 is designed to provide high interaction rates (>1 MHz) but the energy coverage is extremely limited (with a maximum energy of $\sqrt{s_{NN}}$ =4.5 GeV). FAIR SIS/300 might provide a larger coverage (up to $\sqrt{s_{NN}}$ =8 GeV) but it is not approved at present and its operation would in any case only start well beyond 2030.

Consequently, no presently approved experiment is able to cover such a large energy interval and to collect at the same time the large statistics required for truly quantitative measurements. NA60+ would then represent a unique opportunity to investigate the QCD phase diagram in the next decade.

2. Experiment

The experimental program requires to collect data at different energies lower than full SPS energy of 160 GeV/nucleon (example 20-30-40-80-120-160 GeV/nucleon).

The objectives for the thermal radiation studies are:

- the isolation of dilepton spectrum from hadronic phase up to M$\sim$ 2 GeV;
- measurements of temperature with an accuracy at the MeV level (more than $5 \cdot 10^7$ reconstructed pairs from thermal radiation per energy point);
- proton-nucleon collisions at some energy point.

This basic physics program might be accomplished in about 5 years of data-taking.

The apparatus couples a traditional muon spectrometer to a silicon vertex tracker placed before the hadron absorber, as pioneered by NA60 [3, 5, 8, 9]. NA60+ will increase the statistics by a factor $\sim$ 100 over NA60 while retaining a very good signal-to-background ratio even in central Pb-Pb collisions (Fig. 3) [1].

The muon spectrometer is composed of 4 tracking stations and a toroid magnet placed in the middle of the tracking stations. It provides a field integral of 0.75 T·m at R=1 m (simulations for the lowest energies were also performed assuming a reduced field integral of 0.3 T·m at R =1 m).

Placed before the muon spectrometer, the hadron absorber allows muons to be selected. The idea is to have a similar signal-to-background ratio at all energy, implying an absorber system with a scalable thickness. At present an absorber was investigated for measurements at 20-40 GeV/nucleon. It consists of BeO-graphite sections, tuned in the best way to increase the hadron absorption, limiting the multiple scattering. A graphite wall is placed at the end before the trigger stations.

Thus, the hadron absorber provides the muon identification, but at the cost of degrading the kinematics of the muons, because of energy loss fluctuations and multiple scattering. This problem is overcome by measuring particle tracks also before the hadron absorber with a silicon tracker (vertex spectrometer), which is the key element for a precision measurement of muons. Muon tracks are then matched to the tracks measured in the silicon vertex spectrometer in

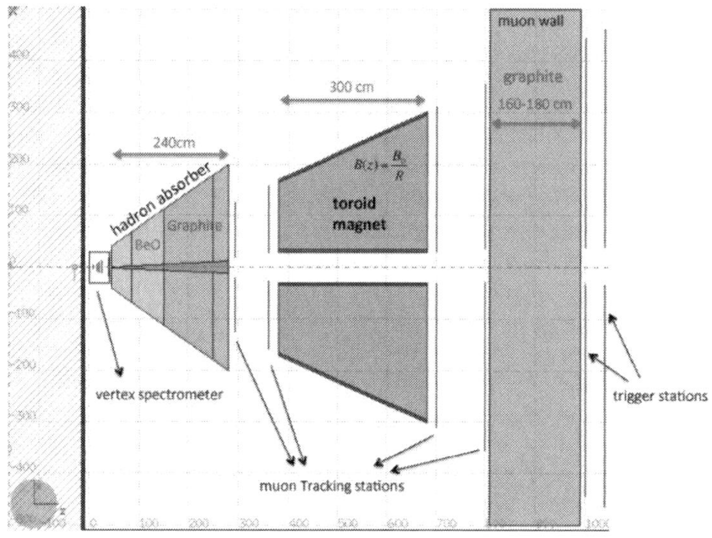

Figure 3. Geometry of the proposed experimental apparatus.

coordinate and momentum space. In addition, the silicon tracker provides also the measurement of charged-particle multiplicity density, which can be used to estimate the energy density.

The vertex spectrometer consists of 5-10 stations of hybrid pixels or monolithic active pixel sensors immersed in a 1.2 T·m dipole field. The results presented in the following are based on the use of hybrid pixels, with a material budget per plane of $\sim 1\%\ X_0$ [10] and space resolution is 10-15 μm (40-50 μm pixel pitch). A disadvantage of this technology is the thickness of the sensors and readout chip. A very interesting alternative might be the monolithic active pixel sensors developed for LHC upgrades [11]: while these sensors have not yet the required readout speed, further developments might lead to competitive sensors in the near future.

The muon trigger is based on two trigger stations placed after the muon wall.

3. Performance study Pb-Pb collisions at 40 GeV/nucleon

First physics performance has been studied for Pb-Pb 0-5% central collisions at 40 GeV/nucleon ($\sqrt{s_{NN}} = 8.8$ GeV, $dN_{ch}/dy = 265$).

The dimuon mass spectrum is simulated as a superposition of various sources:

- the thermal $\mu^+\mu^-$ differential spectra $dN/dM_{p_T}dp_Tdy$ based on the in-medium ρ, ω, 4π spectral functions and the expanding thermal fireball model of [13], with subsequent improvements according to [14];
- the QGP spectrum calculated using a lattice-QCD constrained rate based on the equation of state of [15] with $T_c = 163$ MeV (critical temperature) e $T_{ch} = 148$ MeV (chemical freeze-out temperature);
- a hadron cocktail generator for the 2-body decays of η, ω, ϕ and the Dalitz decays; $\eta \rightarrow \gamma\mu^+\mu^-$ and $\omega \rightarrow \pi^0\mu^+\mu^-$ based on the NA60 ones and on the statistical model of [12];
- Drell-Yan and open charm simulated with the Pythia event generator [16].

In Fig. 4 a data sample composed by $2 \cdot 10^7$ reconstructed pairs in central collisions, corresponding to a total sample of $\sim 5 \cdot 10^7$ integrated in centrality, is shown.

The total reconstructed mass spectrum is shown in black, the combinatorial background coming from muons produced by decays of primary or secondary hadrons is shown by the blue continuous line (simulated using the Fluka package [18, 19]) and the background coming from

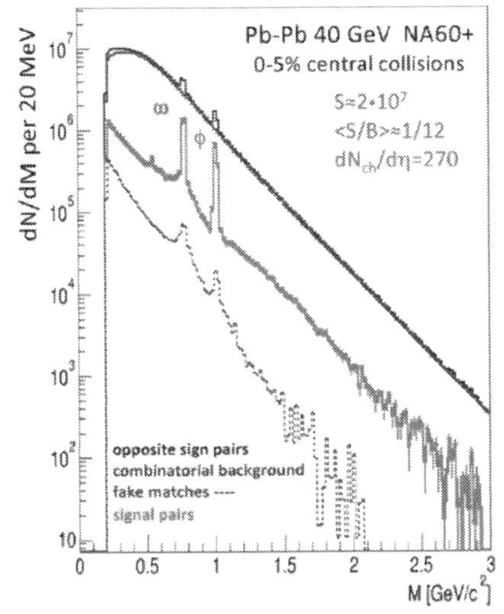

Figure 4. Expected data sample in Pb-Pb 0-5% central collision at 40 AGeV.

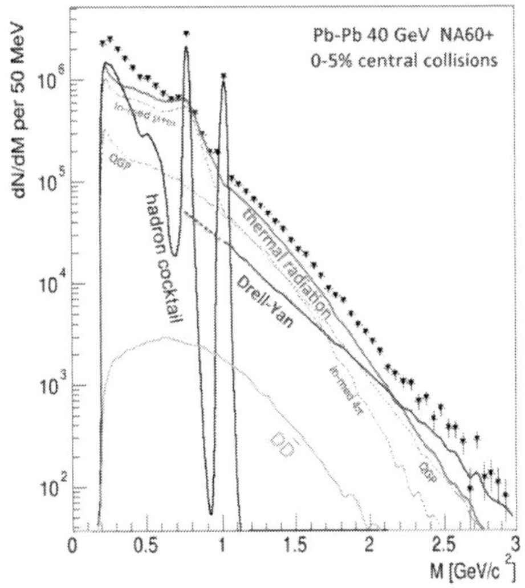

Figure 5. Reconstructed spectrum after background subtraction and different signal components.

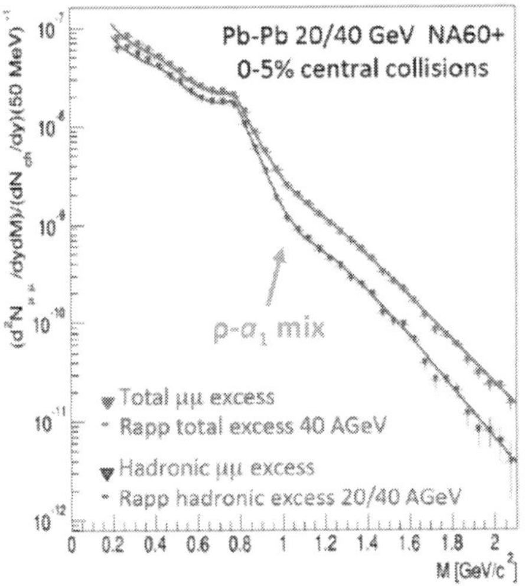

Figure 6. Acceptance-corrected inclusive thermal dilepton spectrum (red) and acceptance-corrected spectrum from hadronic radiation only (green).

wrong hadronic hits in silicon pixel planes, associated to a muon tracks, is shown by the blue dashed line (fake matches). This last contamination was taken into account at reconstruction level including the hadronic hits in the silicon stations according to the pion, kaon and proton multiplicities as measured at 40 GeV/nucleon by the NA49 experiment [17]. The net signal after the background subtraction is shown in red in Fig. 4, the average signal-to-background ratio is 1/12.

All the components of the signal spectrum after subtraction of the combinatorial background are shown in Fig. 5 (the uncertainty from the background subtraction is shown as a yellow band). In the low mass region in-medium $\rho + \omega$ dominates, while the QGP contribution is almost an order of magnitude smaller. The peaks of ω and ϕ mesons are well resolved (with a resolution of (10-15) MeV at the pole mass). This allows the in-medium thermal radiation to be precisely isolated by subtracting the hadron cocktail contributions with the data-driven technique mastered by NA60 [20]. The thermal spectrum is measurable up to (2.5-3) GeV and the QGP yield is still significant, according to the theoretical model considered. Drell-Yan curve exceeds the QGP only above (2.3-2.5) GeV, while the open charm yield is negligible.

The most significant results are summarized in Fig. 6, showing in red the thermal dilepton spectrum obtained after subtraction of the hadronic cocktail, Drell-Yan and open charm contributions (the latter being negligible at this energy).

The precision of the temperature measurement was assessed by fitting the distribution in the range $(1.5 \div 2.5)$ GeV, with $dN/dM \propto M^{3/2}exp(M/T)$. In this way we find $T = (163 \pm 4(stat.) \pm 1(syst.))$ MeV, in perfect agreement with the input value from the theoretical generator of 160 MeV. This shows that a high precision measurement is feasible.

Fig. 6 shows in green the thermal spectrum from hadronic phase $(\rho - a_1)$. The performance for the measurement of the thermal spectrum at the onset of deconfinement has been estimated considering a scenario without QGP and a measured thermal dilepton yield at the level of the in-medium $\rho + \omega + 4\pi$. Moreover, a combinatorial background yield at the level of Pb-Pb central collisions at 20/40 GeV/nucleon was assumed. This shows that a study of the hadronic excess up to M$\sim$ 2 GeV is possible, with a very good sensitivity to $\rho - a_1$ chiral mixing.

4. Summary

NA60+ at the CERN SPS provides a unique opportunity for dilepton measurements with unprecedented precision over the widest possible energy range.

This new experiment will increase the statistics by a factor $\sim$ 100 larger than NA60 and $\sim 10^5$ larger than the present experiments RHIC and LHC, promising a new horizon for a quantitative understanding of chiral symmetry restoration, onset of deconfinement and order transition.

References

[1] Dainese A *et al.* 2016 *Frascati Phys. Ser.* **62**
[2] Muller B 2005 *Nucl. Phys.* A **750** 84
[3] NA60 Collaboration, Arnaldi R *et al.* 2006 *Phys. Rev. Lett.* **96** 162302
[4] NA60 Collaboration, Specht H J *et al.* 2010, *AIP Conference Proceedings* **1322** 1-10
[5] NA60 Collaboration, Arnaldi R *et al.* 2009 *Eur. Phys. J.* C**59** 607623
[6] NA60 Collaboration, Specht H J 2010 *AIP Conf. Proc.* **1322** 110
[7] Csorgo T, Nagy M I, and Csanad M 2008 *J. Phys.* G**35** 104128
[8] Arnaldi R *et al.*, NA60 Coll.2007 *Phys. Rev. Lett.* **99** 132302
[9] Arnaldi R *et al.*, NA60 Coll.2008 *Phys. Rev. Lett.* **100** 022302
[10] Usai G. 2014, *Nucl. Phys.* A**931** 729734.
[11] ALICE Collaboration, Abelev B *et al.* 2014 *J.Phys.* G**41** 087002
[12] Becattini F, Manninen J, and Gazdzicki M 2006 *Phys. Rev.* C**73** 044905
[13] Rapp R and Wambach J 1999 *Eur. Phys. J.* A**6** 415420
[14] Van Hees H and Rapp R 2008 *Nucl. Phys.* A**806** 339387
[15] He M, Fries R J, and Rapp R, *Phys. Rev.* C**85** 044911
[16] Sjostrand T, Mrenna S, and Skands P Z 2006 *JHEP* **05** 026
[17] NA49 Collaboration, Afanasiev S V *et al.* 2002 *Phys. Rev.* C**66** 054902
[18] Bohlen T T, Cerutti F, Chin M P W, Fassò A, Ferrari A, Ortega P G, Mairani A, Sala P R, Smirnov G, and Vlachoudis V 2014 *Nuclear Data Sheets* **120** 211-214
[19] Ferrari A, Sala P R, Fassò A, and Ranft J 2005 *CERN-2005-10 INFN/TC_05/11, SLAC-R-773*
[20] NA60 Collaboration, Arnaldi R *et al.* 2006 *Phys. Rev. Lett.* 96 162302

IOP Conf. Series: Journal of Physics: Conf. Series **841** (2017) 012025 doi:10.1088/1742-6596/841/1/012025

$^{arch}PbMoO_4$ scintillating bolometer as detector to searches for the neutrinoless double beta decay of ^{100}Mo

S Nagorny[1,2], L Pattavina[2], M B Kosmyna[3], B P Nazarenko[3], S Nisi[2], L Pagnanini[1], S Pirro[2], K Schäffner[1], A N Shekhovtsov[3]

[1]INFN – Gran Sasso Science Institute, I-67100 L'Aquila, Italy
[2]INFN – Laboratori Nazionali del Gran Sasso, I-67100 Assergi, Italy
[3]Institute for Single Crystals of NAS of Ukraine, 61001 Kharkov, Ukraine

E-mail: nagorny@lngs.infn.it

Abstract. The $^{arch}PbMoO_4$ scintillating crystal has been produced from archaeological lead for the first time. The advanced technique for deep purification of lead against chemical impurities was used resulting in 99.9995% purity level of final material. The $^{arch}PbMoO_4$ crystal was characterized by means of cryogenics bolometric measurements and demonstrates excellent performances as a scintillating bolometer. The energy resolution (0.3% at 1462 keV of ^{40}K), the high light yield (5.2 keV/MeV for γs, and 1.2 keV/MeV for α particles) and the highly efficient particle identification achieved with this detector, representing the high quality of the crystal. As a final proof for the feasibility of the $^{arch}PbMoO_4$ crystal as a promising detector to search for the neutrinoless double β-decay of ^{100}Mo, the crystal should be produced using the LTG Czochralski technique to prevent the possible contamination during the crystal growth and to increase the production yield.

1. Introduction

The claim of the Heidelberg-Moscow group in 2001 [1] on the neutrinoless double β-decay (0ν2β) observation on ^{76}Ge, impelled numerous experimental efforts in order to probe this result using different nuclei as well as various experimental techniques [2–9]. ^{100}Mo is one of the most promising isotopes to search for the 0ν2β decay due to its high energy transition ($Q_{\beta\beta}$ = 3035 keV [10,11]), the comparably high natural isotopic abundance (9.67% [12]), the possibility to be highly enriched by the ultra-speed gas-centrifuge technology (up to 99.5%) and the reasonable price for such type of enrichment (in the scale of $100/g). The high $Q_{\beta\beta}$ leads to a large phase space integral, which leads to a higher probability of the 0ν2β decay. From an experimental point of view, the $Q_{\beta\beta}$ value higher than of 2615 keV line of ^{208}Tl results in a lower background induced by the natural radioactivity. The use of crystals embedding the isotope of interest operated as scintillating bolometers is one of possible way to realize a very sensitive experimental approach. This technique provides a high experimental sensitivity due to the combination of an excellent energy resolution with a high detection efficiency and the active particle discrimination.

In last decade a different crystals containing molybdenum were tested, demonstrating their feasibility as scintillating bolometers, for example Li_2MoO_4 [13], $CaMoO_4$ [14], $ZnMoO_4$ [15], $Li_2Zn_2(MoO_4)_3$ [16]. The $PbMoO_4$ crystals were also considered as a possible target-material for a next generation experiment aimed to detect the 0ν2β decay of ^{100}Mo [17,18]. Whereas the high quality $PbMoO_4$ crystals can easily be grown by well-known technologies, its application in this field is so far limited by the presence of a high activity of ^{210}Pb in freshly produced lead, with values that range from few [19] to thousands Bq/kg [20]. ^{210}Pb is a decay product of ^{238}U decay chain, which is naturally present in all rocks and ores. During the melting of the lead-containing ore, ^{210}Pb concentrates in the lead metal, while all the other radioactive nuclides belonging to ^{238}U chain (^{234}U, ^{230}Th and ^{226}Ra) are segregated to the slag, since they are chemically very different with respect to Pb. This results in a high purity of lead with respect to natural radioactive nuclides, but ^{210}Pb isotope remains embedded. Eventually ^{210}Pb activity decreases with a half-life of 22.3 years, thus in a metal which was melted hundreds years ago its activity is negligible. In fact, the extremely low level of intrinsic radioactive contamination of archaeological lead, with a limit less than few mBq/kg of ^{210}Pb, has demonstrated by several works [21–23]. So, only archaeological lead can be a suitable material for production of the low-background $PbMoO_4$ crystals.

The main target of the present work is to investigate the performance of the $^{arch}PbMoO_4$ crystal produced using of archaeological lead, as a promising scintillating bolometer to search for the 0ν2β decay of ^{100}Mo.

2. Crystal production

The general requirements on chemical purity of the initial materials to produce high-quality $PbMoO_4$ crystals impose limitations on transition metals concentration (Ti, V, Cr, Mn, Fe, Co, Ni) less than 1 ppm and on some other elements (Mg, K, Mn, Bi, Sn, Mg, Al, Cu) less than 5 ppm. Higher concentration of these impurities will significantly degrade the scintillating properties of the crystal (light output, energy resolution, and intensity of afterglow). With respect to bolometric characteristics this particular requirement is necessary to achieve the extremely low concentration of point defects (vacations and interstitial atoms) and others kind of defects like a color centers and its associates, dislocations and blocks. The specific type of defect in the crystal will determine the dominant mechanism of phonon scattering and consequently will change the heat capacity that will spoil the energy resolution in heat channel of the bolometer.

As a starting material was used a small piece of the lead ingot from the load carried by a Roman Empire age ship sunk near Sardinia [24]. A new advanced technology of vacuum distillation, which described in details in [25], was used for the deep purification of this lead metal. This purification technique is allowed to achieve purity grade of lead more than 99.9995% starting from 99% in the initial metal. The lead oxide (PbO) was produced by means of several sequential chemical transformations: the dissolution of purified lead in a weak nitric acid solution followed by the neutralization of the acid solution by ammonia, and the lead hydroxide $Pb(OH)_2$ precipitated at the end. This substance was washed several times by ultra-pure water and centrifuged. Final product was obtained by annealing of the lead hydroxide in a quartz crucible at 550°C for more than 24 hours.

The $^{arch}PbMoO_4$ single crystal was grown by the conventional Czochralski technique in a platinum crucible (∅40×40 mm) under argon atmosphere. In order to prevent the imbalance of the $PbMoO_4$ charge components due to MoO_3 evaporation during the crystal growth, a small excess of MoO_3 (1.0%) was added to stoichiometric mixture of PbO and MoO_3 powders before the charge synthesis. The total mass of loaded charge into crucible was 350 g, resulting in a single crystalline boule of 175 g (∅20×90 mm). It indicates that the production yield is about 50% from the charge. Finally, a 57.2 g cylindrical crystal with dimensions about ∅20×30 mm was cut from the boule and used as a detector.

3. Low temperature measurements

The $^{arch}PbMoO_4$ crystal was operated over 19 hours as a scintillating bolometer at mK-temperatures, in a ^{3}He/^{4}He-dilution refrigerator at Laboratori Nazionali del Gran Sasso (Assergi, Italy). When an interacting particle deposits energy in the crystal volume, a large fraction of energy is converted to heat, while the rest produces scintillation light. The heat and light signals are simultaneously recorded as temperature increasing by the Germanium Neutron Transmutation Doped thermistors (Ge-NTD) on the main crystal and on the light detector (LD), which is also operated as a bolometer. The main crystal and the LD were housed in a copper structure described in [26]. A reflecting foil (3M VM2002) was used to surround the detector setup, leading to an increase in the overall light collection efficiency. More details on the light detector, which consists of a high purity Ge wafer (∅44.5·0.3mm), can be found in [27]. According to the nature of the interaction, α or β/γ particles will generate different amount of light for the same energy released in the volume of main crystal. This peculiar feature allows to strongly reducing the background, therefore improving detector sensitivity.

The heat and light signals were converted into voltage pulses by Ge-NTD thermistors, were amplified and fed into a 18-bit ADC unit. The trigger of the $^{arch}PbMoO_4$ crystal was software generated while the LD is independently acquired in coincidence with the former. The entire waveform of each voltage pulse is sampled and acquired. The time window has a width of 512 ms sampled with 1024 points. The amplitude and the shape of the voltage pulse were then determined off-line making use of the Optimal Filter technique [28]. Detailed information on the electronic system and data acquisition can be found in [29–31].

The amplitude of the heat and light signals is energy-calibrated by means of known calibration sources. The light detector was calibrated using a ^{55}Fe X-ray source permanently facing the LD. An energy resolution (FWHM) of 590 eV was evaluated for the X-ray doublet of 5.9 and 6.5 keV. While the heat channel is energy-calibrated with α peaks of internal radioactive contamination by the daughter nuclides of U/Th natural decay chains, using a second order polynomial with zero intercept as a calibration curve.

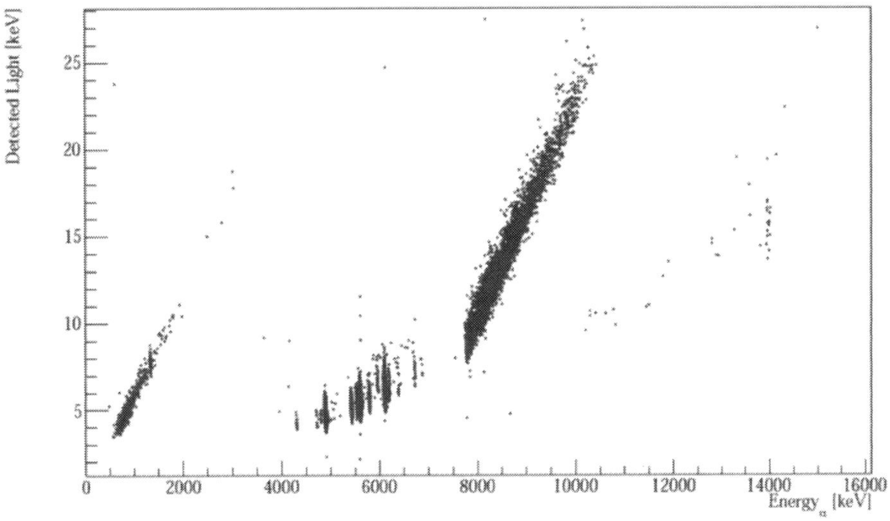

Figure 1. Light-vs-Heat scatter plot corresponding to 19 hours of background measurement with the 57.2 g $^{arch}PbMoO_4$ scintillating crystal, working as a scintillating bolometer. One can recognize several well separated events distributions caused by different type of particles (see description in text below).

The total acquired statistics is shown in Fig. 1, where the detected light is shown as a function of the energy detected by the $^{arch}PbMoO_4$ crystal. The nature of the interacting particles is clearly understood by the different Light Yield (LY), defined in this context as the ratio of the energy measured in the light detector (in keV) to the energy measured in the heat channel of the $^{arch}PbMoO_4$ crystal absorber (in MeV). From Fig. 1, one can see several easily distinguishable distributions induced by detection of different type of particles: a) band in energy range 0.5-3.0 MeV related to β/γ induced interactions (single β/γ events); b) structure between 4 MeV to 7 MeV represent α peaks of internal contamination from U/Th natural decay chains (single α events); c) highly populated band from 8 to 11 MeV is caused by ^{212}Bi-^{212}Po coincidence events (sum of β and α events); d) pure populated distribution from 10 to 15 MeV induced by coincidence of two single α events due to internal contamination from U/Th natural decay chains. The pure β/γ events are characterized by LY of about 5.2 keV/MeV, while pure α's have lower LY on the level of 1.2 keV/MeV.

This $^{arch}PbMoO_4$ crystal demonstrates the outstanding bolometric characteristics. Offering perfect discrimination between α and β/γ events due to big difference in LY, the detector also has an extremely good energy resolution in heat channel both, for α particles (FWHM = 8.4 keV or 0.2% at E_α = 4870 keV of ^{226}Ra) and for γ's (FWHM = 5.1 keV or 0.3% at E_γ = 1462 keV of ^{40}K). Such excellent energy resolution clearly indicates the high perfection of the crystalline structure and the low defects concentration.

4. Internal α contamination

The "pure" α events were selected from entire statistics by applying cut on the amount of emitted light (i.e. LY < 2.0 keV/MeV) to reject β/γ events, and on deposited energy (i.e. < 7.0 MeV) to select only single α events, with the overall efficiency close to 100%. The spectrum of α events obtained applying these cuts is presented in Fig. 2.

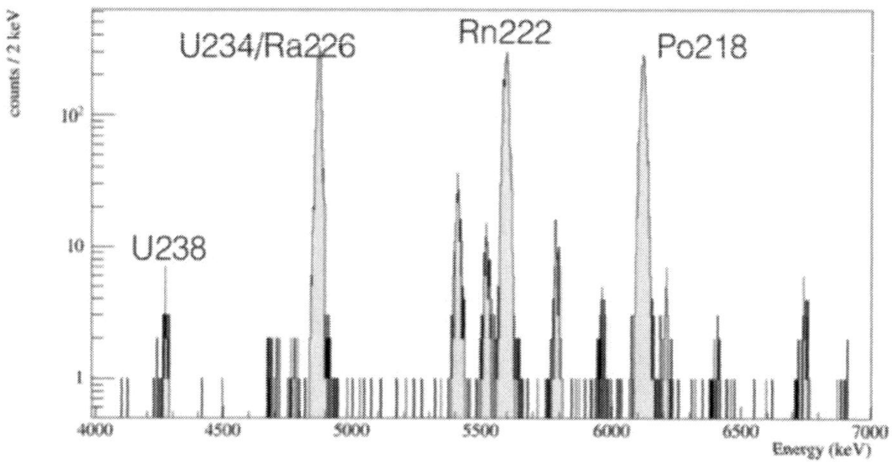

Figure 2. The selected "pure" α events from entire statistics collected with the $^{arch}PbMoO_4$ crystal 57.21 g of mass, during 19 hours of background measurements. The events are attributed to internal contaminations of the crystal by daughter α active nuclides of U/Th decay chains.

As one can see from Fig. 2 and values listed in Tab. 1, the $^{arch}PbMoO_4$ crystal is extremely highly contaminated by the radioactive nuclides from ^{226}Ra sub-chain of ^{238}U chain, as well as with daughter nuclides from ^{235}U and ^{232}Th chains. Such level of internal contamination is absolutely unexpected, considering all efforts dedicated to the production of the high chemical and radioactive purity PbO and MoO_3 powders. We can only suppose that contamination incident is occurred during the crystal growth.

As was mentioned above (see Sec. 2), the $^{arch}PbMoO_4$ crystal was grown by the conventional Czochralski technique, where an opened Pt-crucible with loaded charge was covered by several layers of the aluminum oxide ceramics to provide a thermal insulation. In our opinion, the huge radioactive contamination of the crystal can be the result of the contact between the melted charge and the small piece of the aluminum oxide ceramics. This assumption matches with our experience gained in the framework of the LUCIFER/CUPID-0 project [32,33], devoted to the development of the low-background $Zn^{82}Se$ crystals starting from the high purity Zn and the enriched ^{82}Se isotope. A huge activity of nuclides from ^{226}Ra sub-chain (about 0.2 kBg/kg) and ^{40}K (1 kBq/kg) was induced by a 0.3 g piece of aluminum oxide ceramics incorporated in a certain batch of the ZnSe powder. Assuming the same radiopurity of aluminum oxide ceramics in both cases (i.e. $Zn^{82}Se$ powder production and $^{arch}PbMoO_4$ crystal growth), the internal contamination of the $^{arch}PbMoO_4$ crystal is compatible with activity caused by a piece of ceramics with mass of about 30 mg, fallen into the crucible during the crystal growth.

Table 1. Internal contamination of the 57.21 g $^{arch}PbMoO_4$ crystal.

Chain	Nuclide	Activity [mBq/kg]
^{232}Th	^{232}Th	10±7
	^{228}Th	99±9
	^{224}Ra	86±8
	^{220}Rn	21±5
	^{212}Bi	49±7
^{238}U	^{238}U	38±6
	^{234}U	98±9
	^{230}Th	13±3
	^{226}Ra	2310±42
	^{222}Rn	2370±42
	^{218}Po	2340±42
	^{210}Po	266±16
^{235}U	^{235}U	14±3
	^{227}Th	17±4
	^{223}Ra	30±5
	^{219}Rn	5±2
	^{211}Bi	30±5

5. Discussion

Finally we would like to stress several aspects of the presented results. First of all, the technological issue of the radiopure crystal production. A possible solution to produce a low-background high-quality crystals is to use the Low Thermal Gradient (LTG) Czochralski technique [34,35], which operates with the closed Pt-crucible, that completely eliminates risks of contamination incident. Another very important advantage of this technique is the small temperature gradient at the level of 1 K/cm, which is one-two order of magnitude lower in comparison to the conventional Czochralski method. It leads to an absence of intensive evaporation of charge component from the melt and the losses did not exceed the level of 0.5% of the initial charge. Moreover the LTG Czochralski technique allows to crystallize up to 90% of the loaded charge (already achieved with BGO [36], $CdWO_4$ [37,38] and $ZnMoO_4$ [15] crystals), which is crucial in case of costly material, like enriched isotopes or highly purified archaeological lead. It should be also stressed that, despite the low molar fraction of Mo in the $PbMoO_4$ compound (26%) the number of Mo nuclei in 1 cm^3 of crystal is the same ($0.11\cdot10^{23}$) as for the Li_2MoO_4 crystal, that has highest molar fraction of Mo (55%). This is the result of higher density of the $^{arch}PbMoO_4$ crystal (6.83 g/cm^3), in comparison to the Li_2MoO_4 crystal (3.03 g/cm^3). Moreover, the highest Z and density of the $^{arch}PbMoO_4$ crystal will increase the detection efficiency of the high energetic electrons produced in the $0\nu2\beta$ decay of ^{100}Mo, taking in consideration the detector with the same dimensions. Finally, we do not see any drawbacks in the application of the $^{arch}PbMoO_4$ crystal produced from the archaeological lead as detector to search for the $0\nu2\beta$ decay of ^{100}Mo, especially taking into consideration its excellent bolometric performances.

As a next step we are going to grow the $^{arch}PbMoO_4$ crystal from the highly purified archaeological lead and molybdenum oxide, using the LTG Czochralski technique to prevent possible contamination incident during crystal growth and to increase significantly the production yield. It will allow us to produce a high-quality large-volume $^{arch}PbMoO_4$ crystal with high performance as a low-background scintillating bolometer.

References

[1] Klapdor-Kleingrothaus H V *et al.*, 2001 *Mod. Phys. Lett.* A **16** 2409
[2] Agostini M *et al.*, 2016 *Nucl. Part. Physics Procs.* **273 – 275** 1876
[3] Alfonso K *et al.*, (CUORE Collaboration) 2015 *Phys. Rev. Lett.* **115** 102502
[4] Ebert J *et al.*, 2016 *Phys. Rev. C* **94** 024603
[5] Kim G B *et al.*, 2015 *Adv. High Ener. Phys. **2015** 1*
[6] Iida T *et al.*, 2014 *Nucl. Phys. B Proc. Suppl.* **00** 1
[7] Danevich F A *et al.*, 2016 *J. Phys.: Conf. Ser.* **718** 062009
[8] Artusa D R *et al.*, 2016 *Eur. Phys. J.* C **76** 364
[9] Gando A *et al.*, 2013 *Phys. Rev. Lett.* **110** 062502
[0] Audi G *et al.*, 2012 *Chin. Phys.* C **36** 1287
[11] Wang M *et al.*, 2012 *Chin. Phys.* C **36** 1603
[12] Berglund M and Wieser M E 2011 *Pure Appl. Chem.* **83** 397
[13] Cardani L *et al.*, 2013 *J. Instrum.* **8** 10002
[14] Yoon W S *et al.*, 2015 *Nucl. Instr. Meth* A **784** 143
[15] Barabash A S *et al.*, arXiv:1405.6937v2, accepted by *Eur. Phys. J.* C
[16] Bashmakova N V *et al.*, 2009 *Func. Mater.* 16 266
[17] Minowa M *et al.*, 1992 *Nucl. Instr. Meth.* A **322** 500
[18] Danevich F A *et al.*, 2010 *Nucl. Instr. Meth.* A **622** 608
[19] Laubenstein M *et al.*, 2004 *Appl. Rad. and Isotopes* **61** 167
[20] Bunzl K and Kracke W 1985 *Nucl. Instr. Meth.* A **238** 191
[21] Alessandrello A *et al.*, 1991 *Nucl. Instr. Meth.* B **61** 106
[22] Alessandrello A *et al.*, 1993. *Nucl. Instr. Meth.* B **83** 539
[23] F Danevich F A *et al.*, 2009 *Nucl. Instr. Meth.* A **603** 328
[24] Alessandrello A *et al.*, 1998 *Nucl. Instr. Meth.* B **142** 163
[25] Boiko R S *et al.*, 2011 *Inorg. Mater.* **47** 645
[26] Cardani L *et al.*, 2013 *J. Instrum.* **8** 10002
[27] Beeman J *et al.*, 2013 *J. Instrum.* **8** 05021
[28] Gatti E and Manfredi P F 1986 *La Rivista Nuovo Cimento* **9(1)** 1
[29] Pirro S *et al.*, 2000 *Nucl. Instrum. Meth.* A **444** 331
[30] Arnaboldi C *et al.*, 2002 *Nucl. Instrum. Meth.* A **520** 578
[31] Piperno G *et al.*, 2011 *J. Instrum.* **6** 10005
[32] Beemann J W *et al.*, 2013 *Adv. High Energy Phys.* **2013** 237973
[33] Beeman, J W *et al.*, 2016 *J Low Temp Phys* **184** 852
[34] Borovlev Yu A *et al.*, 2001 *J. Cryst. Growth* 229 305
[35] Galashov E N *et al.*, 2010 *Func. Mater.* **17** 504
[36] Grigoriev D N *et al.*, 2014 *J. Instrum.* **9** 09004
[37] Belli P *et al.*, 2010 *Nucl. Instr. Meth.* A **615** 301
[38] Barabash A S *et al.*, 2011 *J. Instrum.* **6** 08011

The first pion-Ar cross-section measurement with the LArIAT experiment

I Nutini [1,2] on behalf of the LArIAT Collaboration

[1] GSSI - Gran Sasso Science Institute, Viale F. Crispi, 7, 67100 L'Aquila, Italy
[2] Fermi National Accelerator Laboratory, U.S. Department of Energy - Neutrino Division

E-mail: irene.nutini@gssi.infn.it

Abstract. A complete understanding of neutrinos properties requires a study and a characterization of the interactions of the daughter particles created in a neutrino-nucleus interaction. The Liquid Argon In A Testbeam (LArIAT) experiment is a small-scale liquid argon detector situated in the Fermilab Test Beam Facility. The LArIAT experiment is exposed to a tertiary beam comprised of mostly pions along with a mix of muons, protons, kaons, and electrons. LArIAT's goal is to characterize the response of the Liquid Argon Time Projection Chamber (LArTPC) to known incoming charged particles and measure their interactions in Argon, in order to understand their cross-sections and to help developing and tuning simulations and reconstruction algorithms for LArTPC neutrino experiments. The world's first measurement of a pion cross-section on an Argon target, made with the LArIAT detector, is presented here.

1. Introduction

The LArIAT experiment [1, 2] is part of the international US Neutrino Program.
The aim of the accelerator neutrino physics program at Fermilab is to investigate neutrino flavor oscillations and the existence of sterile neutrinos. A beam of neutrinos is fired into the detector and the secondary particles produced by their interactions are detected and utilized to reconstruct the info about the neutrino.
Liquid Argon Time Projection Chambers (LArTPC) technology has been utilized for many accelerator neutrino detectors. In a TPC, planes of angled wires collect the electrons that are produced by ionization of charged particles in the active medium and drifted along a uniform electric field. Liquid argon is chosen as the active medium because it is relatively dense, ensuring higher scattering centers density. It has an high electron lifetime - electrons are not absorbed while crossing the medium - and it has also an high scintillation light yield, being also transparent to its own scintillation light. The LArTPCs scalability, low energy threshold, calorimetric resolution, and ability to reconstruct particle interactions in 3D with mm-resolution make them ideal for the detection of the charged products from neutrino interactions [3].

The US Neutrino Program is based on the use of the liquid argon TPC technology. TPCs of varying scales are planned or already in operation for the Fermilab Short-Baseline (MicroBooNE, SBND, ICARUS) and Long-Baseline (DUNE) neutrino program. Considering this long-term plan for neutrino-argon interactions studies with LArTPCs, a dedicated calibration of this emerging technology is invaluable. The LArIAT experiment aims to characterize LArTPC performance in the energy range relevant to these neutrino experiments by operating the refurbished ArgoNeuT detector [4] in a tunable testbeam at Fermilab. It explores particle response calibration, technical R&D, and several physics topics relevant to current and future liquid argon detectors.

2. The LArIAT Experiment

The LArIAT experiment is a 0.26 ton active mass Liquid Argon Time Projection Chamber (LArTPC) located at the Fermilab Test Beam Facility (FTBF). The TPC is exposed to a tertiary charged particles beam comprised of pions ($\pi^{\pm}$) along with a mix of muons ($\mu^{\pm}$), protons (p), kaons ($K^{\pm}$) and electrons ($e^{\pm}$) in the range between 200 MeV and 2 GeV. The beamline is equipped with detectors that aid to identification and selection of particle species and momenta, in order to characterize the response of the LArTPC to known incoming particles. These results can then be used to tune simulation and reconstruction algorithms for future LArTPC experiments.

2.1. The Fermilab Test Beam

Primary protons (120 GeV) from the Fermilab Main Injector are shot onto an Al target. A tunable secondary pion beam is produced (8 to 64 GeV). Those pions hit a Cu target that is placed at the entrance of the test beam enclosure and a tertiary beam of charged particles, mainly pions, is produced.

2.2. The beamline auxiliary detectors

Several auxiliary detectors are disposed along the beamline to monitor the tertiary beam impinging on the TPC, see Figs. 1 and 2. The 4 Multi Wire Proportional Chambers (MWPCs) and the bending magnets allow the momentum reconstruction before the particle enters the TPC. The dipole magnets allow to have a tunable momentum in 0.2 - 2 GeV/c range and to select positive and negative polarities. There are also two Time Of Flight (TOF) counters at the start and at the end of the beam line. Utilizing the TOF in conjunction with the reconstructed momentum it can be possibile to discriminate among several particle species.

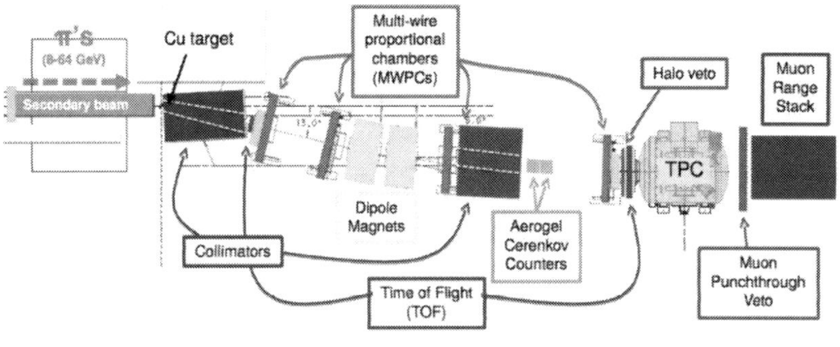

Figure 1: Schematic view of the LArIAT beamline.

2.3. The LArIAT TPC and light collection system

The LArIAT experiment uses preexisting components from the ArgoNeuT experiment, such as the cryostat and the TPC. The cryostat hosts the 550 liters of liquid argon (LAr) and it is vacuum insulated. The LArIAT TPC has an active volume of 90 cm $\times$ 47 cm $\times$ 40 cm (length $\times$ width $\times$ height) with the width serving as the drift direction and the length is along the

IOP Conf. Series: Journal of Physics: Conf. Series **841** (2017) 012026 doi:10.1088/1742-6596/841/1/012026

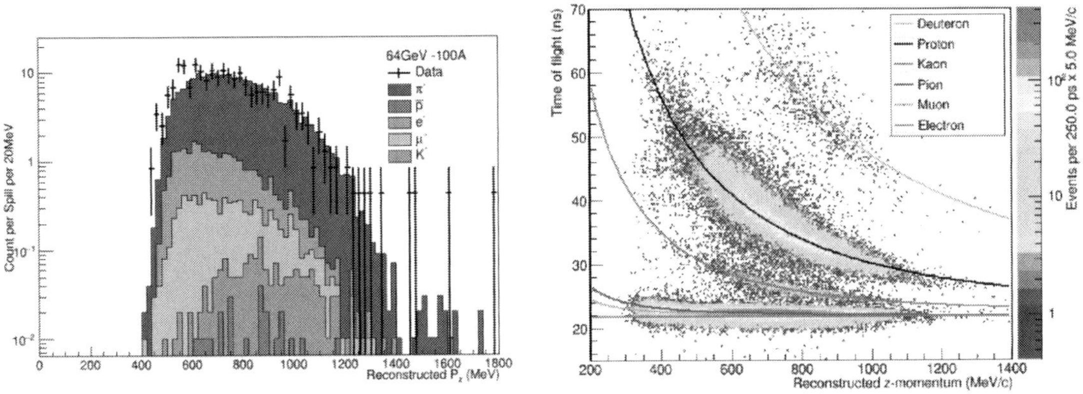

(a) LArIAT's tertiary beam momentum spectrum. **(b)** LArIAT's tertiary beam TOF vs momentum plot.

Figure 2: (a) Reconstructed momentum (P_z, z being the beam direction) profile of the LArIAT's tertiary beam in the high energy tune (negative polarity run), showing a good agreement between simulation and data. (b) Scatter plot of particle time of flight (TOF) against reconstructed momentum P_z, showing the separated populations of muons/pions/electrons, kaons, and protons (positive polarity run).

beam axis. There are three planes of wires (Shield, Induction and Collection) of which two are instrumented with dedicated cold front-end electronics to collect the electrons created during the charged particle interactions in the argon, which drift across the argon volume under a 500 V/cm nominal drift field. Each of the two planes consists of 240 wires oriented at $\pm 60°$ relative to the beam axis, with 4 mm anode wire spacing. Pulse signals on the wires are processed and analyzed through a custom version of LArSoft [5], a dedicated software for LArTPCs automated signals and tracks reconstruction.

In addition to the collection of the ionization charge from the TPC, the detection of the scintillation light produced in LAr is an actual point of interest, both for its trigger function for the TPC events collection and for the possibility of using also this information to improve the energy resolution of this detector technology [6]. The LArIAT light collection system consists of an array of two high quantum efficiency cryogenic photomultiplier tubes (PMT) - a 3" Hamamatsu R-11065 and a 2" ETL D757KFL - and three Silicon PhotoMultiplier Detectors (SiPMs) with custom made cold pre-amplification boards - one SensL single-channel SiPM and two Hamamatsu SiPM Arrays - which are deployed in liquid argon and mounted behind the wire planes of the TPC. The PMTs and the SiPMs are mostly sensitive to optical photons, while Ar scintillation wavelength is 128 nm. Wavelength shifting tetraphenyl butadiene (TPB) coated foils covering the TPC walls have been utilized to shift the argon scintillation light into the visible spectrum (blue). This technique provides enhanced light collection efficiency and uniformity.

In Fig.3 the LArIAT's TPC and light collection system are shown.

3. LArIAT's data taking and analyses

The LArIAT experiment started taking data in May 2015: Run I lasted 3 months. Run II lasted from February to July 2016: 5 times more statistics than Run I was acquired and the beam tuning was improved, ensuring higher-quality data. Run III is planned for next year (2017) and it will be more R&D focused.

LArIAT offers analyses covering detector calibration as well as physics topics relevant to current and upcoming LArTPC-based neutrino experiments. Each analysis relies on software-

(a) LArIAT's TPC inside the cryostat.

(b) LArIAT's light collection system.

Figure 3: (a) LArIAT cryostat hosting the TPC. The TPC structure and electronic readout is outlined. (b) Cryostat flange hosting the LArIAT's light collection system.

based automated reconstruction of signals from the beamline detectors, TPC, and light collection system. An example of a real LArIAT data event showing a candidate kaon decay is reported in Fig. 4, 2-D views from the wire-planes and 3-D reconstruction.

Calibration studies include measuring the drift electron lifetime, the electronics response calibration (improvement in signal-to-noise ratio using cold front-end electronics for the TPC), and the charge recombination as a function of electric field. The primary higher-level physics analyses on the collected data are the measurement of charged hadron-Ar interaction cross sections and the study of exclusive channels, e/γ shower identification, μ sign determination in the absence of a magnetic field, study of nuclear effects in argon, and Geant4 validation for hadrons interaction models. Regarding the detector technical characterization, other studies are ongoing: the optimization of the event reconstruction algorithms, the light collection system characterization, and the improvement of the energy resolution through a combined study of ionization and scintillation light.

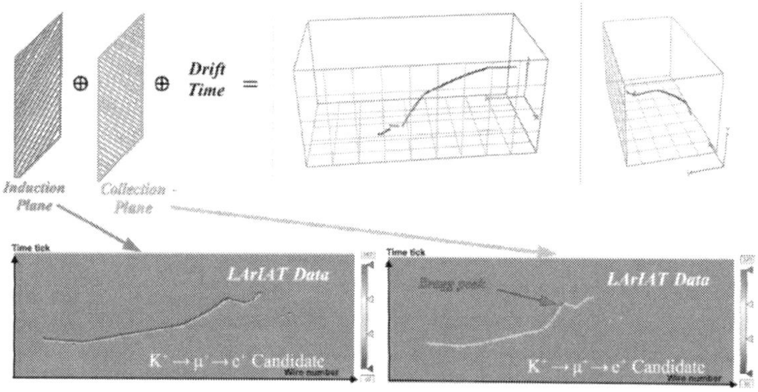

Figure 4: Example of LArIAT event imaging: kaon decay candidate. Event views on the Collection and Induction TPC planes and the reconstructed 3-D tracks.

4. Total π^- - Ar cross section measurement in LArIAT

One of the goals of the LArIAT experiment is the experimental measurement of charged pion cross section on Ar in a wide energy range. Prior to this analyses, no experimental measurement of this cross section on argon had been preformed. Predictions for only few energies come from interpolation between lighter and heavier nuclei.

4.1. Cross-section definition

The total π^-- Ar interaction cross section σ_{tot} measured in this analysis is defined as the one related to strong interaction processes (including both elastic and reaction channels). In case of the "thin target" approximation, the measured rate of interactions for a given slab of Ar can be used to calculate the total cross section as a function of energy (see Eq.1).

$$\sigma_{tot}(E) \simeq \frac{1}{nz}\frac{N_{interacting}(E)}{N_{incident}(E)} \tag{1}$$

where z is the target thickness (in cm) along the incident pion direction, and n is the scattering centers density in the target, $n = \frac{\rho N_A}{A}$ (in cm^{-3}).

The LArIAT TPC active volume (90-cm thick LAr) can not be considered a thin target for pion interactions in the energy range of interest. Therefore a new technique called the "sliced TPC" method was developed which makes possible an accurate measurement of the pion cross section dependence on energy in the "thick target" LArTPC environment [7], as shown in Figure 5. Using the granularity of the LArTPC, it is possible to treat the wire-to-wire spacing as a series of "thin-slab" targets of $\Delta z \approx 4.5$ mm thickness along the z axis, i.e. direction of the incident particles (pions), if the energy of the pion incident to each slice is measured. Each slice can be considered as a "thin target" and it is possible apply the cross section calculation from Eq.1 iteratively, evaluating the actual kinetic energy of the pion as it enters each slice, exploiting the precise calorimetric reconstruction along the track.

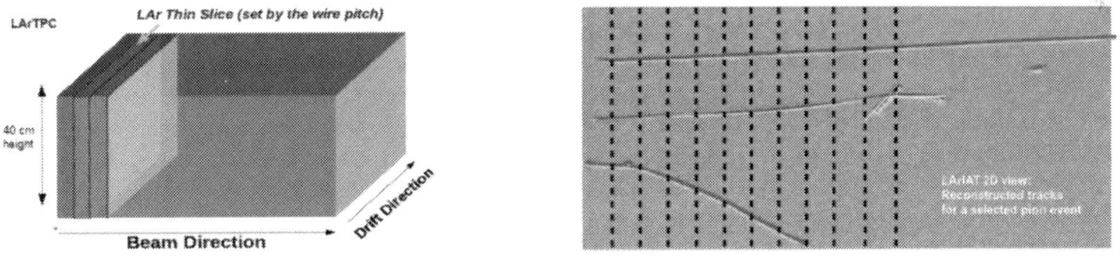

Figure 5: Representation of the analysis of reconstructed tracks with "sliced TPC" technique.

4.2. Analysis on LArIAT's Data

This analysis uses data collected during LArIAT's Run I - Negative Polarity runs.

Selection criteria on beamline detectors response and on TPC events topology were applied to select π^- candidate events. Separation of $\pi/\mu/e$ from (anti-)protons and late bunches from the beam was obtained through the TOF vs momentum measurement. A clean and unique match between extrapolated Wire Chamber track (from the MWPCs) and the start of TPC track was required. Electron-initiated shower events in the TPC were rejected in order to remove the electron contamination from the pion sample. The muon contamination was estimated through Monte Carlo simulation. The performance of the other auxiliary beam line detectors to improve the sample purity is being investigated (e.g aerogel and muon range stack to tag μ^-/e backgrounds).

In the preliminary analysis, the traditional strong processes as well as background processes are currently included in the cross section result presented below in Fig. 6. The background processes involve: negative pion capture at rest and pion decay events. Further studies on data and Monte Carlo to remove π^- capture and decay events from the sample are currently underway. The result shows a relatively good agreement between data and Monte Carlo in a wide energy range except in the first energy bin, in which deficiencies in the Monte Carlo simulation are suspected. A revised simulation of the LArIAT beamline may resolve this discrepancy.
The main systematic uncertainties on the cross section are related to: calibration of the energy deposition (dE/dx) in the TPC (5%), energy loss prior to entering the TPC (3.5%), through-going muons contamination (3%), and wire chamber (initial) momentum reconstruction (3%) [8].

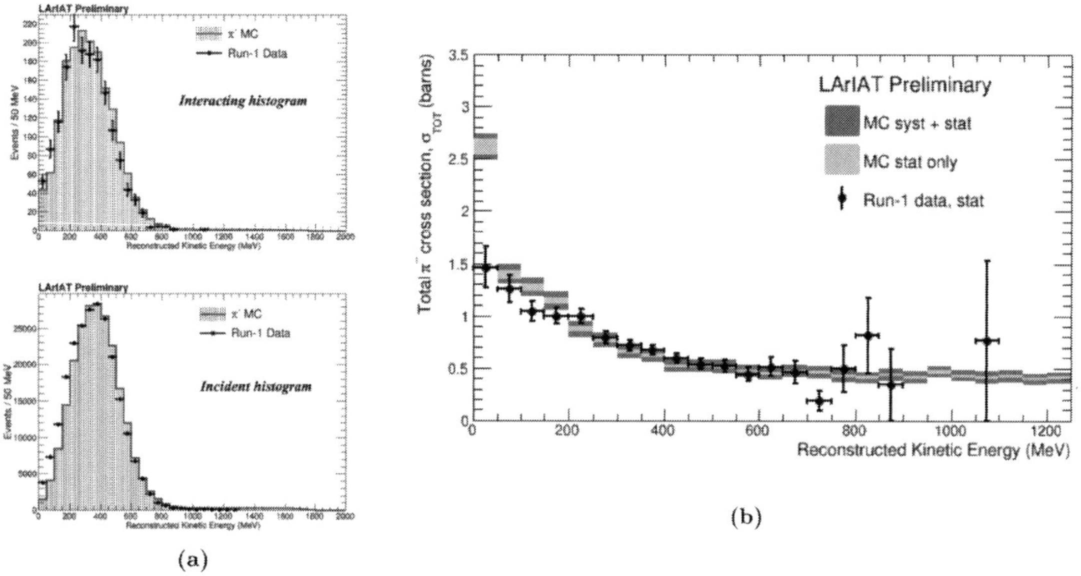

Figure 6: LArIAT Preliminary results on the π^-- Ar cross section analysis [8]. (a) Interacting $N_{interacting}(E)$ and Incident $N_{incident}(E)$ energy distributions. (b) Total π^-- Ar cross section.

5. Conclusions

The data sample from LArIAT's initial run is utilized to make the first measurement of the π^- - Ar cross section and is compared to the expectation from Monte Carlo. More analyses are forthcoming from the LArIAT collaboration utilizing additional data collected in the second beam run which took place between February and August 2016.

References

[1] Cavanna F, Kordosky M, Raaf J, Rebel B and LArIAT collaboration, 2014, arXiv:1406.5560 [physics.ins-det]
[2] LArIAT/T-1034: Liquid Argon TPC In A Test beam, TSW - Technical Scope of Work, February 2013
[3] Rubbia C, 1977, *CERN EP Internal Report* pp 77-8
[4] Anderson C *et al.*, 2012, *JINST* **7** P10019
[5] LArSoft collaboration, Church E D, 2013, arXiv:1311.6774 [physics.ins-det]
[6] Szelc A M, 2013, *JINST* **8** C09011
[7] Nutini I, 2015, FERMILAB-MASTERS-2015-03
[8] Asaadi J on behalf of the LArIAT Collab., FNAL Joint Experimental-Theoretical Physics Seminar, 2016.

Assess the neutrino mass with micro and macro calorimeter approach

A Giachero [1,*]

[1] INFN Milano-Bicocca, Piazza della Scienza 3, Milano I-20126 - Italy
[*] Invited Speaker.

E-mail: `andrea.giachero@mib.infn.it`

Abstract. Thanks to oscillation experiments it is now an established fact that neutrinos are massive particles. Yet, the assessment of neutrinos absolute mass scale is still an outstanding challenge in particle physics and cosmology as oscillation experiments are sensitive only to the squared mass differences of the three neutrino mass eigenstates. The mass hierarchy is not the only missing piece in the puzzle. Theories of neutrino mass generation call into play Majorana neutrinos and there are experimental observations pointing toward the existence of sterile neutrinos in addition to the three weakly interacting ones. Three experimental approaches are currently pursued: an indirect neutrino mass determination via cosmological observables, the search for neutrinoless double β-decay, and a direct measurement based on the kinematics of single β or electron capture decays.

Bolometers and calorimeters are low temperature detectors used in many applications, such as astrophysics, fast spectroscopy and particle physics. In particular, sensitive calorimeters play an important role in the neutrino mass measurement and in the search for the neutrinoless double β-decay. There has been great technical progress on low temperature detectors since they were proposed for neutrino physics experiments in 1984. This general detector paradigm can be implemented in devices as small as a micrometer for sub eV radiation or as large as 1 kg for MeV scale particles. Today this technique offers the high energy resolution and scalability required for leading edges and competitive experiments addressing the still open questions in neutrino physics.

After 30 years of observations, oscillation experiments have firmly established neutrinos are massive fermions whose three weak interaction flavor eigenstates (ν_e, ν_μ, ν_τ) are a quantum superposition of three mass eigenstates (ν_1, ν_2 , ν_2). Thanks to the accurate characterization of the neutrino oscillations, observed for solar, atmospheric, accelerator and nuclear reactor neutrinos, the neutrino mixing matrix $U_{\ell i}$ is now precisely determined. Neutrino oscillations are sensitive only to the difference of the squared masses $\Delta m^2_{ij} = |m^2_i - m^2_j|$. Two mass square differences have been determined reasonably well: $\Delta m^2_{12} \simeq (7.37 \pm 0.17) \cdot 10^{-5}\,\mathrm{eV}^2$ and $|\Delta m_{23}|^2 \simeq |\Delta m_{13}|^2 \simeq (2.50 \pm 0.04) \cdot 10^{-3}\,\mathrm{eV}^2$, but the sign of Δm^2_{13} is however unknown [1]. This sign settles the neutrino mass hierarchy: if it is positive the hierarchy is normal ($\mathcal{NH}$) with $m_1 \ll m_2 \ll m_3$ on the contrary if it is negative the hierarchy is inverted ($\mathcal{IH}$), with $m_3 \ll m_1 \lesssim m_2$. If $m_1 \lesssim m_2 \lesssim m_3$ the pattern is quasi-degenerate. Past oscillation experiments were not sensitive to the neutrino mass hierarchy. Currently, optimistic efforts to determine the neutrino mass hierarchy are basically accelerator based long-baseline (i.e. NOνA [2] and T2K [3]) and reactor (i.e. JUNO [4]) experiments.

As alternative to oscillation experiments three experimental methods can assess the neutrino masses and the neutrino mass scale: the analysis of Cosmic Microwave Background (CMB) temperature fluctuations [5], the search for the neutrinoless double β-decay ($0\nu\beta\beta$), and the analysis of the single β or electron capture (EC) decay. The first method can give important indirect information on the neutrino mass by using space-borne instruments to precisely measure the CMB thermal fluctuations. These bounds depend on the delicate interplay between CMB and galaxy power spectra. This approach depends critically on cosmological and astrophysical assumptions and requires therefore independent checks. The observable of these measurements is the sum of the neutrino masses: $\Sigma = \Sigma_{i=1}^{3} m_i$, where m_i are the mass eigenvalues of the three neutrino mass eigenstates. The most stringent limit is $\Sigma < 0.12\,\mathrm{eV}$ @ 95% C.L. [6]. The second method has been suggested since long time as a powerful tool to measure electron neutrino effective mass, although its very existence depends on the assumption that the neutrino is Majorana particle ($\overline{\nu} \equiv \nu$). The critical problems of this approach are the theoretical uncertainties in the calculations of the nuclear matrix elements (NME) and phase space factors (PSF), that enter the $0\nu\beta\beta$ lifetime expressions [7]. The observable is the effective neutrino Majorana mass: $m_{\beta\beta} = |\Sigma_{i=1}^{3} m_i \eta_i |U_{ei}|^2|$, where η_i are the CP Majorana phases ($\eta_i = \pm 1$ for CP conservation) and U_{ei} are the elements of the electron sector of the neutrino mixing matrix. The most stringent limits depend on the isotope and up-to-dately are: $m_{\beta\beta}(^{76}\mathrm{Ge}) \leq (160 \div 260)$ meV [8], $m_{\beta\beta}(^{130}\mathrm{Te}) \leq (260 \div 760)$ meV [9, 10], $m_{\beta\beta}(^{136}\mathrm{Xe}) \leq (61 \div 165)$ meV [11]. In the last method the neutrino mass can be directly assessed by studying the kinematics of low energy nuclear β or EC decays, where the neutrino is not directly observed but the energy of the decay products is precisely measured [12]. In these experiments the observable is the electron neutrino mass: $m_\beta = \sqrt{\Sigma_{i=1}^{3} m_i^2 |U_{ei}|^2}$. Experiments based on the classical spectrometer technique provided the most stringent limits of $m_\beta < 2\mathrm{eV}$ @ 90% C.L. [13, 14].

The investigation of fundamental neutrino properties, like its mass and its nature, calls for the design of a new generation of experiments. Different detection techniques and isotopes are being actively promoted by experimental groups across the world. Low temperature detection (LTD) has been suggested as high resolution soft X-ray detectors in 1984 by D. McCammon and collaborators [15]. The same year E. Fiorini and T.O. Niinikoski proposed LTD for neutrino physics experiments [16]. After more than 20 years of developments, the low temperature techniques are now ready for challenging the most pressing open questions in neutrino physics. Today LTD are widely used for neutrinoless double β-decay searches, for calorimetric direct neutrino mass experiments, but they play also an important role for dark matter searches [17] and X- and γ-ray spectroscopy in general [18].

1. Low temperature micro- and macro-calorimeters

In general, a low temperature calorimeter consists of: an *energy absorber*, in which the energy deposited by a particle is converted into phonons, a *temperature transducer* (sensor) that converts thermal excitations into a readable electric signal variation, and a weak *thermal link*. The temperature rise ΔT is related to the energy release E and, denoting by C the heat capacity of the absorber, can be written as $\Delta T = E/C$. The accumulated heat flows then to the heat bath through the thermal link and the absorber returns to the base temperature with a time constant $\tau = C/G$ where G is the thermal conductance of the link: $\Delta T(t) = E/C \cdot e^{t/\tau}$. The decay constant τ should be long enough to allow the development of a high resolution signal but short enough to avoid pile-up. In order to obtain a measurable temperature, the heat capacity and the base temperature of the absorber crystal must be very small. For these reasons low temperature detectors are usually operated at a temperature around 10 mK and are made of superconductor or dielectric materials, so that, by the Debye law ($C \propto (T/\Theta_D)^3$), only the lattice heat capacity plays a role. The temperature variation measured by a low temperature calorimeter provides a precise estimation of the released energy, limited only by the so-called thermodynamic limit: $\Delta E_{rms} = \sqrt{k_B T^2 C}$, where T is the equilibrium operating temperature and k_B is the Boltzmann

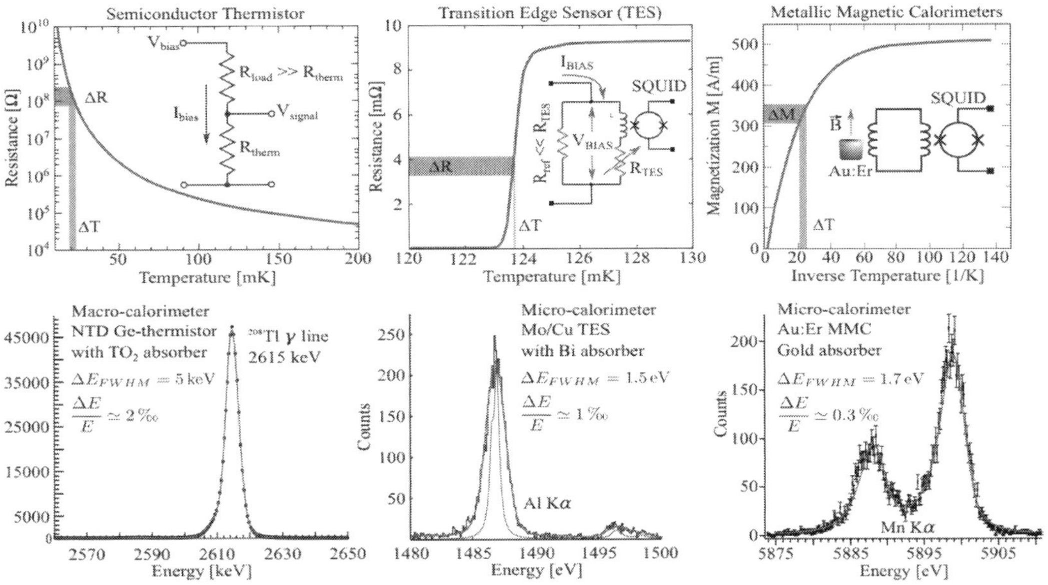

Figure 1. Different sensors with temperature behaviour and energy resolutions. The TES spectrum (middle column) is courtesy of NIST. The thermistor (left column) and MMC (right column) spectra are edited from [9] and [19], respectively.

constant, independent of the weak link G. The absorber linear dimensions can range from few hundred micron (micro-calorimeters) to few centimeters (macro-calorimeters) with rise time from few μs to few ms, respectively.

The performance of the sensor matters significantly to achieve a high-energy resolution with low temperature calorimetric measurements. In recent years, a lot of technologies have been developed, but three of them play a major role in neutrino experiments: Si- or Ge-thermistors, Transition edge sensors (TESs), and Magnetic Micro Calorimeters (MMCs). All of them provide a very high efficiency and energy resolution of the order of ‰ in the keV or in the MeV energy range.

In semiconductor thermistors working at cryogenics temperature, the resistivity dependence on temperature is governed by an exponential law $\rho(T) = \rho_0 \, e^{(T_0/T)^{1/2}}$, where T_0 and ρ_0 are parameters controlled by the doping level (Fig. 1, left column). Arrays of micro- and macro-calorimeters based on these devices have been widely used. Micro-calorimeters for X-ray spectroscopy achieved energy resolutions lower than 5 eV @ 6 keV with tin or HgTe absorbers [20], while macro-calorimeters with Neutron Transmutation Doped Ge-Thermistor for neutrinoless double β-decay achieved energy resolutions of 5 keV at the 2615 keV ^{208}Tl γ line [10].

TESs are superconducting thin films, operated just below its transition temperature (T_c). The energy deposition can drive the superconducting TES to a normal state, hence a sharp change in resistance can be sensed in TES even for a small change in temperature (Fig. 1, center column). TESs can be two orders of magnitude more sensitive than the semiconductor thermistors but, since the resistance is very low (tens of mΩ), they cannot be readout by conventional FET amplifier and they must be coupled with Superconducting Quantum Interference Devices (SQUID) current amplifiers. TESs are widely used in large microcalorimeter arrays for X-ray spectroscopy achieving record resolutions below 2 eV at 1.5 keV, with Gold and Bismuth absorber [18]. Their high energy and time resolutions make TESs ideal detectors for developing microcalorimeters array for the direct measurements of the neutrino mass [21]. Moreover, since

TESs have a very low energy threshold, that is key requirement for light Dark Matter research, they are used in several cryogenic dark matter experiments.

MMC sensors are quite different from the previous two. They are paramagnetic sensors exposed to a small magnetic field. The temperature rise causes a change in the sensor magnetization, which is sensed by a SQUID magnetometer (Fig. 1, right column). The MMCs use the paramagnetic susceptibility of gold doped with a low concentration of erbium ions (Au:Er) when placed in a DC magnetic field. In a paramagnet, the magnetization is inversely proportional to temperature, making it very sensitive to small changes at low temperatures. MMCs provide very high energy resolutions, i.e. 1.7 eV @ 6 keV, and they are used as micro-calorimeters for X-ray spectroscopy [19] and neutrino mass measurements [22] and as macro-calorimeters for the search of the neutrinoless double β-decay [23].

2. Macro-calorimeter for neutrinoless double β-decay

Double β-decay is a rare process in which a nucleus changes its atomic number by two units. It is possible only in nuclei with an even number of neutrons and protons, in which the single β-decay is energetically forbidden. In neutrinoless double β-decay only the two electrons are emitted [7]. The decay violates the lepton number conservation by two units and it is possible only if the neutrino is a massive Majorana particle. Its transition width is proportional to the square of the effective Majorana mass $m_{\beta\beta}$. The searches for a $0\nu\beta\beta$ signal rely on the detection of the two emitted electrons: a monochromatic line at the decay $Q_{\beta\beta}$-value.

Since the two-neutrino double β-decay ($2\nu\beta\beta$) allowed by Standard Mode is an unavoidable source of background for $0\nu\beta\beta$ searches, detectors with excellent energy resolutions are fundamental. From simple statistical considerations, the sensitivity $S^{0\nu}$ of $0\nu\beta\beta$ decay search is given by $S^{0\nu} \propto \varepsilon\,\eta\sqrt{(M\,t)/(B\,\Delta E)}$ where ε, η, M, t , ΔE and B are the detector efficiency, the active isotope abundance, the source mass, the measuring time, the energy resolution and specific background at $Q_{\beta\beta}$, respectively.

There are two experimental approaches for $0\nu\beta\beta$ searches: active source external to the detector and active source internal to the detector (*calorimeter*). Low temperature macro-calorimeters are particularly suitable for the second approach when the decaying isotope under study can be embedded in the absorber material. They provide high energy resolution and large flexibility in the choice of the sensitive material. Several interesting calorimetric candidates have been proposed and tested. By far the most advanced results are obtained with TeO_2 low-temperature macro-calorimeters. With this compound, it is possible to grow large crystals with reasonable mechanical and thermal properties together with a very large content of the candidate $0\nu\beta\beta$ ^{130}Te (isotopic abundance: η = 34.167%). The $Q_{\beta\beta}$-value of the decay is around 2528 keV, falling between the peak and the Compton edge of the 2615 keV gamma line of ^{208}Tl, the highest-energy gamma from the natural decay chains; this leaves a relatively clean window to search for the signal.

The CUORE experiment is based on TeO_2 macro-calorimeters and represents the largest thermal detector array ever built [24]. The detector is composed by 988 calorimeter, arranged in 19 towers of 52 crystals, for total mass of 741 kg. Each tower consists of 13 floors of 4 crystals each. The tower structure is made of ultra pure copper and the crystals are coupled to it by mean of small PTFE supports. The absorber is a TeO_2 cubic crystal, $5 \times 5 \times 5\,\mathrm{cm}^3$ in size, with an NTD-Ge thermistor glued onto the crystal surface. The CUORE technology was already tested during the last ten years. In particular CUORE-0 was a single CUORE-like tower built using the low-background assembly techniques developed for CUORE and running at LNGS from 2013 to 2015 [9]. The average energy resolution in the Region Of Interest, defined as the FWHM of the 2615 keV γ-ray peak, resulted 4.9 keV, with a corresponding RMS of 2.9 keV. This demonstrates that the CUORE goal of 5 keV of energy resolution is feasible. The background in the ROI resulted $B = 0.058$ cnts/keV/kg/yr. With these data, the collaboration sets a 90% C.L. lower bound on the decay half-life of $T^{0\nu} > 2.7 \cdot 10^{24}$ yr. Combining this with the exposure of ^{130}Te from the past Cuoricino experiment, the collaboration sets a limit of $T^{0\nu} > 4.0 \cdot 10^{24}$ yr (90% C.L.). This corresponds to a limit range on the effective Majorana mass of $m_{\beta\beta} < 270 - 760$ meV, using the most up-to-date NME calculations. The expected upper limit on surface-related backgrounds in the ROI in CUORE, extrapolated from test results and the performance of CUORE-0, is around 10^2counts/keV/kg/y. The projected $T^{0\nu}$ sensitivity after five years of live-time is $T^{0\nu} > 9.5 \cdot 10^{25}$ y at the 90% C.L., which corresponds to an upper limit on the effective Majorana mass in the range $m_{\beta\beta} < 50 - 130$ meV, approaching the Inverted Hierarchy region [24]. After successful commissioning of the CUORE cryostat, the 19 towers have been deployed in the cryostat in August of 2016. The begin operation at base temperature are currently in progress and the first data are foreseen for mid-2017.

3. Micro-calorimeter for direct neutrino mass measurements

Experiments based on kinematic analysis of electrons emitted in single β-decay, are the only ones dedicated to effective electron-neutrino mass determination. The method consists in searching for a tiny deformation caused by a non-zero neutrino mass to the spectrum near its end point. The most stringent result is $m_\beta < 2$ eV @ 90% C.L. and comes from electrostatic spectrometers on tritium decay ($Q = 18.6$ keV) [13, 14]. KATRIN, a next generation experiment, is designed to reach a sensitivity of $m_\beta < 0.2\ eV$ in five years [25]. With KATRIN, this experimental approach reaches its technical limits and it is therefore mandatory to develop alternative and complementary experimental methodologies for the direct neutrino mass measurements. The micro-calorimetric measurement, where the β-source is embedded in the detector absorber, is a promising alternative.

Two experiments investigated the feasibility of micro-calorimeter based on the ^{187}Re isotope: MANU [26] and MIBETA [27]. The first used one detector with NTD-Ge thermistor glued on a 1.6 mg single crystal of metallic rhenium while the second used an array of eight Si:P implanted Si-thermistors with $AgReO_4$ for a total mass of 2.2 mg. Both collected a statistic of around 10^7 β-events, corresponding to a neutrino mass sensitivity of $m_\beta < 20$ eV. To improve this limits by a factor 100, so that to reach the sub-eV sensitivity needed to approach the Inverted Hierarchy, a statistic increase of 10^8 is required. One of the most interesting candidate as an alternative to ^{187}Re is ^{163}Ho. The calorimetric measurement of the ^{163}Ho was suggested in 1982 by De Rujula and Lusignoli [28] but only in recent times low temperature microcalorimeters reached the necessary maturity to be used in a large scale experiment with the potential to reach the sub-eV neutrino mass sensitivity. In a ^{163}Ho calorimetric experiment the statistical sensitivity on the neutrino mass is proportional to $\sqrt[4]{1/N_{ev}}$, where N_{ev} is the number of the detected events. Considering micro-calorimeters with ^{163}Ho-embedded absorber, with high energy and time resolutions ($\simeq$ 1 eV and $\simeq$ 1 μs, respectively) and a pile-up fraction within the range $f_{pp} = 10^{-3} - 10^{-6}$, a sub-eV neutrino sensitivity can be reached collecting a total number of events around $N_{ev} > 10^{17} - 10^{19}$ [29]. To achieve these requirements a Megapixel arrays experiment is needed, where the multiplexing factor (number of multiplexed detector signals) and the read-out bandwidth play a crucial role for future developments. Currently, three experiments explore the approach of using electron capture on ^{163}Ho to assess the neutrino mass: ECHo [22], HOLMES [30], and NuMECS [31].

ECHo employs MMCs sensor with microwave multiplexing. The holmium source is enclosed in gold absorbers. With their first prototype, ECHo could demonstrate excellent energy resolution of 7.6 eV @ 6 keV and fast rise times of $\tau_r = 130$ ns. A larger detector array with 16 pixels, increased purity and activity (0.1 Bq) is being tested at the moment. In the next step ECHo-1k will exploit 100 pixels to collect 10^{10} events in 1 year, pushing the neutrino mass sensitivity of below 10 eV. In the final step ECHo-1M will be composed by 10^4 pixels with the aim to reach a sensitivity below 1 eV in 1 year. HOLMES is making use of the TES technology with microwave multiplexing. The collaboration is currently performing the detector optimization and the read-out R&D development. A custom ion-implanter is being assembled to embed the ^{163}Ho in the detectors. First result on the developed detectors showed an energy resolution of about 6 eV @ 6 keV and a rise time of $\tau_r \simeq 10\,\mu s$, fulfilling the HOLMES requirement [30]. The first tests with first implanted sub-arrays will start in 2017, while the commissioning of the final 1024 pixels array will be performed in 2018. The final goal is to reach a neutrino mass statistical sensitivity below 2 eV. NuMECS is also pursuing the TES technology. The collaboration focus is on ^{163}Ho production via proton activation of dysprosium, as opposed to the most common neutron irradiation on ^{162}Er. In their prototype the source nuclei are enclosed as liquid drop in a nanoporous gold absorber. NuMECS successfully measured a ^{163}Ho spectrum with an energy resolution of about 40 eV FWHM. NuMECS future plans include the deployment of four 1024 pixel arrays, aiming at a statistical sensitivity of about 1 eV.

4. Conclusion

The study of frontier subjects in neutrino physics requires the use of frontier detectors. Experiments with large number of pixel, large mass and very high efficiency and energy resolutions, of the order of ‰ in the MeV or in the keV energy range, are a fundamental tool to study and understand the neutrino properties. The goal of the next future experiments is to approach and cover the Inverted Hierarchy region. Low-temperature micro and macro calorimeters today are mature devices which often represent the only technical opportunity to satisfy these challenging demands.

References

[1] Capozzi F *et al.* 2016 *Nucl. Phys. B* **908** 218–234
[2] Adamson P *et al.* (NOvA) 2016 *Phys. Rev. D* **93** 051104
[3] Abe K *et al.* (T2K) 2013 *Phys. Rev. Lett.* **111** 211803
[4] An F *et al.* (JUNO) 2016 *J. Phys. G* **43** 030401
[5] Ade P A R *et al.* (Planck) 2016 *Astron. Astrophys.* **594** A13
[6] Palanque-Delabrouille N *et al.* 2015 *J. Cosmol. Astropart. Phys* **1511** 011
[7] Dell'Oro S *et al.* 2016 *Adv. High Energy Phys.* **2016** 2162659
[8] Agostini M 2016 First results from GERDA Phase II presentation at NEUTRINO 2016
[9] Alfonso K *et al.* (CUORE) 2015 *Phys. Rev. Lett.* **115** 102502
[10] Alduino C *et al.* (CUORE) 2016 *Phys. Rev. C* **93** 045503
[11] Gando A *et al.* 2016 *Phys. Rev. Lett.* **117**(8) 082503
[12] Drexlin G *et al.* 2013 *Adv. High Energy Phys.* **2013** 293986
[13] Lobashev V *et al.* 2001 *Nucl. Phys. B - Proceedings Supplements* **91** 280–286
[14] Kraus C *et al.* 2005 *Eur. Phys. J. C* **40** 447–468
[15] Moseley S H, Mather J C and McCammon D 1984 *J. Appl. Phys.* **56** 1257–1262
[16] Fiorini E and Niinikoski T 1984 *Nucl. Instrum. Meth. A* **224** 83–88
[17] Agnese R *et al.* (SuperCDMS) 2013 *Appl. Phys. Lett.* **103** 164105
[18] Ullom J N and Bennett D A 2015 *Superconductor Science and Technology* **28** 084003
[19] Porst J P *et al.* 2014 *J. Low. Temp. Phys.* **176** 617–623
[20] McCammon D 2005 *Topics in Applied Physics* **99** 35–62
[21] Nucciotti A 2016 *Adv. High Energy Phys.* **2016** 9153024
[22] Gastaldo L *et al.* (ECHo) 2014 *J. Low. Temp. Phys.* **176** 876–884
[23] Kim G B *et al.* (AMoRE) 2015 *Adv. High Energy Phys.* **2015** 817530
[24] Artusa D R *et al.* (CUORE) 2015 *Adv. High Energy Phys.* **2015** 879871
[25] Angrik J *et al.* 2004 *Tech. Rep., Forschungszentrum, Karlsruhe, Germany* 447–468
[26] Galeazzi M, Fontanelli F, Gatti F and Vitale S 2000 *Phys. Rev. C* **63**(1) 014302
[27] Sisti M, *et al.* 2004 *Nucl. Instrum. Meth. A* **520** 125–131
[28] Rújula A D and Lusignoli M 1982 *Phys. Lett. B* **118** 429–434
[29] Nucciotti A 2014 *Eur. Phys. J. C* **74** 3161
[30] Alpert B *et al.* 2015 *Eur. Phys. J. C C* **75** 112
[31] Croce M P *et al.* 2016 *J. Low. Temp. Phys.* **184** 958–968

Heterogeneous nucleation of ice in the atmosphere

A Nicosia[1,2,3], **M Piazza**[1], **G Santachiara**[1] **and F Belosi**[1]

[1] Institute of Atmospheric Sciences and Climate (ISAC-CNR), Bologna, Italy
[2] Department of Physics and Earth Sciences, University of Ferrara, Ferrara, Italy
[3] Laboratoire de Météorologie Physique (LaMP-CNRS), Université Blaise Pascal, Clermont-Ferrand, France

E-mail: a.nicosia@isac.cnr.it

Abstract. The occurrence of ice-nucleating aerosols in the atmosphere has a profound impact on the properties of clouds, and in turn, influences our understanding on weather and climate. Research on this topic has grown constantly over the last decades, driven by advances in online and offline instruments capable of measuring the characteristics of these cloud-modifying aerosol particles. This article presents different aspects to the understanding of how aerosol particles can trigger the nucleation of ice in clouds. In addition, we present some experimental results obtained with the Dynamic Filter Processing Chamber, an off-line instrument that has been applied extensively in the last years and that circumvents some of the problems related to the measurement of Ice Nucleating Particles properties.

1. Introduction

Laboratory experiments have shown that the freezing of very pure water droplets occurs at about -41 °C for droplets of 1 μm in diameter and at about -35 °C for droplets of 100 μm in diameter [1]. Hence, in the atmosphere, homogeneous freezing (i.e. the formation of ice crystals from pure water and haze droplets) is relevant only for high clouds, as for example cirrus. For mixed phase clouds, the formation of ice crystals is mainly related to the presence of a low concentration of aerosol particles called Ice Nucleating Particles (INP). These nuclei provide a substrate capable to lower the energy barrier between the supersaturated water vapour or the supercooled droplet (metastable liquid) and the thermodynamically stable crystalline phase. These particles may act as "deposition nuclei" (direct passage from the vapour state to the solid state) or as "freezing nuclei" (passage from the liquid intermediate phase to the solid phase). Three additional sub-cases are distinguished for the freezing nuclei: (i) the direct freezing of supercooled water by a nucleus contained within the droplet (immersion case); (ii) the condensation of water vapour on the nucleus surface and the subsequent freezing (condensation-freezing mode); (iii) the collision between a nucleus and a supercooled droplet with the subsequent freezing (contact-freezing mode).

Which process -among deposition, immersion freezing, condensation freezing and contact freezing- is the most relevant, may depend on the meteorological conditions and the specific nucleus, its chemical composition and dimension.

Particles with a crystallographic structure similar to that of ice crystal lattice (which has a hexagonal arrangement) tend to be effective as ice nuclei, although this condition is neither necessary nor sufficient. Traditionally, it has been thought that good heterogeneous ice nuclei should be insoluble solids, but extensive researches have shown that also soluble species can act in the formation of the solid phase in both the immersion [2, 3] and the deposition mode [4].

The capability of an INP species to nucleate ice in a specific mode is influenced mainly by the temperature T and the supersaturation with respect to water S_w or ice S_i.

A parameter used to classify the ice nucleating capabilities of a species is the activation fraction, f_{IN}, calculated as the ratio of the number of ice crystals detected, to the total particle number concentration. The activated fraction can be used to identify the activation region which is peculiar to a certain species in relation to the parameters T, S_w and the specific nucleation mode.

Some inorganic particles (e.g. clay particles, volcanic ash, soil dust) present the highest activation fraction in a region of temperatures above -25 °C and for this reason they are considered to play an important role in cloud ice phase. Some recent studies have shown that also organic materials (such as bacteria, pollen, lichen, fungi, decaying vegetation material, marine bacteria associated with plankton) can be effective ice nucleators at fairly high temperatures (below -4 °C), but the activation fraction seems to be lower compared to clay particles. In recent years, focus on anthropogenic aerosol has asserted that also soot and metallic oxides can play a role in the INP population [5]. Eventually, further data are needed to clarify the significance of biogenic aerosols (as pollens, bacteria and cellulose), marine aerosol contributions or new possible INP sources. More generally, it is essential to investigate the composition and concentration of INP in different regions in order to predict the impact of natural and anthropogenic aerosol particles on cloud properties and to improve our understanding on weather and climate.

Research on this topic has grown constantly over the last decades but the study still presents some critical aspects that make its analysis and parameterization difficult. First of all, the small number concentration of INPs. It should be considered that no existing ice nucleus measuring system is capable of detecting ice formation by all known mechanisms simultaneously, then different techniques should be considered. Lastly, processes in the atmosphere, as the coating, may reduce the ice nucleating capabilities of INP species and these aging effects should not be neglected.

The instruments commonly used to investigate INP are: mixing chambers [6], expansion cloud chambers [7], continuous flow diffusion chambers [8], filter processing methods [9-11], immersion mode cooling chambers [12] and immersion mode offline methods [13].

Among these techniques, the filter processing is the most versatile one and has been largely used for its convenience and simplicity. Actually, aerosol particles can be easily collected on membrane filters, everywhere, and then conserved and transported. Inside the laboratory, a number of methods and instruments exist to process the filter, induce the activation of the INPs present, and distinguish them from the remaining aerosols collected.

The Dynamic Filter Processing Chamber (DFPC) is one of these instruments and it allows the control of the supersaturated water vapour S_w and the temperature T, in order to explore either the deposition or the condensation-freezing activation mode. We applied this instrument to a number of field campaigns [14-16] and we show the benefit obtained with respect to other techniques, and the INP concentrations obtained.

2. The dynamic filter processing chamber

The DFPC is a replica of the Langer dynamic developing chamber [11], housed in a refrigerator and used to detect and determine the concentration of aerosol particles active as INP at different supersaturations S_w and temperatures T.

The experimental method used to measure the INP activated fraction comprises three main steps. First the aerosol is collected on a filter through a common sampling line which combines a filter holder, a pump and eventually a cyclone or a particulate monitoring sampling head (as PM10 or PM1). The sampling is done in parallel with the monitoring of aerosol properties through an optical counter or a Scanning Mobility Particle Sizer (SMPS). Secondly, the filter is pretreated at the laboratory, by filling its pores with paraffin. More specifically, the filter is placed on a metallic support covered with a smooth layer of paraffin. Then the paraffin is slightly heated and rapidly cooled in order to penetrate deeper inside the filter pores. The main goal of this pretreatment is to enhance the thermal conductivity between the filter and the supporting substrate. Lastly the filter is processed inside DFPC.

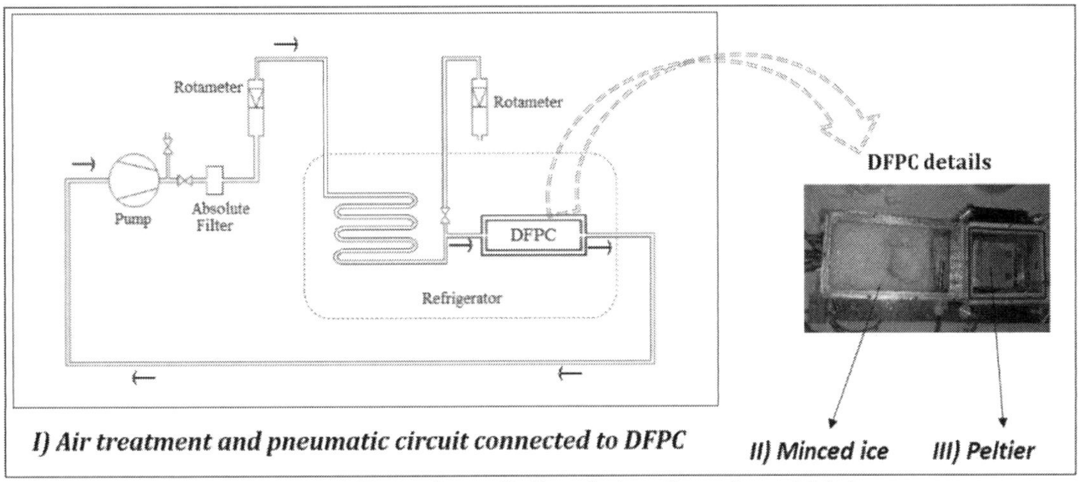

Figure 1. DFPC scheme (left) and view from above (right).

Fig. 1 reports the instrumental set-up which can be divided into three sections: the air pneumatic circuit (section I), the supersaturation control level (section II) and the filter temperature control (section III).

In section I, the air is filtered and forced by a pump to flow through the remaining sections in a closed loop. Section II is filled with minced ice, which is cooled by the base plate, while section III allocates a Peltier plate, where the filter is placed. When air spreads into the ice bed (section II), it becomes saturated with respect to ice, and then enters in the following section (III) and borders on the filter. The values of T and S_w are measured before approaching the filter with a thermo-hygrometer (model HD 9216, DeltaOhm). Supersaturations are calculated theoretically from vapour pressures of ice and water at the considered temperatures [17]. By varying the different temperatures inside the chamber and the vapour supersaturation with respect to liquid and/or ice, it is possible to activate the INP both in the deposition and in the condensation-freezing modes. Ice can form by deposition provided that the air is supersaturated with respect to ice and the temperature is low enough [1] such as, for example, at S_w = 0.92 and T = -18 °C. Ice can form by condensation-freezing if the air is supersaturated both with respect to ice and water such as, for example, at S_w = 1.02 and T = -18 °C.

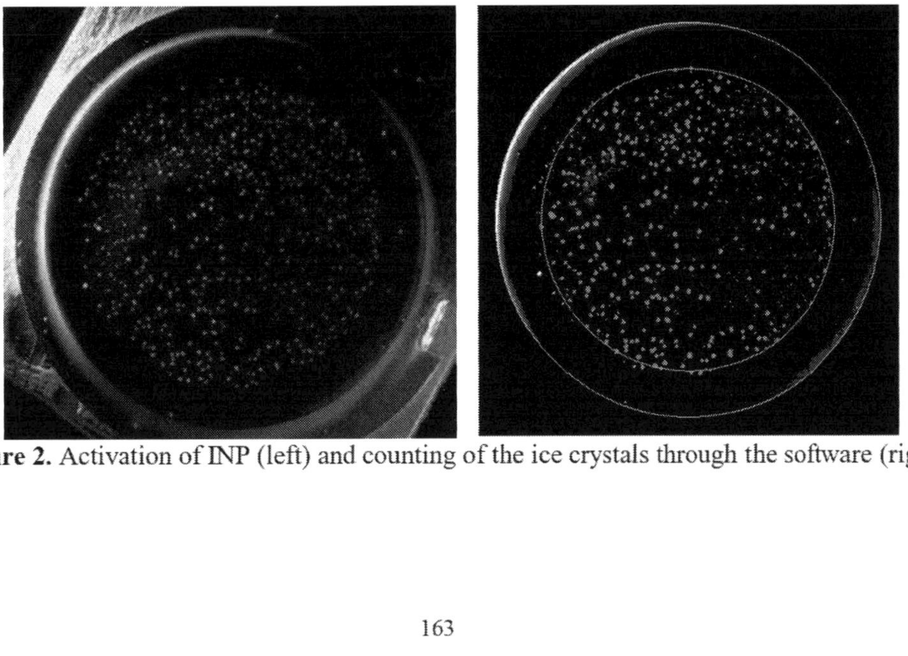

Figure 2. Activation of INP (left) and counting of the ice crystals through the software (right).

Finally, the INP activated fraction can be derived from the number of crystals grown on the filter and the initial sampling conditions (i.e. the volume of sampled air and the concentration measured by an optical counter or a SMPS). To count the number of ice crystals, a photograph of the filter is taken with a high resolution camera sensor (model DP-M17 USB digital microscope from Conrad electronic international) and analysed with an algorithm that runs under the Image Processing Toolbox of Matlab. The code has been conceived at ISAC-CNR and it excludes the external corona of the filter (which is pressed by the filter holder during the aerosol sampling and in which the air does not pass through). Fig. 2 reports an example of the ice crystals grown on the filter and the subsequent image processing and counting of the software.

3. Ice nucleating particles concentrations

The DFPC has been employed for the measurements of Ice Nucleating Particle concentrations [14-16] in the framework of BACCHUS (FP7-603445) and AIR-SEA LAB (CNR funded) projects between 2014 and 2016 at the following sampling sites: San Pietro Capofiume (SPC) in the north of Italy; Mace Head (MH) in Ireland; and the Global Atmospheric Watch station of Monte Cimone (MC) in Italy.

The first site, SPC, is considered a rural background site and the measurements were performed during a winter campaign (SPC1) and a spring campaign (SPC2). The second site, MH, is exposed to the North Atlantic Marine Boundary Layer and the campaign was performed during summer. The third site, MC, is the highest peak of the north Italian Apennines (2165 m a.s.l.) and it hosts the Italian Climate Observatory "O. Vittori". MC is a mountain site and it is considered representative of the free tropospheric conditions of the Mediterranean basin/southern Europe (MB/SE) during the cold months. In this last site two campaigns were performed, one in spring (MC1) and another one in autumn (MC2).

Aerosols were collected using a parallel PM1 - PM10 sampling system and the filters were then processed inside the DFPC at -22 °C (for the MH site) and -18 °C (for the other sites). The supersaturation level with respect to water S_w was fixed to 0.96 and to 1.02 to explore both the deposition and the condensation freezing activation mode. Fig. 3 and 4 show the box chart plot of the INP concentration values obtained at the different sites for respectively PM1 and PM10 aerosol size fraction (S_w of 1.02). The boxes indicate the 25th-75th percentiles, while lower and upper bars show the minimum and the maximum, respectively.

The coastal site MH appears with the lowest median INP concentration value (both in the PM10 and PM1 analysis). Nevertheless the INP concentration variability could reflect not only the different site but also the particular air masses which are affected by short and long-term transport of crustal and anthropogenic components.

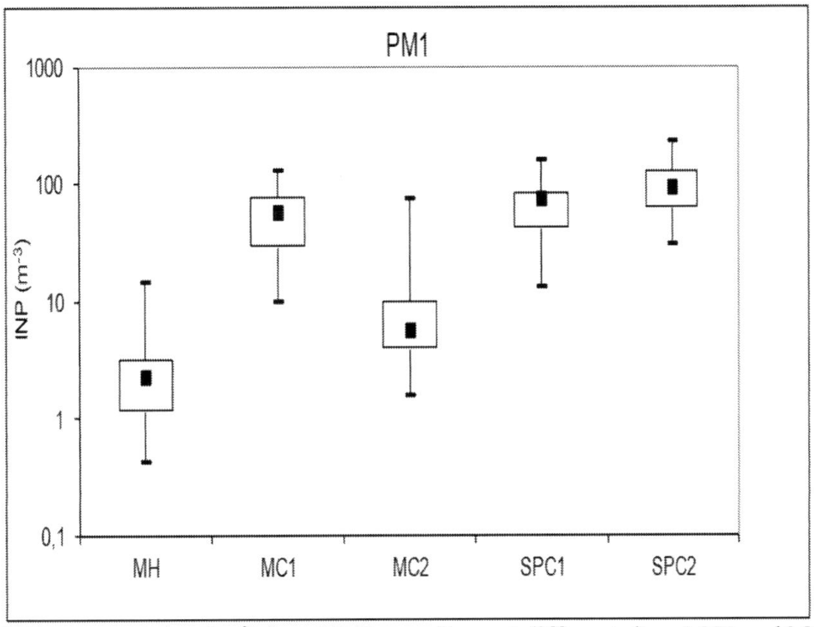

Figure 3. INP concentration (m^{-3}) from PM1 samplings at different sites: MC1 and MC2 refer to Monte Cimone (spring 2014 and autumn 2015); SPC1 and SPC2 refer to San Pietro Capofiume (winter 2014 and spring 2014, respectively); MH refers to Mace Head (summer 2015).

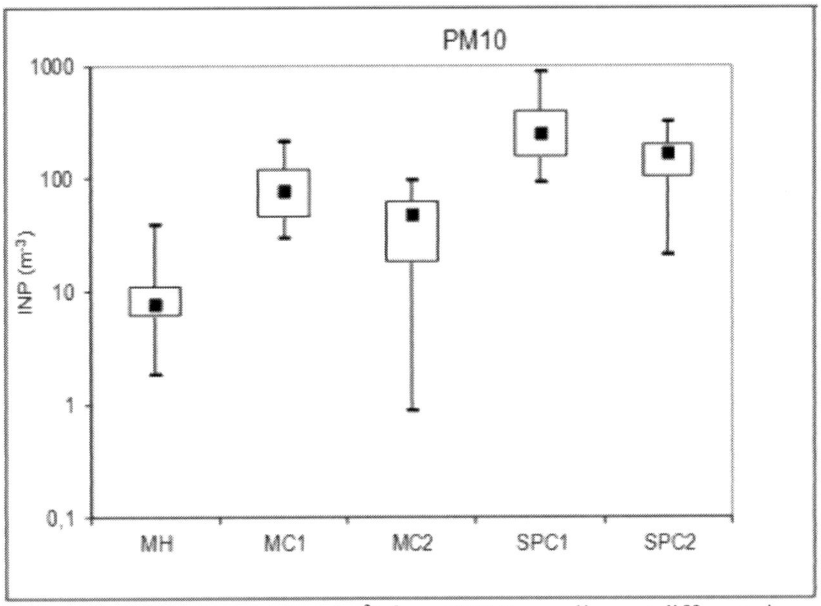

Figure 4. INP concentration (m^{-3}) from PM10 samplings at different sites.

Saharan dust transport episodes were observed during the campaigns of SPC2 and MC1. These episodes are generally characterized by increased INP concentration in the PM10 size range, particularly at S_w of 1.02, and an enhanced concentration of particles in the fine and coarse fractions. However, our observations show a decrease of the activated fraction during these events. This could depend on the aging processes (e.g. coating with sulphate or sea salts) during the long range transport of aerosol, as explained by [18]. Considering SPC1 in Fig. 3, the reduced nucleation activity of PM1 particles -with respect to SPC2- can be reasonably attributed to the enhance contribution of carbonaceous particles from biomass burning (which are poor ice nuclei), characterizing the site during the cold months. On the other hand, the higher activation of super-micrometer particles observed in winter (SPC1 in Fig. 4) suggests different nucleating properties for the coarse aerosol populations in the two campaigns. Unfortunately, the lack of chemical/mineralogical composition data does not allow testing this hypothesis.

INP PM1/PM10 ratios are included between 0.2-0.3 at SPC1 (winter campaign) and 0.6 at MC and SPC2 (spring campaign). Then, not only different locations but also various periods of the year should be monitored to obtain reliable INP concentrations. Our results highlight also the need of measuring the freezing activity of the total aerosol population, and not only the fine fraction, as reported by [19].

4. Conclusions

We presented some aspects related to the characterisation of INP and their measurement. The DFPC is one of the instruments that are used to investigate the ice nucleation capabilities of aerosol particles collected onto a filter. Its goal is mainly the characterisation of INP when the deposition and the condensation-freezing modes are explored. We applied this instrument to evaluate the INP concentrations of several field campaigns and we showed the benefit obtained with respect to other techniques. In particular, the DFPC can investigate all the size spectrum of particles up to 10 μm, while on-line measurement devices allow measurements of particles up to about 2 μm. Results show that coarse aerosol particles (size range 1-10 μm) contribute significantly to the total INP concentration. We could find a case with the prevalence of INP in the coarse fraction -with respect to PM1 size fraction- during winter in San Pietro Capofiume. In addition, Saharan dust transport events were observed in a few measurement campaigns and related to the variability of the INP concentration. The measurement of INP properties achieved with the DFPC instrument could be crucial in the search for a better understanding of the aerosol-cloud interactions.

References

[1] Mason B J 1971 *The physics of clouds* (Oxford Univ. Press: Oxford) p 160
[2] Zuberi B, Bertram A K, Koop T, Molina L T and Molina M J 2001 *J. Phys. Chem. A* **105** 6458
[3] Zobrist B et al 2006 *Atmos. Chem. Phys.* **6** 3115
[4] Shilling J E, Fortin T J and Tolbert M A 2006 *J. Geophys. Res.* **111** D12204
[5] Salam A, Lesins G and Lohmann U 2008 *Air Qual. Atmos. Health* **1** 135
[6] Bundke U, Nillius B, Jaenicke R, Wetter T, Klein H and Bingemer H 2008 *Atmos. Res.* **90** 180
[7] Tajiri T, Yamashita K, Murakami M, Orikasa N, Saito A, Kusunoki K and Lilie L *2013 J. Meteor. Soc. Japan* **91** 687
[8] Rogers D C, DeMott P J, Kreidenweis S M and Chen Y 2001 *J. Atmos. Oceanic Technol.* **18** 725
[9] Bigg E K, 1996 *Tellus* **48B** 223
[10] Klein H, Haunold W, Bundke U, Nillius B, Wetter T, Schallenberg S and Bingemer H 2010 *Atmos. Res.* **96** 218
[11] Langer G and Rodgers J 1975 *J. Appl. Meteor.* **14** 560
[12] Kohn M, Lohmann U, Welti A and Kanji Z A 2016 *J. Geophys. Res. Atmos.* **121** 4713
[13] Budke C and Koop T 2015 *Atmos. Meas. Tech.* **8** 689
[14] Santachiara G, Di Matteo L, Prodi F and Belosi F 2010 *Atmos. Res.* **96** 266
[15] Belosi F, Santachiara G and Prodi F 2014 *Atmos. Res.* **145–146** 105
[16] Belosi F, Rinaldi M, Decesari S, Tarozzi L, Nicosia A and Santachiara G 2017 *Atmos. Res.* **186** 116-126
[17] Perry J H 1963 *Chemical Engineers' Handbook* (McGraw-Hill Company: New York)
[18] Levin Z, Teller A, Ganor E and Yin Y 2005 *J. Geophys. Res.* **110** D20202
[19] Mason R H et al. 2016 *Atmos. Chem. Phys.* **16** 1637

IOP Conf. Series: Journal of Physics: Conf. Series **841** (2017) 012029 doi:10.1088/1742-6596/841/1/012029

Detection of terrestrial gamma-ray flashes with the AGILE satellite

A Ursi[1], M Marisaldi[2,3], M Tavani[1,4], P Sanò[5], D Casella[5] and S Dietrich[5]

[1] INAF-IAPS, National Institute for Astrophysics, Roma, Italy
[2] INAF-IASF, National Institute for Astrophysics, Bologna, Italy
[3] Birkeland Centre for Space Science, University of Bergen, Norway
[4] Department of Physics, University of Roma Tor Vergata, Roma, Italy
[5] ISAC-CNR, Institute for Atmospheric Science and Climate, Roma, Italy

E-mail: alessandro.ursi@iaps.inaf.it

Abstract.
Terrestrial gamma-ray flashes are brief submillisecond gamma-ray emissions, produced during thunderstorms and strictly correlated to lightning and atmospheric electric activity. Serendipitously discovered in 1994 by the Compton Gamma Ray Observatory, these elusive events have been further investigated by several missions and satellites devoted to high-energy astrophysics, such as RHESSI, AGILE and Fermi. Terrestrial gamma-ray flashes are thought to be bremsstrahlung gamma-rays, produced at the top of thunderclouds by avalanches of electrons accelerated within thunderstorm strong electric fields and abruptly braked in the atmosphere. Exhibiting energies ranging from few keV up to several tens of MeV, terrestrial gamma-ray flashes are the most energetic phenomenon naturally occurring on Earth and they can represent a severe risk for airplanes and aircraft transports, both for the crew and the on board electronics, that should be carefully investigated and understood.

The AGILE (Astrorivelatore Gamma ad Immagini LEggero) satellite is an entirely Italian mission, launched in 2007 and still operational, aimed at investigating gamma-ray emissions from cosmic sources. The wide energy range and the unique submillisecond trigger logic of its on-board instruments, together with the narrow quasi-equatorial orbit of the spacecraft, make AGILE a very suitable instrument to detect and investigate terrestrial gamma-ray flashes. Recent improvements rose up the terrestrial gamma-ray flashes detection rate and lead to the observation, for the first time, of multiple events occurring within single thunderstorm processes.

1. Introduction

Terrestrial Gamma-ray Flashes (TGFs) are brief and intense gamma-ray emissions coming from the lower part of terrestrial stratosphere ($\sim$ 15 km a.s.l.), strictly correlated with thunderstorms and atmospheric electric activity. All observed TGFs exhibit durations from tens of microseconds to few milliseconds and energies ranging from few MeV to several tens of MeV. TGFs take place at high altitudes, at thundercloud tops, making them very difficult to observe directly from the ground, due to the gamma-ray absorption in the high-density lower atmospheric layers: hence, despite being a completely terrestrial phenomenon, the only way to study these elusive objects is from space, by using high-energy detectors placed on-board satellites for astrophysics, on-board the International Space Station or on-board airplanes and stratospheric balloons. As space experiments specifically dedicated to TGFs are yet to come, most of the studies about this phenomenon have been carried out by using high-energy astrophysics satellites. TGFs are a sort of borderline phenomenon between atmospheric physics and astrophysics, representing a really attractive challenge for both the scientific fields.

TGFs were unexpectedly observed for the first time in the early 90s by the BATSE experiment [1], one of the four main instruments on-board the Compton Gamma-Ray Observatory, and subsequently more deeply investigated by the NASA Reuven Ramati High-Energy Solar Spectroscopic Imager (RHESSI). In the last seven years, wide contributions to their study have been brought by the AGILE and by the Fermi space telescope. Since the beginning, almost all of the observed events took place within the same geographic region (i.e., within the tropical region and over the continents) and the same time interval (i.e., mostly in the afternoon) of a thunderstorm, suggesting a correlation between the two phenomena, as shown in Fig. 1. Successive investigations pointed out that individual radio atmospheric Extremely Low Frequency and Very Low Frequency waves, briefly called sferics, which represent a sort of electromagnetic signature of lightning strokes, were associated with TGFs, highlighting a direct correlation between TGFs and the single lightning discharges of a thunderstorm.

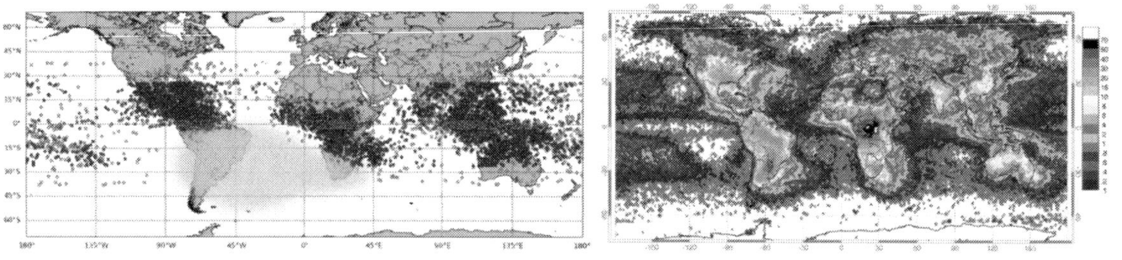

Figure 1. (a) Geographic distribution of all TGFs detected by BATSE (yellow), RHESSI (green), AGILE (red) and Fermi (blue). The green region corresponds to the South Atlantic Anomaly, where the enhanced charged particle background prevents any detection. (b) Lightning rate distribution acquired by the NASA Lightning Imaging Sensor.

The production mechanism of these events is still matter of debate. To date, the most accepted hypothesis suggests the upper part of Earth's troposphere behaves as a particle accelerator, under thunderstorm conditions: free electrons in air can be accelerated to relativistic energies by these intense electric fields and produce hard X- and gamma-rays via a bremsstrahlung process, against atoms and nuclei in the atmosphere. If the thundercloud electric field is strong enough, a huge number of electrons can reach ionization energies, producing secondary electrons, which may undergo the same acceleration process, giving rise to an avalanche process of so-called runaway electrons, named Relativistic Runaway Electron Avalanche [2]. Such a model needs an external seed, to initiate a process, such as a cosmic ray or some other energetic particle coming from outside the system. Nevertheless, two alternative mechanisms have been proposed, both providing avalanche processes without the necessity of external seeds: the Thermal Breakdown and the Relativistic Feedback [3,4]. These models require very strong electric fields and large avalanche distances, that make them impossible to be reproduced in laboratories, but not so hard to be achieved in typical thunderstorm conditions (especially near the lightning source region).

2. The AGILE satellite

AGILE is a mission of the Italian Space Agency, with coparticipation of the Istituto Nazionale di Astrofisica and the Istituto Nazionale di Fisica Nucleare, together with several Italian research Institutes, Universities and industrial partners [5]. A schematic view of AGILE is given in Fig. 2. The satellite was launched on 23 April 2007 from the Satish Dhawan Space Center (India) on a PSLV-C8 rocket and delivered into a low-inclination (2.5°) orbit, at 540 km altitude: the orbit was chosen near-equatorial in order to undergo a reduced background, being the lowest inclination orbit ever achieved by a high-energy astrophysics mission, to date. The nominal scientific observation started December 1st 2007. Aim of the AGILE program is to provide very effective detection and imaging capabilities, both in the hard X- and gamma-ray energy range [18 keV $\div$ 50 GeV], providing coverage of the $>$ 100 MeV window, left unobserved for 7 years after the end of the EGRET mission, in 2000. The AGILE energy range, together with an excellent timing resolution and large field of view, allows to investigate high-energy emissions from cosmic sources: main targets of the mission are compact objects, active galactic nuclei, galactic molecular clouds and gamma-ray bursts, soft gamma-ray repeaters, pulsars and pulsar wind nebulae, supernova remnants, the Galactic center and TeV sources, providing a diagnostic of particle acceleration and radiation processes in extreme conditions. The AGILE program and scientific objectives partially overlap and are complementary to those of other high-energy astrophysics missions (CHANDRA, INTEGRAL, RXTE, XMM-Newton, SWIFT, SUZAKU, RHESSI and the Fermi space telescope) and ground-based instrumentation (radio telescopes, optical observations and TeV observatories), providing fast response to new detected gamma-ray transients and alert for multiwavelength observations. Quicklook data analysis, rapid reaction and fast communication of new high-energy transients is a crucial part of the AGILE program and one of the key assets of the mission.

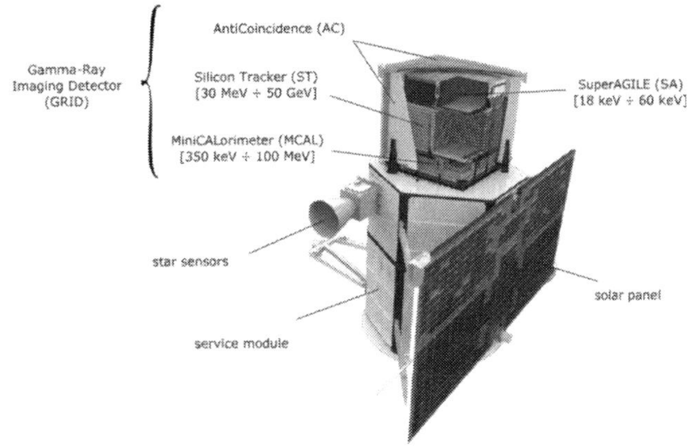

Figure 2. Schematic view of the AGILE satellite and its the scientific instruments and electronics: a silicon tracker, a minicalorimeter, a coded mask detector and an anticoincidence module, plus an on-board data handling unit.

3. The AGILE TGF detection algorithm

AGILE turned out to be a very suitable satellite for the detection of TGFs. Most of the events are detected by its MiniCALorimeter (MCAL) instrument, as its on-board trigger logic works on different trigger timescales, consisting in four dynamic software logics (8 s, 1 s, 256 ms, 64 ms) and three static hardware logics (16 ms, 1 ms, 293 μs): given the shortness of TGFs, most of the detected events are observed in the 293 μs and in the 1 ms time windows. The hardware logics do not depend upon the background variation and require a settable fixed threshold, kept as low as possible, to increase the detection efficiency. A large number of triggers in the submillisecond and in the 1 ms logics are produced by electronic noise in the instruments, affecting the detection: as a consequence, in order to be classified as genuine TGFs, the triggered events are carefully analyzed, by means of an off-line selection strategy. For each trigger acquisition, whose duration depends on the trigger logic and has been modified several times during the mission lifetime, due to telemetry requirements, all photon-by-photon data are scanned by means of a 1 ms dynamic time window: whenever a count is encountered, a secondary 300 μs wide time window is opened, starting from the count time. If at least 6 counts fall within this second time window, the count group is defined as a cluster, providing a preliminary estimate of the TGF duration. At this point, a selection algorithm is applied, that selects all clusters fulfilling the following requirements: on-board trigger fired is within 20 ms from the count cluster start time, a total number of counts $N \geq 9$, a hardness ratio HR ≥ 0.5 (with HR defined as the ratio between the number of counts with $E \geq 1.4$ MeV over the number of counts with $E < 1.4$ MeV), at least one count is released in each half MCAL detection plane and a maximum photon energy of 30 MeV.

This technique allows to reject spurious triggers, due to charged particles releasing counts in the MCAL planes or due to electronic noise, not requiring a successive visual inspection. This procedure allowed the publication of the first AGILE TGF catalog [6], consisting in 308 events, acquired in about 28 months of observation and available online at *http://www.asdc.asi.it/mcaltgfcat/*. The reliability of the obtained TGF sample is shown by the typical geographic pattern of the detected events, clustering over land regions, and by the local time distribution, that reflects the lightning activity: a relaxing in the algorithm selection criteria translates into a contamination of the selected sample, with both the TGF spatial and time distributions smoothed towards more uniform pattern, due to a probable contamination by erratic events, such as cosmic rays and electronic noise in the instruments.

4. Detection of high-energy TGFs

The cumulative energy spectrum of 130 TGF events, satisfying all the illustrated selection criteria, but no constraints on the photon energy, showed a high-energy spectral component, besides the well-known power law [7]: this component exhibits no exponential attenuation and extends up to 100 MeV, challenging the current accepted models for the TGF production mechanism.

5. Localization of TGFs

The AGILE satellite is the only mission that performed localization of TGFs, by means of an imaging instrument [8]: events have been searched with the Gamma-Ray Imaging Detector (GRID), sensitive between 30 MeV and 50 GeV, in correspondence (within 200 ms) to TGFs detected by the MCAL instrument. By considering a set of 119 TGFs, a total of 8 correspondences was obtained. The GRID photon directions are obtained by exploiting the standard analysis for gamma-ray photons in the AGILE field of view and are localized with an accuracy of $5^\circ - 10^\circ$, at 50 MeV. All events turned out to be completely compatible with a terrestrial incoming direction, from a production site close to the satellite footprint less than 400 km. The AGILE silicon tracker detected events provide both the first direct localization from space of TGFs and an evidence that these phenomena emit a significant fraction of their energy in the energy range well-above 20 MeV.

6. Enhanced detection of TGFs

A common issue affecting all TGF detectors is dead time, given the large fluence (~ 0.1 ph cm^{-2}), released in a very short timescale. During dead time, an instrument is busy in processing previous events and not sensitive to successive input pulses: such an effect increases with the event rate to which it is exposed. For what concerns AGILE, MCAL detections of TGFs were heavily affected by dead time, mostly induced by the AC (Anti-Coincidence) units, aimed at rejecting signals due to background charged particles, preventing as well the detection of events with duration shorter than $\sim 100\ \mu s$. The net result is, the time distribution of the AGILE TGFs is biased toward longer values, with respect to other satellites: the $\sim 100\ \mu s$ average time duration observed by RHESSI and Fermi is shifted to about 300 μs for the AGILE sample, preventing the finding of correlated radio sferics [9,10]. A new enhanced configuration was set, by disabling the AC veto for the MCAL, maintaining the same selection criteria used in the previous catalog: the enhanced configuration increased the TGF detection rate of about one order of magnitude, allowing the observation of $60 \div 70$ TGFs per month and of events with shorter durations (down to few tens of μs) and the finding of 39 simultaneous sferics, detected by the World Wide Lightning Location Network, occurring within $\pm 200\ \mu s$ from the associated event [11]. The absence of the AC veto also increased the number of spurious triggers, but the selection criteria adopted in the algorithm ensure the reliability of the new TGF sample, whose phenomenological features (geographic and local time distribution) are consistent with the TGF sample collected in the previous standard configuration. The high detection rate of events, concentrated within a narrow equatorial latitude belt, provides a large number of TGFs observed per unit area and per unit time: this constitutes a unique large TGF data set, unbiased by dead time effects, which can be exploited for understanding the relationship between TGFs and lightning activity and to test theoretical production models.

7. The AGILE TGF meteorological pipeline system

A fast-processing pipeline system was recently established for AGILE, aimed at retrieving meteorological data acquired by geostationary meteorological satellites and at reconstructing the meteorological scenario associated to the regions where the events occurred. All meteorological data are continuously downloaded from the EUMETSAT portal and stored into a 1 week buffer memory at the Istituto di Scienze dell'Atmosfera e del Clima in Rome: this allows the investigation of the initial stages of the evolution of the thunderstorm associated to the TGF, together with its on-set and successive development. The products processed by the pipeline are particularly aimed at highlighting the presence of convection in the atmosphere, by means of cloud top altitude estimates and of dedicated algorithms, such as the severe storm RGB composite or the Global Convective Diagnostics (GCD). Data from other non-geostationary satellites, microwave sensors and lightning detection networks can be implemented in the pipeline in a successive moment. By exploiting this pipeline and the associated meteorological archive, it is possible to perform both precise studies on individual events and associated storms, as well as statistical analyses on large scale samples. This data archive can be made available to the community in the near future and could help shedding light on the thunderstorm type, stage and degree of association correlated to the TGF production.

8. Detection of multiple TGFs from thunderstorms

A strong point of the AGILE satellite relies in its quasi-equatorial orbit, which represents a unique feature among the satellites currently detecting TGFs: such a low inclination allows AGILE to monitor the same geographic region at each orbital passage, without experiencing significant latitudinal shifts and making the satellite able to observe a large number of multiple TGFs from the same convective region, even at successive overpasses. The study of the waiting times distribution between successively detected events highlighted repeated enhancements, corresponding to the same passage and to each of the following orbital resonances. Such feature made the AGILE satellite observe 79 cases of convective systems producing multiple TGFs, in the period from March 2009 to November 2015: in particular, 28 of these systems produced more TGFs, detected by AGILE during the same passage, and 51 of these systems lasted several hours, producing events detected up to 4 overpasses after. Moreover, 3 mixed cases of repeated TGFs from the same thunderstorm system detected either during the same passage and at successive passages have been observed. Meteorological images acquired by geostationary satellites included in the AGILE meteo-retrieving pipeline show the presence of convective systems in the region of interest, frequently lasting several hours, in order to produce more TGFs detected at successive passages. The underlying thunderstorms exhibited very different extensions and structures, suggesting there are no privileged types of thunderstorm for TGF production, as already pointed out by previous studies [12-14]. A statistical study on the variation in time of the GCD algorithm in the region of interest was carried out: the cumulative distribution of the multiple TGFs occurrence times shows that these events tend to take place in the leading part of the thunderstorm development, when the system is increasing its height and size. This not only represents a further confirmation of the efficiency of the AGILE satellite as a TGF detector, but it is also the first reported evidence that single thunderstorm systems can produce multiple TGFs throughout their entire lifetime.

9. Conclusions and future perspectives

After almost 10 years since its launch, AGILE is still capable of producing high-quality scientific results, representing a successful and still promising mission both for high-energy astrophysics and geophysics. TGFs have driven an increasing attention from the scientific community, for their still poorly understood meteorological contest and their possible source of danger for planes flying nearby their production region; moreover, the small number of missions currently detecting these events make TGF science even more challenging. In this perspective, the AGILE satellite plays a major role: the recently increased TGF detection rate makes AGILE the mission with the highest TGF detection rate to date and the recently established meteorological pipeline allows for the development of a wide database of high-quality data, that can be exploited to perform improved analyses on the relationship between TGFs and other meteorological parameters, as well as to test the current theoretical production models. Finally, the narrow equatorial orbit of AGILE puts the satellite in a privileged condition, for what concerns the observation of repeated TGFs, produced by the same thunderstorm, further underlining the connection between these events and lightning discharges.

Acknowledgments

AGILE is a mission of the Italian Space Agency, with coparticipation of the Istituto Nazionale di Astrofisica), the Istituto Nazionale di Fisica Nucleare, and several Italian research Institutes, Universities and industrial partners.

References

[1] Fishman G J *et al.*, 1994 *Science* **264**, 1313–1316
[2] Gurevich A V *et al.*, 1992 *Phys. Lett.* A **165**, 463–468
[3] Dwyer J R *et al.*, 2008 *J. Geophys. Res.* D **113**, 10103
[4] Dwyer J R *et al.*, 2012 *J. Geophys. Res.* A **117**, 2308
[5] Tavani M *et al.*, 2008b *Astron. Astrophys.* **502**, 995–1013
[6] Marisaldi M *et al.*, 2010a *J. Geophys. Res.* **115**, A00E13
[7] Tavani M *et al.*, 2011b *Phys. Rev. Lett.* **106**, 18501
[8] Marisaldi M *et al.*, 2010b *Phys. Rev. Lett.* **105**, 128501
[9] Dwyer J R and Cummer S A 2013 *J. Geophys. Res.* **118**, 3769–3790
[10] Connaughton V *et al.*, 2013 *J. Geophys. Res.* **118**, 2313–2320
[11] Marisaldi M *et al.*, 2015 *Geophys. Res. Lett.* **42**, 9481–9487
[12] Smith D M *et al.*, 2010 *J. Geophys. Res.* **115**, A00E49
[13] Splitt M E *et al.*, 2010 *J. Geophys. Res.* **115**, A00E38
[14] Chronis T *et al.*, 2016 *BAMS* D **97**, 639-653

Review of techniques for 2D camera calibration suitable for industrial vision systems

G D'Emilia[1], D Di Gasbarro[1]

[1]Dipartimento di Ingegneria Industriale, dell'Informazione e di Economia, Università degli Studi dell'Aquila, ITALY

david.digasbarro@graduate.univaq.it

Abstract. In camera calibration the goal is to determine a set of camera parameters that describe the mapping between 3-D reference coordinates and 2-D image ones. Correction for image distortion in cameras is an important topic in order to obtain accurate information from the vision systems. A review of calibration techniques for the vision systems is presented among the different methods for camera calibration found in literature. The advantages and limitations of these techniques are also discussed. This work analyzes the techniques that can be useful in modern industrial contest with the aim of knowing the effective distortion of the lens and the possibility to determinate the measurement uncertainty of the vision system.

1. Introduction

The industrial vision systems are widely used in many different applications: part identification of complex systems, defect inspection, Optical Character Recognition (OCR) reading, 2D code reading, pieces counting and dimensional measurement. If the system is used for measurement applications, it is necessary to calibrate preliminary the vision system. In industrial applications the most of the calibration procedure is generally devoted to the identification of pixel/mm ratio. In fact, many other optical problems can occur, like chromatic aberration, aberration of axial points, astigmatism, coma and distortion of the image. In particular, the distortion of the image is the most diffused error in image acquisition and elaboration. Usually, the top class industrial vision systems have in their software suite some tools to correct the effect of distortion of the images. The results of their application generally are satisfactory, but often it is impossible to know about the method for the characterization of the vision system, about the characteristics of the optics or about the uncertainty of the system. This problem is very important in case of critical situations (validation of very tight geometrical tolerances or research applications). In literature some methods for the camera calibration are described, which are different from the point of view of the methodology and of the parameters they take into account.

The aim of this paper is to review literature techniques for 2D camera calibration and to discuss some comments on these methods and on their suitability for industrial vision systems. All the most important methods will be described and explained, with reference to the optics features and their capability of facing optical problems. In the paper the comparison and evaluation criteria of the methods will be experimentally validated with reference to the Zhang method, applied with a commercial high performance Camera Calibration tool, acting on images of a Omron industrial vision system. Finally, the guidelines for an on-line industrial implementation of algorithms based on these methods will be also analyzed.

2. Industrial vision systems applications

During the years the number of applications in industrial contests is more and more increasing. This trend has been made easy by the combination of many factors: the reduction of the costs of the vision systems, the easier software implementation and maintenance of vision systems, the revamping and optimization of the production line, in compliance with the "zero-defect" approach. All the present time, most applications are still based on grayscale systems. Furthermore, for a large number of kind of inspection tests the algorithms are based on binary images. In these cases, it is necessary to convert the RGB image, acquired by the vision system, into a grayscale image and then to a binary one. For the transition from grayscale to binary image a threshold value of the gray level should be set, that divides the pixels that will be converted in white pixels from the pixels that will be converted in black ones. These functions are generally included in the software tool of the vision system. The dimensional measurements are usually based on the identification of the pixels describing the contour of the piece, then calculating the difference between the pixels' position on the borders of the object under inspection. Anyway, the most common, simple and accurate technique of inspection is based on binary images.

3. Review of the calibration methods

In cases the vision systems are used for measurement applications, to calibrate preliminary the vision system is mandatory. Generally, in industrial context the calibration procedure is based on the identification of the pixel/mm ratio. This parameter is fundamental for the correct evaluation of the dimension of an object under inspection, but it is not the only parameter to check. In effect many optical problems can occur, in particular, the distortion of the image is the most diffused in image acquisition and elaboration and remarkable measurement errors can arise.
In literature some methods for the camera calibration are described. Due to the lack of information, it is likely that the tools of the industrial vision system used are among these. In the following the main literature techniques for 2D camera calibration will be reviewed, putting in evidence any comments and observations useful to discuss their suitability for industrial vision systems; aspects concerning the optics features and their capability of facing optical problems on-line are also discussed. In order to make the examination of the existing methods more complete, in the following section a procedure to obtain the calibration parameters is presented, based on a high performance commercial tool (Matlab).

3.1. Plumb line method

Brown [1] proposed a first experimental method to calibrate an optical system used for measurement aims. This approach, named *plumb line method*, assumes the lenses are perfectly symmetrical only at the design stage, but not in their manufacturing. The previous method like the stellar calibration, explained in the same reference, does not allow for a precise calibration because it considers in the mathematical model the lenses to be perfectly symmetric. This more realistic assumption causes the necessity to implement a mathematical model and then a practical model for camera calibration. The method needs to determine at least 10 parameters: 8 parameters of the inner cone of the lens and 2 parameters for each of the calibrated lines. In reality the lines used to calibrate the system are generally 7. Although this method is the first to be proposed for calibration of optical systems most of the aspects to be taken into account (lens geometry and characteristics, camera model) are considered; therefore, the uncertainty calibration according to this method is very satisfactory. In his paper, Brown defines a formula which could accurately model radial distortion at a range of different focal length settings. If the radial distortion parameters at two separated focal distances are known, the values for all settings between them may be accurately found. Furthermore, the method specifies the formulas for radial distortion within the photographic field of view. This is only of significance for close range camera applications. In effect, if a camera is focussed for a particular finite distance, targets at different distances will display slightly different quantities of radial distortion. The advantage of this method is the high measurement accuracy achievable (±0.5 mm with a camera at 2 m of distance from the target), that it is comparable to the accuracy of the best technique proposed in this review. The disadvantages of this method are the occurrence of a manual determination of the points necessary for the calibration (this could be a problem in an automated system), the needs of computer-intensive full-scale nonlinear search and the complexity to determine the real principal point as offset from the ideal axis of the lens. Some similar methods are proposed in [2-6] and the results are similar to [1]. A revolution in the close range photography was made by the introduction of cameras based on Charged-Coupled Device (CCD) sensors that allow to have directly digital images: the new devices are much smaller respect the full frame film and then the resolution of the images is less accurate. Brown's idea was developed by Clarke, Fryer [7] and Chen [8] in the 90's. In fact, these papers, apply the plumb line method in a new context. The principal innovation in these works is that only the first term of the polynomial function of the radial distortion is considered because the resolution of the images is linked to the pixel dimension and it is not necessary to study all the terms of the calibration function. This assumption is in agreement to another branch of study in camera calibration named Direct Linear Transformation (DLT) proposed by Abdel-Aziz and Karara [9].

3.2. Two stage method

Tsai [10] proposed a new algorithm for camera calibration named *two stage method*. The work starts from the same camera model considerations theorized by Brown [1], but the aim of the paper is the characterization of a real time procedure for camera calibration. The calibration procedure needs a calibration pattern composed by some black squares on white background. The procedure needs to identify the corner of the black squares. Others type of pattern can be used with the same procedure. The accuracy of the pattern is a contribution of uncertainty. After the acquisition of one image, it is possible to apply the two stage method. The first stage is useful to compute the 3D orientation and position of the pattern and to compute the scale factor. In the second stage it is possible to compute the effective focal length, the distortion coefficients and the position of the pattern on the z axis. The limit of the two stage method is based on some simplification in the camera model (called pinhole model) and the experimental results are in several cases less accurate than for others methods. The achievable uncertainty of the system can be in order of ± 1 mm with a camera at 2 m of distance from the target. An extension of this method was developed by Heikkila [11] that combines the two stage method with the DLT ones. The results of this method are discussed in the following subsection.

3.3. Direct Linear Transformation methods

Abdel-Aziz and Karara [9], Clarke and Fryer [7] and Chen [8], Sturm [11], Heikkila [12-13] proposed several methods for camera calibration based on the DLT assumption. That hypothesis presents a simplified model based only on linear equation for radial and tangential distortion and is in practical equivalent for real camera calibration to the complete model. Clarke, Fryer [7] and Chen [8] use the DLT method to simplify the algorithm of the Plumb Line Method. Sturm [11] proposed an algorithm similar to the plumb line method but based on a particular 3D calibration pattern composed of 3 different planes with calibrated white dots on black surfaces. This method reaches a very low uncertainty also without knowing the principal point of the image, the focal length, the aspect ratio and without considering the skew factor. This method is very simple to implement and very general too, but it is necessary to have a specific and very accurate 3D calibration pattern. Some similar methods are proposed in [14-16], and the results are similar to [9]. The most accurate methods based on the DLT simplification are [8,11] and the achievable uncertainty of the system is in the order of ±0,6 mm with a camera at 2 m of distance from the target. Heikkila [12-13] combines the two stage idea with the DLT and proposed a new method called "four step calibration procedure". Also in this case a specific 3D calibration object with 256 white dots on each face and a black background has to be provided. In the first step the camera parameters are calculated using a DLT method and the lens distortion is assumed to be insignificant; the focal length and the image center are known. In the second phase, the camera model is applied in compliance with the two stage method and a least square estimation technique is used to calculate the camera parameters. In the third phase, the error caused by the asymmetry of the dot projections is compensated and an estimate of the camera parameters is recalculated. The last step is the "image correction"; in other words, all the others unknown parameters, not calculated due to simplification in the camera model or in the DLT method, are now synthesized in a couple of values used to correct the center point of the camera. These values are calculated with a specific formula based on the inverted camera model. The achievable uncertainty of the system is in order of ±0.07 pixels that correspond at ±0.02 mm with a camera at a distance of 0.5 m from the target. The empirical correction allows to find the camera parameters for the calibration with a very good approximation but the method appears very difficult to be applied in industrial context being based on a 3D calibration pattern.

3.4. Unknown orientations plane

In the end of the 90s, Zhengyou Zhang studies a different technique of camera calibration [17]. He proposed a procedure that requires only a simple planar pattern (cheaper respect to a 3D pattern) based on black squares on white surface, which can be also printed with a laser printer. The quality of the pattern is of concern in the uncertainty budget, but in a lot of cases a good calibration can be achieved also with a printed pattern. The only request of this technique is the necessity to observe the pattern at different orientations (at least two, but in our experience it is better to use almost 15 orientations because this is also a contribution of uncertainty). The most interesting aspect of this method is that is not necessary to know the original distance or orientation of the pattern from the camera neither the motion during the calibration. The found calibration parameters are very similar to the ones calculated by Brown [1] or Tsai [10] methods, but the calculation procedure is based on the maximum likelihood estimation. This statistical approach is possible because there are some different views of the same pattern. This way the approach introduces another contribution of uncertainty. Therefore, this method isn't the most accurate, in fact the achievable uncertainty of the system is in order of some tenths of pixel but it results to be very quick and flexible; the pattern is cheap and then can be easily applied in industrial context. For these reasons in the authors' opinion these methods can be evaluated the best in compliance with the aim of this paper.

3.5. Others

The problem of camera calibration was analysed by others researchers that find several solutions also with a large amount of experimental results. These solutions are related to a specific problem of camera inspection in certain situation like high-speed tensile testing machine [18], guide unmanned vehicles [19], surface profiling applications [20] or older joint analysis [21].

4. Camera calibration in Matlab

In the paper the comparison and evaluation criteria of the methods will be experimentally validated with reference to the Zhang method, applied with a Camera Calibration toolbox for Matlab [22] acting on images of an Omron industrial vision system.

This toolbox is available on the web and recently has been included in Matlab Computer Vision System Toolbox [23]. The last version has been developed with a specific app to guide the user in a camera calibration. The application can use online camera or some images acquired previously. The steps in compliance with Zhang method are acquisition of several images with the calibration pattern in different positions, extraction of the corner of the black squares of the pattern, calculation of the calibration parameters. In output the system gives back all the information with their uncertainties like focal length, principal point, skew, radial and tangential distortions. By this way, it is possible to use the results of camera calibration to automatically correct future images acquired with a self-generated Matlab script. Our experimental equipment (Fig.1) are composed by Omron FZ-SC2M camera and VS-1214V optics, 2D calibration pattern and the elaboration are made with Camera Calibration toolbox for Matlab [22]. The acquired images (Fig. 2) are imported to Matlab and elaborated with Camera Calibration toolbox. The corner extraction is made manually in compliance with the toolbox procedure. The automatic elaboration gives back us the results synthesized in the bottom of Fig. 3. In the upper part of Fig. 3 the effect of radial distortion on the CCD pixels is shown. For simplicity only the radial distortion has been taken into account.

The Omron FZ-SC2M camera has a CCD resolution of 1600 x 1200 pixels being a high-end quality camera for industrial applications. The optics used has also a very good quality and the calibration results

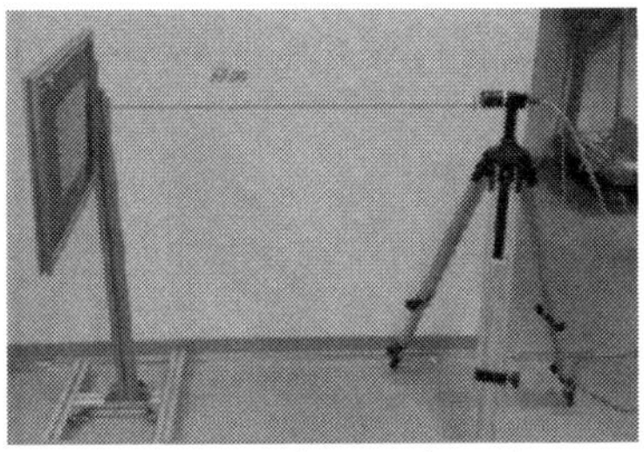

Figure 1. Experimental equipment.

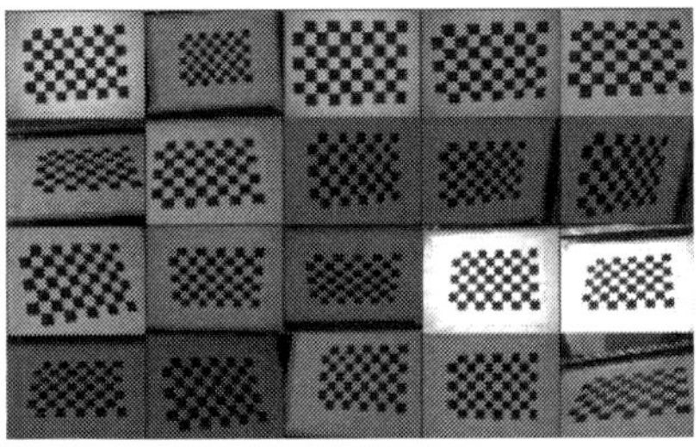

Figure 2. The acquired images.

confirm the expectation. Nevertheless, each optics is a single piece, because the lens is very difficult to manufacture. The optics under calibration has a theoretical centre (also named principal point) in the pixel 800 x 600 (at the centre of the CCD), in fact the real principal point is in pixel 818 x 599. This error, if not correctly evaluated, can generate a measurement error in order of 1.1% on the horizontal axis. Another very interesting result for our purpose is the radial coefficient; this figure represents the mean pixel error due to the curvature of the lens. The mean value is quite contained (in order of -0.1 pixels) thanks to the high manufacturing quality of the lens, but at the edge of the lens it can be three orders of magnitude higher (in order of 10 pixels), in particular on the left edge. This error probably is a combination of two different situations: the distortion of the lens and the error on the principal point. The sum of these two error, if not correctly evaluated, can generate a measurement uncertainty in order of ± 1.5%. Otherwise with calibration of the lens the uncertainty contribution of these factors is limited to the uncertainty measurement of the radial coefficient, that is one order of magnitude lower (0.01 pixels). Another important contribution in the uncertainty budget is the pixel error or in other words the maximum error of discrimination on one surface, due to the pixel dimension. In this case the pixel error is 0.35 pixels on the horizontal axis and 0.39 pixels on the vertical one. Generally, in a high quality calibrated vision system this is the main contribution to the uncertainty budget of the system.

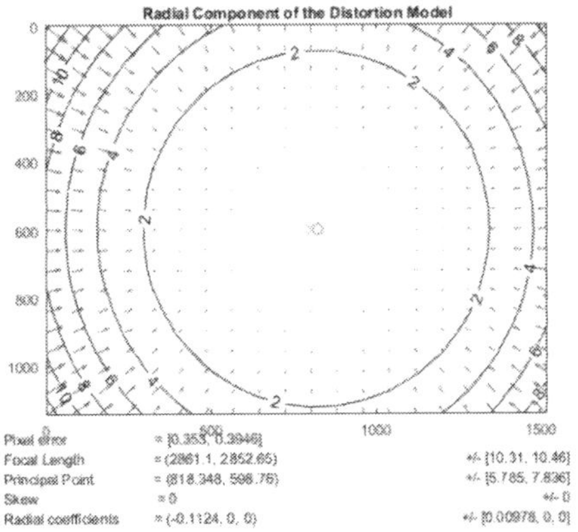

Figure 3. Radial distortion and calibration results.

5. Discussion

The existing activities of camera calibration have been discussed. For each method advantages and limitations have been identified and analysed. Some comments have been provided, concerning the suitability of using the methods in industrial context. The Zhang method has been extensively investigated and appears to be a good choice for camera calibration in the industrial context. A possible practical solution based on the Matlab toolbox is presented with an experimental test. The main advantage of applying this method is the possibility, after the calibration, to assign an uncertainty to the measure obtained by the vision system. This is very appealing in customer-supplier relationships for companies operating with certified quality management system. The guidelines for an on-line industrial implementation of algorithms based on these methods have been also analysed. The results are encouraging and foster further research.

6. Future work

As a natural follow up of this work, an implementation in an industrial context will be carried out. Generally, the limitation of Matlab algorithms is the difficulty of using them in real applications. However, in this case a "black box" tool, receiving in input the raw image and providing in output the image corrected with the calibration parameters, can be devised. A Matlab Compiler SDK, capable to convert the Matlab code in C/C++, has been recently released, and this will allow implementing this method in the common industrial software based on C/C++ languages.

References

[1] Duane C B 1971 *Photogramm. Eng*, ***37*** 855-866
[2] Faig W 1975 *Photogramm. Eng. Remote Sensing*, ***41*** 1479-1486
[3] Gennery D B 1979 *Proc. Image Undest. Workshop*, 101-108
[4] Isaguirre A, Pu P, Summers J 1985 *Proc. Int. Conf. Robotics and Automation*, 74-79
[5] Ziemann H 1986 *Proc. ISPRS Symposium*, 41-48
[6] Shortis M R, Snow W L, Goad W K 1995 *Int. Arc. Photogramm. Remote Sensing*, ***30,*** 53-59
[7] Clarke T A, Fryer J G 1998 *The Photogrammetric Record*, ***16*** 51-66
[8] Fryer J G, Clarke T A, Chen J 1994 *International Archives of Photogrammetry and remote sensing*, ***30*** 97-101
[9] Abdel-Aziz Y, Karara, H M 1971. *Urbana-Champaign*, **1** 18
[10] Tsai R 1987 *IEEE Journal on Robotics and Automation*, ***3*** 323-344
[11] Sturm P F, Maybank S J 1999 *Computer Vision and Pattern Recognition*, 1999 IEEE Computer Society Conference on. **1**
[12] Heikkila J, Silvén O 1996 IEEE *Pattern Recognition, 1996., Proceedings of the 13th International Conference on* **1** 166-170
[13] Heikkila J, Silvén O 1997 *Computer Vision and Pattern Recognition, 1997 Proceedings.*, 1997 IEEE Computer Society Conference on 1106-1112
[14] Malhotra V M 1971 *Proc. Symp. on Close Range Photo. Sys.* 62-80
[15] Okamoto A 1984 *Photogramm. Eng. Remote Sensing*, ***50,*** 705-711
[16] Wong K W 1975 *Photogrammetric Eng. Remote Sensing* ***41***, 1355-1373
[17] Zhang Z 1999 *Computer Vision, 1999 The Proceedings of the Seventh IEEE International Conference on* **1** 666-673
[18] Guiqin L I, Guo L, Wang Y, Guo Q, Jin Z 2012 *Transaction on Control and Mechanical Systems* **1** 99-103
[19] Wang J, Li Z, Ren F, Liu Z, Shen A 2016 *IEEE OCEANS 2016-Shanghai* 1-4
[20] Abu-Nabah B A, ElSoussi A O, Al Alami A E K 2016 *Optics and Lasers in Engineering* **84** 51-61
[21] Heinemanna D, Knabnerb S, Baumgartena D 2016 *ISPRS-International Archives of the Photogrammetry, Remote Sensing and Spatial Information Sciences*, 27-30
[22] Bouguet J Y 2004 *Camera calibration toolbox for Matlab*
[23] *Matlab Computer Vision System Toolbox User's Guide* R2016a edition, chapter 5 pp 14–40

IOP Conf. Series: Journal of Physics: Conf. Series **841** (2017) 012031 doi:10.1088/1742-6596/841/1/012031

Cryocooled Josephson standards for AC voltage metrology

P Durandetto, A Sosso, E Monticone, B Trinchera, M Fretto and V Lacquaniti

I.N.Ri.M. - Istituto Nazionale di Ricerca Metrologica - Strada delle Cacce, 91 - 10135 Torino (Italy)

E-mail: p.durandetto@inrim.it

Abstract. The Josephson effect is worldwide used as a basis for constant reference voltages in national metrological institutes and in calibration laboratories of industry. Research on Josephson voltage standards is aiming at a fundamental change also in the metrology of the volt for AC and arbitrary waveforms: programmable Josephson voltage standards converting a digital code into a quantum-accurate stepwise waveform are already available in primary laboratories and even more advanced standards for converting sub-nanosecond binary coded pulses into any arbitrary signal with quantum accuracy are now actively developed and tested. A new experimental setup based on a two-stage Gifford-McMahon cryocooler has been developed at INRiM for the operation of AC-Josephson voltage standards. Among its distinct features, the possibility of employing both the aforementioned techniques (programmable and pulsed Josephson voltage standards) is particularly interesting. Quantum-based AC voltage sine waves have been synthesized with both programmable and pulse-driven arrays, although their accuracy is still limited by thermal oscillations due to the cryocooler piston motion.

1. Introduction

The application of superconducting devices based on Josephson effect for the synthesis of AC voltage waveforms has been investigated for many years. Indeed, arrays of many thousands of Josephson junctions for DC voltage standards are, to date, among the most important achievements of superconducting electronics.

Josephson Voltage Atandards (JVSs) are employed for high precision DC voltage calibrations since the early years after the phenomenon was predicted by Brian Josephson in 1962 [1]: he describes how it can be used to relate, in some conditions, the constant voltage across a junction to the frequency of the alternate bias current flowing through it.

For voltage metrology, a Josephson junction is essentially described by the equation expressing the relationship between the frequency of the current and the voltage across the junction:

$$V_n = n \cdot \frac{h}{2e} \cdot f \qquad (1)$$

where n is an integer, e is the electron charge and h is the Planck constant. Voltage V_n and frequency f are related by a direct proportionality and, since the coefficient is given in terms of the two fundamental constants e and h, its value is universal. Eq. (1) tells us that driving a Josephson junction by a microwave source of frequency f, constant-voltage steps at integer

multiples of $h/2e$ appear. Junction voltages of about 150 μV are obtained by radiating it with microwave bias frequencies around 70 GHz, for $n = 1$. To achieve the output voltage required for a practical quantum standard, hysteretic Superconductor-Insulator-Superconductor (SIS) technology was successfully adopted, following an idea by Levinsen [2] to fabricate arrays with tens of thousands of highly uniform junctions and output up to 10 V. Arrays with several thousands of DC Josephson junctions have been used for many years as voltage standards with ultimate accuracy (up to few parts in 10^{11} [3]). Lately, voltage standard research has focused on the metrological application of Josephson effect for AC voltages in a wide range of frequencies and amplitudes. Quantum AC voltage standards are based on non-hysteretic Josephson junctions that can be realized with SNS (N: normal metal) self-shunted junctions, where voltage is easily adjustable by varying the bias current and Shapiro steps are intrinsically stable.
The main techniques employed for the generation of quantum-based waveforms are programmable and pulse-driven AC-JVSs. Programmable arrays are subdivided in sub-circuits with series-connected junctions to generate voltages following a power of 2-sequence (1, 2, 4, 8, 16, ...). Independent bias currents are used to activate/deactivate the array's subsections, thus controlling the output voltage as in DACs of semiconductor electronics. Programmable arrays can generate stepwise quantum waveforms in the kHz range and amplitudes up to 20 V [4]. On the other hand, it is possible to use short pulses instead of a sinusoidal rf (radiofrequency) signal, allowing to modulate the rf period while keeping junctions phase-locked over a wide range of frequencies (up to the MHz range) [5]. The principle of operation of pulse-driven JVS (or Josephson Arbitrary Waveform Synthesizer, JAWS) is based on the quantization of magnetic flux, since the time-integrated area of every voltage pulse of each junction is exactly equal to the magnetic flux quanta $\phi = h/2e$. Fundamental accuracy follows from the control of the flux quanta transferred through the junctions by the pulse signal and the output voltage is exactly calculable from fundamental constants if the pulse repetition rate is known [6].
Cryogen-free operation of JVSs is now regarded as crucial to widen the range of applications of quantum standards, due to its greater ease of use with respect to liquid helium and for possible shortages and high cost of the latter. Furthermore, in cryocooler the length of the cables can be at least halved with respect to liquid helium system, allowing to minimize the systematic errors that affect high-frequency voltage measurements [7].
In this paper we present last results obtained with both programmable and pulse-driven Josephson standards in cryocooler setup.

2. Measurement setup

2.1. Cryogenic system

The cryocooler in use for the operation of programmable and pulse-driven arrays is a two-stage GM type with top loading sample assembly, 1 W cooling power at 4.2 K, 8 kW compressor input power and minimum temperature achievable of about 2.9 K with no load. The cold finger top was fitted with an additional copper disk (coldplate) that contains a germanium resistance temperature sensor and a heater for temperature monitoring and control (Fig. 1). A second thermometer (a silicon diode, b) in Fig. 1) is positioned close to the chip in order to better estimate the array operating conditions and detect unwanted thermal gradients [8]. Indeed, when a Josephson array is irradiated by rf signal, it warms up and consequently induces the rise of the local temperature and the worsening of the array's operating margins. All heat has to be dissipated through a series of suitable thermal links with high thermal conductivity [9]. All measurements performed make use of a cryopackage structure which is already described in [10]. Finally, GM coolers are affected by temperature oscillations of the coldplate that are caused by the cryocooler piston periodic motion and lead to the worsening of AC-JVSs operation. Interventions aimed at reducing these effects are under study.

IOP Conf. Series: Journal of Physics: Conf. Series **841** (2017) 012031 doi:10.1088/1742-6596/841/1/012031

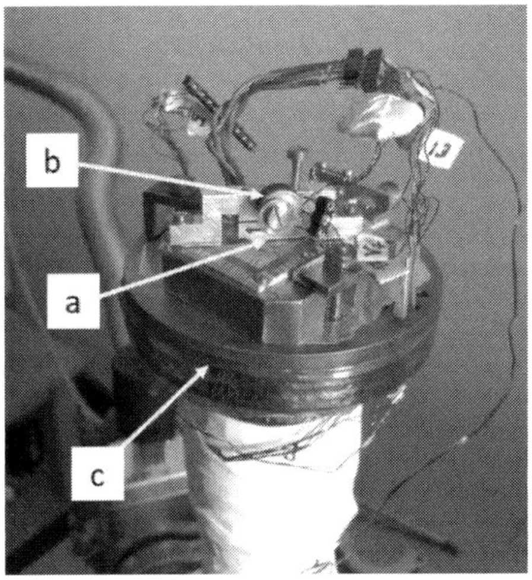

Figure 1. Picture of the cryocooler's coldplate with a Cu sample holder positioned on it. a) is the Josephson chip, b) is a silicon-diode thermometer and c) is the heater wire wound around the coldplate, suitably guarded with kapton tape.

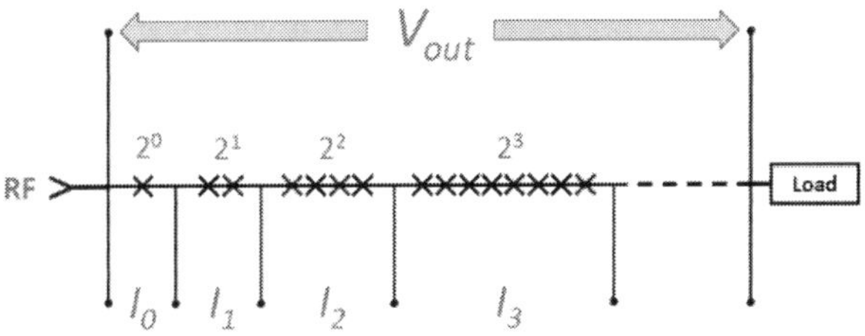

Figure 2. Scheme of programmable JVS.

2.2. Programmable Josephson standard

As mentioned before, Programmable-JVS (PJVS) is made up by series-connected subarrays whose size follows a power of 2-sequence (Fig. 2). Each subarray is individually current-biased on the nth Shapiro step (where n is typically 0 or ±1). The arrays under test were realized in cooperation with PTB (Physikalisch-Technische Bundesanstalt, Germany) and are made up by 8192 Nb/Al-AlO$_x$/Nb (SNIS) intrinsically-shunted Josephson junctions. The SNIS technology is interesting for application to a cryocooled standard, allowing to set a compromise between device and refrigerator requirements. Indeed, an important requirement for cryocooler operation is the temperature stability of the electrical parameters above 4.2 K, as explained in reference [11]. While SIS and SNS junctions show temperature dependence of the electrical parameters V_c and I_c below or above 4.2 K respectively, SNIS junctions can minimize it over an extended range by optimizing some intrinsic properties. Moreover PJVS operation requires rf signal transmission through a microwave guide (WR-12, 60÷90 GHz) that can convey significant heat to the coldplate. We adopted a stainless steel guide, with gold-coated internal surface to reduce the dramatic signal attenuation of the stainless steel [8].

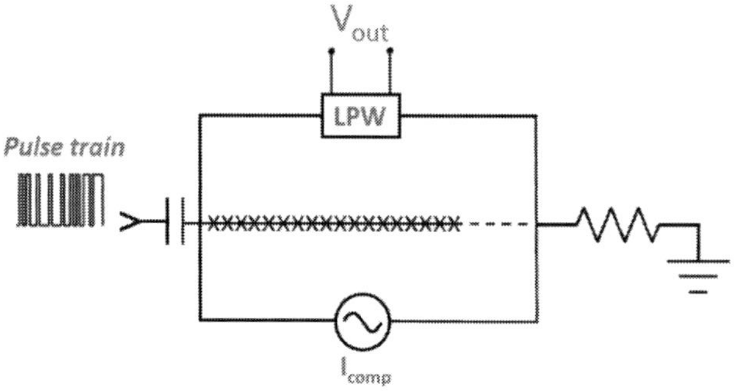

Figure 3. Scheme of pulse-driven JVS.

2.3. Pulse-driven Josephson standard

With pulse-driven technique AC and arbitrary voltages can be obtained by modulating the repetition frequency of pulses, allowing the synthesis of very pure signals [12]. The desired waveform is encoded into a 0/1 or 1/0/-1 bitstream through $\Sigma\Delta$ modulation techniques employed for ADC in electronics (Fig. 3). The work presented here uses arrays of SNS junctions realized by PTB with Nb_xSi_{1-x} as barrier material. Our pulse-driven samples consist of two arrays, with 4000 junctions each, integrated on a 10 mm×10 mm Si chip. For each array, junctions are arranged in double stacks and are embedded into the center line of a 50 Ω coplanar-wave-guide (CPW) which ensures a suitable propagation of pulses [13]. The chip is mounted onto a chip carrier 23 mm×40 mm made on Rogers RO3006. Two conducting CPW paths made of copper with a 2 μm gold layer on top (without nickel) are used for pulse transmission to the arrays. The CPW lines of this carrier are connected to two PCB-SMA launchers for connection to coaxial cables [14]. For the transmission of the pulse train we employed a 50 cm long stainless steel coaxial cable, suitably anchored to the cryocooler first and second stages for reducing the direct thermal link with the external environment at room temperature. A I/O DC-block is needed for the electrical decoupling of the sample and to avoid common-mode voltage errors [15].

3. Results

The main measurements performed with our programmable chip were concentrated on the optimization of operating margins on the 1st and the 2nd Shapiro steps by adjusting coldplate's temperature and microwave input power. The output voltage of the Josephson array was observed and measured with a compact wideband digital modular system for dynamic measurement with fast frame rate [16]. The flatness of voltage steps was determined under microwave radiation by recording the current-voltage (IV) characteristics of all the PJVS subarrays (inset of Fig. 4) and was deduced by a linear fit within 100 μA of current operating margins: for the entire array (8192 junctions) we obtained 0.8 μV/μA at 1.28 V and 0.4 μV/μA at 2.56 V. These results suggest the use of these quantized voltage steps for synthesis of stepwise sine waves. Indeed, as can be seen in Fig. 4, we synthesized sine waves at a frequency of 1 kHz and with amplitude of about 1 V_{rms} at $T \simeq 6.5$ K using the 1st step and of about 2 V_{rms} at $T \simeq 4.7$ K using the 2nd step.

As regards the pulse-driven JVS, to avert any common-mode voltage on the termination resistor, the AC-coupling technique was used [15], [17], where the low-frequency component of the pulse train is removed through a DC-block with the scope of canceling the common-mode signal at the array output. Then, in order to ensure the proper operation of the array, the low-frequency component has to be re-injected as a compensation current signal I_{comp} (see Fig. 3) at the same frequency of the desired waveform and has to be phased with the pulsed signal. The Pulse

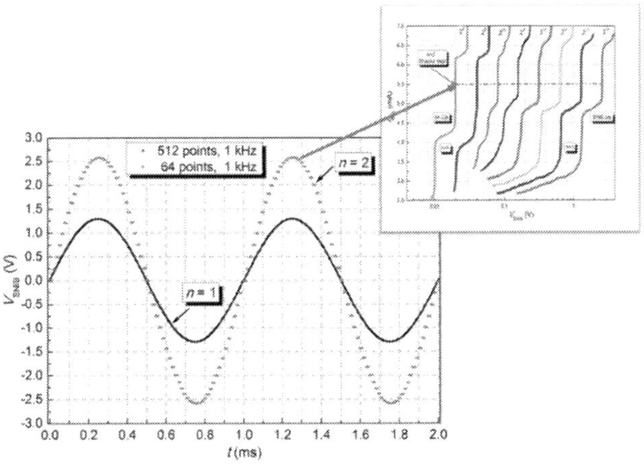

Figure 4. Stepwise sine waves at 1 kHz synthesized with the PJVS array in cryocooler. The black curve (1 V_{rms}) is obtained at $T \simeq 6.5$ K with all the 1st Shapiro steps of the PJVS subsections, while the red curve (2 V_{rms}) is obtained at $T \simeq 4.7$ K with all the 2nd Shapiro steps. The inset shows some of the overlapping quantized steps employed for the synthesis of the 2 V_{rms} waveform, where the horizontal red line represents the common bias current ($\simeq 5.5$ mA) for each section.

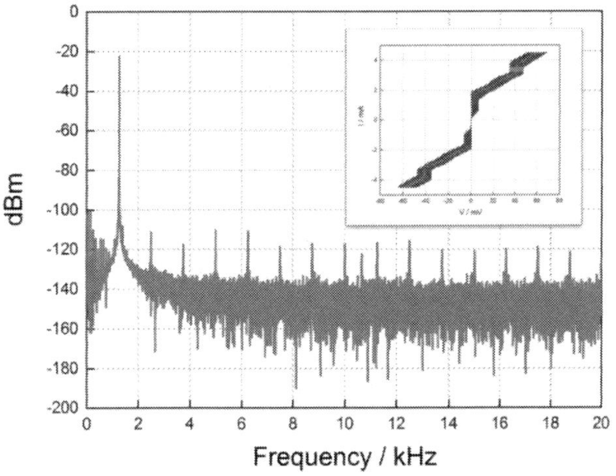

Figure 5. Observed FFT spectrum of a unipolar sine wave at 1.25 kHz at 11.7 mV_{rms} obtained with pulse-driven JVS at about 5 K in cryocooler. The inset shows the IV characteristic of the JVS when radiated by the 0/1 pulse pattern of a $\Sigma\Delta$-modulated sine wave. The red area on the 1st step represents the operating margins exploited for the waveform synthesis.

Pattern Generator (PPG) we employed delivered unipolar non-return-to-zero (NRZ) pulses with a maximum clock frequency of 12.5 GHz. The output voltage was observed and measured with the aforementioned digital modular system, appropriately reconfigured for this experiment. By optimizing five parameters (temperature, pulse amplitude, offset, amplitude and phase of I_{comp}), we observed FFT spectra of unipolar sine waves at 1.25 kHz and amplitudes from 4.3 mV to 11.7 mV, (Fig. 5). The PPG clock frequency was set to 10 GHz. The compensation current

I_{comp} was provided by an additional AC synthesizer locked to the PPG 10 MHz clock reference. System tuning for optimal operating margins has been performed by observing the frequency spectrum at temperatures around 5 K.

4. Conclusion

In this work we presented our helium-free system based on a GM cryocooler for the operation of AC-JVSs. We have demonstrated the feasibility of synthesizing quantum-based sine waves in the kHz range with the two different techniques presented here. Stepwise sine waves with amplitude of 1 V_{rms} and 2 V_{rms} have been synthesized with the programmable standard by current-biasing on the 1st and the 2nd Shapiro steps respectively, while with the pulsed standard we have generated waveforms with amplitude from 4.3 mV to 11.7 mV and distortion around 80 dBc.

These preliminary characterizations show that the output waveform is affected by temperature fluctuations due to intrinsic vibration of the cryocooler. The use of temperature compensation technique and low-vibration cryocooler systems will further improve the overall performance of the ac-JVS synthesizer.

Acknowledgments

This work was partly carried out with funding by the European Union within the EMPIR project JRP 15SIB04 QuADC. The EMPIR is jointly funded by the EMPIR participating countries within EURAMET and the European Union.

References

[1] Josephson B D 1962 *Physics letters* **1** 251–253
[2] Levinsen M, Chiao R, Feldman M and Tucker B 1977 *Appl. Phys. Lett.* **31** 776–778
[3] Wood B and Solve S 2009 *Metrologia* **46** R13
[4] Müller F, Scheller T, Wendisch R, Behr R, Kieler O, Palafox L and Kohlmann J 2013 *IEEE Trans. Appl. Supercond.* **23** 1101005–1101005 ISSN 1051-8223
[5] Kim J, Sosso A and Clark A F 1998 *J. Appl. Phys.* **83** 3225–3232
[6] Benz S P and Hamilton C A 1996 *Appl. Phys. Lett.* **68** 3171–3173
[7] Landim R P, Benz S P, Dresselhaus P D and Burroughs C J 2008 *IEEE Trans. Instrum. Meas.* **57** 1215–1220
[8] Sosso A, Fretto M, Lacquaniti V, Monticone E, Rocci R, Serazio D and Trinchera B 2015 *J. Supercond. Novel Magn.* **28** 1181–1184
[9] Howe L, Burroughs C J, Dresselhaus P D, Benz S P and Schwall R E 2013 *Applied Superconductivity, IEEE Transactions on* **23** 1300605–1300605
[10] Durandetto P, Monticone E, Trinchera B, Lolli L, Serazio D, Fretto M and Sosso A 2016 arXiv:1609.03384
[11] Lacquaniti V, De Leo N, Fretto M, Sosso A, Müller F and Kohlmann J 2011 *Supercond. Sci. Technol.* **24** 045004
[12] Benz S, Dresselhaus P, Burroughs C and Bergren N 2007 *IEEE Trans. Appl. Supercond.* **17** 864–869
[13] Kohlmann J, Kieler O F, Iuzzolino R, Lee J, Behr R, Egeling B and Müller F 2009 *IEEE Trans. Instrum. Meas.* **58** 797–802
[14] Oliver FO K, Thomas S and Johannes K 2013 *World Journal of Condensed Matter Physics* **2013**
[15] Benz S P, Burroughs C J and Dresselhaus P D 2001 *IEEE Trans. Appl. Supercond.* **11** 612–616
[16] Trinchera B, Sosso A, Pogliano U, Lacquaniti V, Monticone E, Fretto M, Rocci R, Tet L R and Serazio D 2014 *Precision Electromagnetic Measurements (CPEM 2014), 2014 Conference on* (IEEE) pp 114–115
[17] Benz S P and Waltman S B 2014 *IEEE Trans. Appl. Supercond.* **24** 1–7

Metrological traceability of carbon dioxide measurements in atmosphere and seawater

F Rolle[1], **E Pessana**[1], **M Sega**[1]
[1] Istituto Nazionale di Ricerca Metrologica, Torino, Italy
E-mail: f.rolle@inrim.it

Abstract. The accurate determination of gaseous pollutants is fundamental for the monitoring of the trends of these analytes in the environment and the application of the metrological concepts to this field is necessary to assure the reliability of the measurement results. In this work, an overview of the activity carried out at Istituto Nazionale di Ricerca Metrologica to establish the metrological traceability of the measurements of gaseous atmospheric pollutants, in particular of carbon dioxide (CO_2), is presented. Two primary methods, the gravimetry and the dynamic dilution, are used for the preparation of reference standards for composition which can be used to calibrate sensors and analytical instrumentation. At present, research is carried out to lower the measurement uncertainties of the primary gas mixtures and to extend their application to the oceanic field. The reason of such investigation is due to the evidence of the changes occurring in seawater carbonate chemistry, connected to the rising level of CO_2 in the atmosphere. The well established activity to assure the metrological traceability of CO_2 in the atmosphere will be applied to the determination of CO_2 in seawater, by developing suitable reference materials for calibration and control of the sensors during their routine use.

1. Introduction

The accurate and reliable determination of environmental gaseous pollutants is fundamental for the monitoring of the trends of these analytes in atmosphere and to obtain significant information about the current pollution levels and the future trends on a global scale. Hence, the application of the metrological concepts to this field is necessary to assure the reliability of the measurement results and to support the planning of suitable actions to reduce the negative effects on the environment and on the human beings.

In particular, the determination of carbon dioxide (CO_2) levels is important for the role of this gas in the greenhouse effect. CO_2, together with other gases such as water vapour and methane, absorbs the IR radiation and reflects part of it back to the terrestrial surface. This phenomenon contributes to the terrestrial warming, maintaining the temperature of the Earth at the value of 15 °C. Without the atmosphere and without the greenhouse effect the Earth would have a temperature of -15 °C [1] and would be covered in ice. During the last century, this positive effect has become responsible for an excessive global warming, due to the increasing levels of CO_2 and other greenhouse gases in atmosphere and, consequently, the temperature on the Earth's surface raised above 15 °C. Water vapour is responsible for two thirds of this effect, while CO_2 for one fourth.

In addition, the higher level of CO_2 absorbed by the oceans from the atmosphere causes an acidification of the marine waters, with fallouts on the global ecosystem. Although the natural absorption of CO_2 by the oceans helps to mitigate the climatic effects of anthropogenic emissions of CO_2, the resulting decrease in pH will have negative consequences primarily for oceanic calcifying organisms. Since the beginning of the industrial revolution, it has been estimated that surface ocean pH has fallen off by slightly more than 0.1 units on the logarithmic scale of pH. It is expected to drop by a further 0.3 to 0.5 pH units and these changes are predicted to be continuous and rapid as the oceans take up more anthropogenic CO_2 from the atmosphere than they can eliminate.
The choice of addressing the metrological traceability of CO_2 measurement results derives from CO_2 relevance for the World Meteorological Organisation (WMO), an intergovernmental organization that promotes observation of atmospheric environment. WMO operates with programmes as the "Global Atmosphere Watch" (GAW) which is divided into focal areas for different pollutant species. Suggestion of GAW in exposure limits can be implemented in international and national legislation, as in the European Directive 2008/50/EC [2] that is dedicated to ambient air quality and cleaner air in Europe.
As for the marine environment, the partial pressure of CO_2 (pCO_2) in seawater is an Essential Climate Variable which can be monitored in situ with networked sensors. At the European level, the Marine Strategy Framework Directive 2008/56/EC [3] asks for provisions for the adoption of methodological standards to assess the status of the marine environment, to implement its monitoring and to reach environmental targets. In addition, during the last year, a Joint Action called "European Marine Sensors Calibration Network" has been launched within the European Joint Programming Initiative "Healthy and Productive Seas and Oceans" (JPI Oceans - http://www.jpi-oceans.eu/) in order to bring together the competencies of various communities such as the oceanographic, metrological and sensor producers.
In this framework, there is a growing need of assessing the conformity of the instrumentation and the sensors necessary for the determination of CO_2 in atmosphere and seawater, in particular considering their measurement uncertainties.
In this work, an overview of the activity carried out at INRiM for the establishment of metrological traceability of the measurements of CO_2 and in particular for the preparation of primary reference gas standards for the molar fraction of CO_2, is presented. A focus on the uncertainty evaluation and on some applications for the atmospheric monitoring is given.
Metrological traceability is defined, in the International Vocabulary of Metrology [4], as "the property of a measurement result whereby the result can be related to a reference through a documented unbroken chain of calibrations, each contributing to the measurement uncertainty". This definition highlights the necessity of the availability of suitable reference standards for the calibration of the instrumentation and for the development and validation of measurement procedures. INRiM, being the Italian National Metrological Institute, has among its duties to realise and maintain primary standards for the basic and derived units of the International System of Units (SI). In the gas analysis field, suitable primary reference materials are represented by gas mixtures prepared by means of primary methods, such as gravimetry and dynamic dilution. A primary method of measurement in the SI is defined as "a method having the highest metrological qualities whose model (mathematical equation) and realisation are completely described and understood in terms of SI units" [5].

2. Reference gas standards

2.1. Gravimetry

Gravimetry is a primary method which can be used for the preparation of primary reference standards and it can be applied to the preparation of primary gas mixtures in high pressure cylinders. Gravimetry gives direct traceability to mass standards and the purity of the parent gases. The preparation procedure followed at INRiM is divided in two steps: 1) the conditioning and 2) the weighing. During the conditioning step the gas cylinder is evacuated and heated for some hours, then is filled with matrix gas (usually nitrogen or synthetic air). The conditioning cycle is generally repeated for three times. The conditioning step aims to desorb contaminants from internal cylinder walls as the presence of contaminants would lead to fast degradation of the gas mixture. During the second part of the procedure the mass of the gas cylinder is compared with the mass of an empty reference cylinder, and calibrated mass standards are arranged on the top of the lighter cylinder to maintain the difference between the gas cylinder and the reference gas cylinder within 1 g, thus optimising the mass comparator performance. The use of calibrated mass standards also assures the metrological traceability of the mass measurement (Figs. 1(a) and 1(b)). The reference cylinder and the gas cylinder are alternatively weighed according to the double substitution scheme (A-B-B-A), where A represents the sample cylinder and B the empty cylinder used as reference. This comparison is carried out to minimise the correction for the buoyancy effect due to the shape of the cylinder. This correction is calculated taking into account the air density during the weighing process. The sample cylinder is weighted prior and after the addition of each component of the final gas mixture. For a bi-component mixture the weighing procedure would be: 1) empty cylinder, 2) cylinder after the introduction of the analyte gas, which can be present in a pure gas or in a parent gas mixture, 3) cylinder after the introduction of the matrix gas. The mass of the gas introduced inside the cylinder is related to the gas concentration [6–8]. The simplified model equation for the calculation of the molar fraction of a generic analyte in a bi-component mixture starting from a parent gas mixture is the following:

$$\chi_{\text{fin}} = \chi_{\text{in}} \frac{m_2 - m_1}{m_3 - m_1} \tag{1}$$

where:
χ_{fin} = molar fraction of the analyte gas in the final mixture (μmol/mol)
χ_{in} = molar fraction of the analyte gas in the parent gas mixture to be diluted (μmol/mol)
m_i = mass (g) of the gas determined in each of the three weighing steps (1: tare, 2: tare + parent mixture, 3: tare + parent gas mixture + matrix gas).

(a)

(b)

Figure 1. a) Primary gravimetric gas mixtures produced at INRiM; b) Detail of the INRiM facility for the weighing of a primary gas mixture, with calibrated mass standards.

2.2. Dynamic dilution

A dynamic dilution system can mix in real time two or more gas mixtures to generate a new gas mixture. Two gases can be mixed together in order to reach the desired molar fraction, starting from a more concentrated gas mixture and a matrix gas. The reference gases obtained by dynamic dilution can be used for instrumental calibration or validation of calibration curves by changing the parent gases. With respect to gravimetry, dynamic dilution has the advantage that a gas mixture can be diluted in real time on a range of different concentrations and can be more easily used also for reactive gases that are instable in high pressure cylinders. In addition, mixtures generated by dynamic dilutions can also be used to check and validate gas standards prepared by gravimetry. At INRiM a dynamic dilution system composed of two Mass Flow Controllers (MFCs) and a mixing chamber is used (Figs. 2(a) and 2(b)). MFCs are calibrated at the INRiM micro-gas station, which is one of the Italian gas flow rate primary standards. Three different MFCs having full scale range (FSR) of 200 sccm, 500 sccm and 2000 sccm can be used [9]. Two pairs of MFCs of FSR 200 sccm and 2000 sccm or, alternatively, of FSR 500 sccm and 2000 sccm are employed depending on the mixtures to be prepared. A polynomial regression curve was obtained from the calibration results of the MFCs: the regression curve gives a correction for the nominal flow displayed by the MFCs. The molar fraction of a generic gas mixture obtained by dilution with two MFCs [10] is calculated according to the following equation:

$$\chi_{\mathrm{a\_out}} = \frac{\chi_{1\mathrm{a}} Q_1 + \chi_{2\mathrm{a}} Q_2}{Q_1 + Q_2} \tag{2}$$

where:
$\chi_{\mathrm{a\_out}}$ = molar fraction of the analyte gas in the final mixture (μmol/mol),
Q_1 = flow of the MFC used for the analyte gas that has to be diluted (sccm=cm^3/min),
Q_2 = flow of the MFC used for the diluting matrix gas (sccm=cm^3/min),
$\chi_{1\mathrm{a}}$ = molar fraction of the analyte gas to be diluted (μmol/mol),
$\chi_{2\mathrm{a}}$ = molar fraction of the analyte in the diluting gas matrix (μmol/mol).

The uncertainties of Q_1 and Q_2 are obtained during the calibration, while uncertainties of the concentrations $\chi_{1\mathrm{a}}$ and $\chi_{2\mathrm{a}}$ are certified by the gas mixture producers.

(a) (b)

Figure 2. INRiM dynamic dilution system: a) MFCs connected to gas cylinders; b) MFCs and mixing chamber.

3. CO_2 mixtures produced at INRiM

At INRiM, a facility for the production of primary CO_2 standards by gravimetry has been used since 2004. Due to the huge amount of gas mixtures that are needed to cover different ranges of concentration during the calibration of instrumentation, a new system for dynamic dilution was recently set up. In both cases, reference gas mixtures are produced paying particular attention to the traceability of the gas standard, hence primary methods are used for this purpose. Primary reference mixtures of CO_2 in nitrogen or synthetic air matrices are prepared by gravimetry in high pressure cylinders of aluminum alloy, having an internal volume of 5 L. The CO_2 mixtures cover the molar fraction range (50-1000) μmol/mol. The relative gravimetric standard uncertainty associated with these mixtures decreases with increasing molar fraction and is in the range (0.5-0.01)%. The gravimetric uncertainty depends on:

- uncertainty on the molar mass of the parent gases;
- uncertainty on the parent gas mixture composition;
- uncertainty of the weighing process.

For the dynamic dilution, the same molar fraction range and uncertainties can be obtained. The main sources of uncertainty contributing to the uncertainty of the preparation of dynamic mixtures are:

- uncertainty of the parent gas mixture composition;
- uncertainty of the flow rates;
- uncertainty of the purity of the parent gases.

The main uncertainty contribution is related to the parent gas mixture composition for both the preparation techniques. Due to the wide range of concentrations that can be obtained by mixing two gas mixtures, the dilution system results in a powerful tool for situations that request the use of many gas mixtures. For this reason, the dynamic dilution can be employed to calibrate CO_2 analysers and sensors, to validate calibration curves and to assign analytical values in the stability study of CO_2 gas mixtures prepared by gravimetry in high pressure cylinders. The dynamic dilution proved not to introduce any additional contribution to the uncertainty budget, maintaining an uncertainty comparable with the uncertainty of the parent gas mixture to be diluted.

4. Applications for atmospheric monitoring

Different analytical techniques can be used for the monitoring of atmospheric pollutants. Non Dispersive Infrared spectroscopy (NDIR) is suggested as a detection method by the WMO, being a robust and selective technique, easily applicable to portable instruments. As NDIR is selective to CO_2, it cannot be applied to the determination of degradation products and impurities. On the other hand, Fourier Transform Infrared spectroscopy (FTIR) is sensitive to all molecules having a net change in their dipole moment. For this reason, FTIR is used at INRiM to identify the presence of impurities and degradation products inside the gas mixtures which might affect the nominal concentration of the mixtures, their stability and also the response of the analysers. FTIR is a powerful tool also for the identification of the presence of different isotopes of carbon in CO_2 mixtures, an important aspect considering that the determination of the ratio $^{13}CO_2/^{12}CO_2$ ($\delta^{13}C$) helps to understand the global carbon cycle. Typically, the NDIR analysers use narrow band path filters to determine $^{12}CO_2$ in all CO_2 molecules and can ignore the measurement of $^{13}CO_2$ partially or totally. However, if the carbon isotopic abundances of CO_2 samples deviate from those in CO_2 gas standard, the NDIR measurement will be affected by a bias. The discrimination of $^{13}CO_2$ from $^{12}CO_2$ is extremely important not only for obtaining a correction to be used for analytical responses. Carbon isotope fractionation is associated with photosynthesis and the plants, depending on their metabolism, prefer ^{12}C. This fractionation results in lowering the $\delta^{13}C$ by about 20 per thousand for land plants and 10 per thousand for marine plants [11–13]. Moreover when organic carbon is buried, more ^{12}C is locked out of the system in sediments. During summer, the atmosphere contains relatively more ^{13}C, as plants consume $^{12}CO_2$, while in winter plant respiration dominates and atmospheric $^{12}CO_2$ increases while $\delta^{13}C$ decreases. In addition, fossil fuels (originally plants) contain relatively more ^{12}C, so that, as fossil fuels are burned, atmospheric $^{12}CO_2$ increases while $\delta^{13}C$ decreases and the addition of $^{12}CO_2$ by the fossil fuel combustion is more than the amount of $^{12}CO_2$ removed by the terrestrial biosphere and the oceans. Measuring these isotopic tracers helps to understand the global carbon cycle and to assess how the human activity contributes to the increase of CO_2.

5. Conclusions and future perspectives

In this work the activity carried out at INRiM for assuring metrological traceability to the measurement of carbon dioxide (CO_2), included in the group of greenhouse gases, was presented. The aspects related to the metrological traceability are mainly addressed by producing reference gas mixtures by using primary methods, namely gravimetry and dynamic dilution. Gravimetry is a weighing process based on subsequent steps in which the masses of gases introduced in a cylinder are accurately weighted and it gives direct traceability to mass standards and the purity of the parent gases. In order to guarantee the metrological traceability, the weighing process is carried out by comparison with traceable mass standards. The dynamic dilution is a process in which two gases are mixed together by controlling their flows. At this purpose, calibrated Mass Flow Controllers are used, after calibration by the gas flow rate primary standard of INRiM. Future activity will regard the application of the expertise obtained in the preparation of primary reference mixtures and in the establishment of metrological traceability to the calibration of sensors for the monitoring of CO_2 in seawater. Typical sensors for CO_2 measurement in water are based on spectroscopic techniques such as NDIR. The partial pressure of CO_2 in seawater is a variable that can be monitored in situ with networked sensors and there is a growing need of assessing the conformity of these sensors to the user requirements, in particular considering their measurement uncertainties. Establishing metrological traceability for CO_2 in seawater, by developing suitable reference materials for calibration and control of the sensors during their routine use, is a goal to achieve for a better monitoring of the trends of CO_2 in the marine environment and for the comprehension of the global carbon cycle.

References

[1] Ma Q and Tipping R 1998 *The Journal of chemical physics* **108** 3386–3399
[2] *Directive 2008/50/EC of the European Parliament and of the Council of 21 May 2008 on ambient air quality and cleaner air for Europe, OJ L 152, 11.6.2008, p. 1-44*
[3] *Directive 2008/56/EC of the European Parliament and of the Council of 17 June 2008 establishing a framework for community action in the field of marine environmental policy (Marine Strategy Framework Directive), OJ L 164, 25.6.2008, p. 19-40*
[4] *JCGM 200:2012 International vocabulary of metrology Basic and general concepts and associated terms (VIM 3rd edition)*
[5] *BIPM, 1999, Consultative Committee for Amount of Substance (CCQM)-5th Meeting*
[6] *International standard ISO 6142-1:2015 Gas analysis - Preparation of calibration gas mixtures - Part 1: Gravimetric method for Class I mixtures (Geneva, ISO)*
[7] Milton M, Guenther F, Miller W and Brown A 2006 *Metrologia* **43** L7
[8] Alink A and Van der Veen A 2000 *Metrologia* **37** 641
[9] Spazzini P G, Pennecchi F, Pessana E and Piccato A 2015 Analysis of flow meters calibration *9th International Symposium on Fluid Flow Measurement* (CEESI) pp 1–7
[10] Rodrigues N J, Gomes P, Fernandes E S, Ferreira C P and Sampaio J 2009 *19th IMEKO World Congr. on Fundamental and Applied Metrology*
[11] Farquhar G D, O'Leary M H and Berry J A 1982 *Functional Plant Biology* **9** 121–137
[12] Farquhar G D, Ehleringer J R and Hubick K T 1989 *Annual review of plant biology* **40** 503–537
[13] Smith B N and Epstein S 1971 *Plant physiology* **47** 380–384

IOP Conf. Series: Journal of Physics: Conf. Series **841** (2017) 012033 doi:10.1088/1742-6596/841/1/012033

Optimal estimation of parameters of an entangled quantum state

S Virzì[1], A Avella[2], F Piacentini[2], M Gramegna[2], G Brida[2], I P Degiovanni[2], M Genovese[2]

[1] University of Turin, Physics Department, Via Giuria 1, 10125 Turin, Italy
[2] INRIM, Strada delle Cacce 91, 10135 Turin, Italy

E-mail: virz.salvatore@gmail.com

Abstract. Two-photon entangled quantum states are a fundamental tool for quantum information and quantum cryptography. A complete description of a generic quantum state is provided by its density matrix: the technique allowing experimental reconstruction of the density matrix is called quantum state tomography. Entangled states density matrix reconstruction requires a large number of measurements on many identical copies of the quantum state. An alternative way of certifying the amount of entanglement in two-photon states is represented by the estimation of specific parameters, e.g., negativity and concurrence. If we have *a priori* partial knowledge of our state, it's possible to develop several estimators for these parameters that require lower amount of measurements with respect to full density matrix reconstruction. The aim of this work is to introduce and test different estimators for negativity and concurrence for a specific class of two-photon states.

1. Introduction

The amount of entanglement plays a crucial role in quantum information [1 - 3]. Therefore, characterization and quantification of entanglement in a quantum system is a crucial issue for development of quantum technologies.

There exist two ways for measuring the amount of entanglement: the first one is performing a complete quantum state tomography [2] and evaluating parameters like negativity [4] or concurrence [5] by reconstructed density matrix [1, 2], while the second one is based on estimating such parameters with an algorithm based on optimal measurements [6] exploiting some *a priori* knowledge of quantum state.

In our case we implemented a quantum optical system that, by means of spontaneous Parametric Down-Conversion (PDC) [7, 8], generates two entangled photons in the singlet state $|\psi_-\rangle = \frac{1}{\sqrt{2}}(|HV\rangle - |VH\rangle)$, where H is the horizontal component of polarization and V is the vertical one: the notation $|XY\rangle$ is a shortcut to indicate the tensorial product $|X\rangle \otimes |Y\rangle$.

With this *a priori* knowledge we can perform the estimation of the above mentioned parameters. We test two estimators for each parameter: an optimal one and a non-optimal one. An estimator is optimal if the smallest statistical uncertainty associated to it coincides with the limit imposed by the Cramér-Rao bound [9], corresponding to the minimum uncertainty available.

2. Procedure

The experimental setup for singlet-state preparation is shown in Fig. 1: it hosts a 9 W

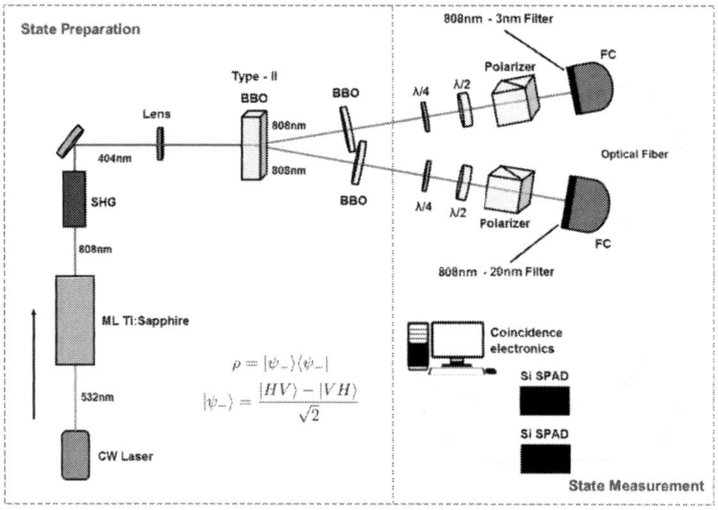

Figure 1: Experimental setup for the two-photon maximally entangled state

CW laser[1] at 532 nm pumping a Ti:Sapphire crystal in an optical cavity. At the exit of this cavity the mode-locked laser generated has a wavelenght of 808 nm with a FWHM[2] of 7 nm. Such laser is frequency-doubled by means of second harmonic generation (SHG) [10] and then injected into a BBO[3] crystal where Type-II PDC occurs. After this outcoming photons make two cones: one with horizontal polarization and the other with vertical polarization. We are interested in photons belonging to the intersections of these two cones, because these photon pairs are in a superposition of singlet state $|\psi_-\rangle$ and triplet state $|\psi_+\rangle$[4]. In order to compensate the walk-off between the two polarizations induced by the birefringence of the BBO and select only the singlet state without any unwanted relative phase, we put another BBO crystal in both photon paths. In the second part of the setup there are in both optical branches a quarter wave plate ($\lambda/4$), a half wave plate ($\lambda/2$), a polarizing beam splitter and a Fiber Coupler (FC) preceded by a bandpass filter: after being projected onto different polarization bases the photons are filtered by the bandpass filters (3 nm and 20 nm FWHM) and fiber coupled.

Fiber-coupled photons are addressed to two Silicon Single-Photon Avalanche Diodes (Si-SPADs), whose outputs are sent to coincidence electronics.

[1] Continuous wave laser
[2] Full Width at Half Maximum
[3] β-Barium borate
[4] $|\psi_+\rangle = \frac{|HV\rangle+|VH\rangle}{\sqrt{2}}$

In order to test the fidelity of the generated entangled state with respect to the expected one, we perform a quantum state tomography and we calculate the Uhlmann's Fidelity [2] between the reconstructed and the theoretical density matrices of the two-photon state.

We also prepare and test the completely decoherent mixture $\rho_{mix} = \frac{1}{2}(|HV\rangle\langle HV| + |VH\rangle\langle VH|)$, obtained by adding a birefringent crystal on one of the two photon paths as shown in Fig. 2. This birefringent crystal has the optical axis orthogonal to the

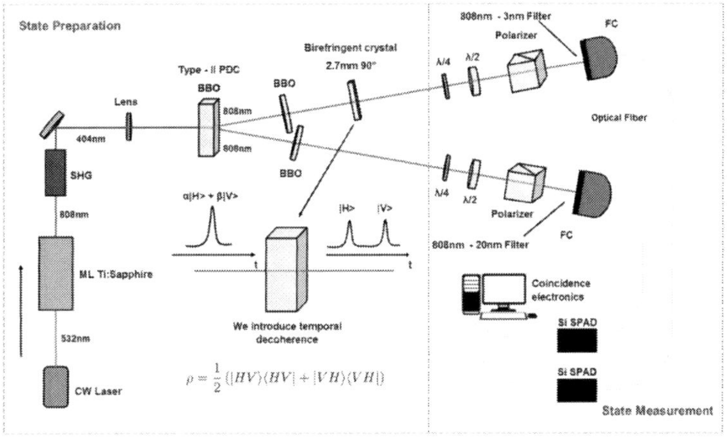

Figure 2: Experimental setup for the two-photon mixed state. We introduced temporal decoherence by putting on one photon path a 2.7 mm thick birefringent crystal with optical axis orthogonal to the photon propagation direction.

photon propagation direction, and its thickness is 2.7 mm, much greater than the photons wavelength, in order to introduce only temporal decoherence into the two-photon state.

By using a fraction p of singlet-state data and a $1-p$ portion of the decoherent-state data, we create a statistical mix of data simulating a generic partially-decoherent state described by the density matrix:

$$\rho = p|\psi_-\rangle\langle\psi_-| + (1-p)\frac{(|HV\rangle\langle HV| + |VH\rangle\langle VH|)}{2} \quad (1)$$

where $p \in [0,1]$.

Quantum tomography validates our *a priori* knowledge of the state: being aware that our state density matrix is in the form (1), we can implement estimator algorithms for the parameters that we want to measure. We choose the $\{|+\rangle, |-\rangle\}^5$ polarization basis for the measurement sets of the estimators, because all the estimators are based on projections onto such basis. With the same setups of Fig. 1 and Fig. 2 we perform coincidences measurements in this polarization basis.

3. Data Analysis

The preliminary results of the density matrix reconstruction with quantum state tomography are shown in Fig. 3 and Fig. 4.

[5] $|+\rangle = \frac{|H\rangle+|V\rangle}{\sqrt{2}}$, $|-\rangle = \frac{|H\rangle-|V\rangle}{\sqrt{2}}$

IOP Conf. Series: Journal of Physics: Conf. Series **841** (2017) 012033 doi:10.1088/1742-6596/841/1/012033

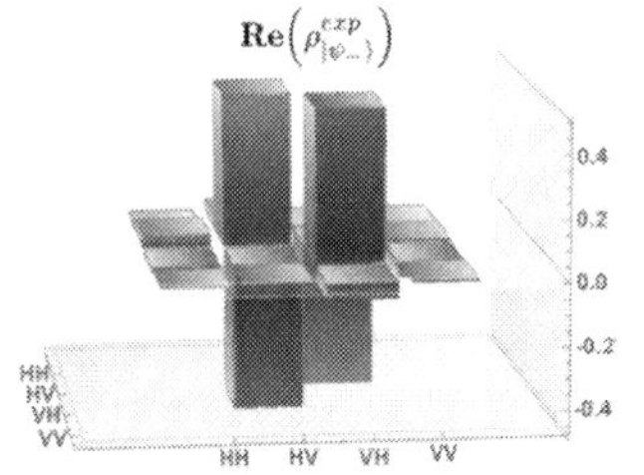

(a) *Real part of* $\rho^{exp}_{|\psi_-\rangle}$.

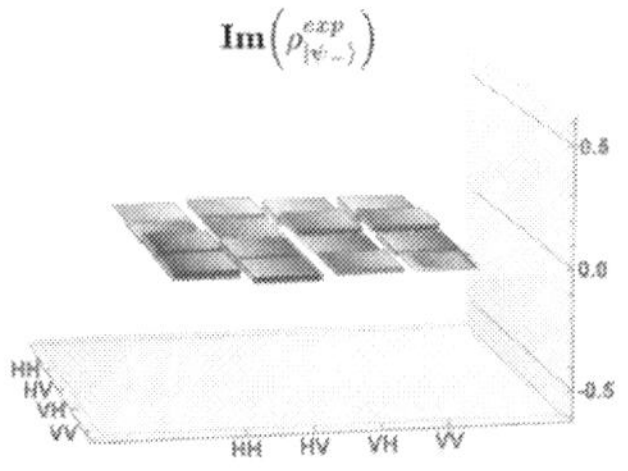

(b) *Imaginary part of* $\rho^{exp}_{|\psi_-\rangle}$.

$$\rho^{th}_{|\psi_-\rangle} = \begin{pmatrix} 0 & 0 & 0 & 0 \\ 0 & \frac{1}{2} & -\frac{1}{2} & 0 \\ 0 & -\frac{1}{2} & \frac{1}{2} & 0 \\ 0 & 0 & 0 & 0 \end{pmatrix}$$

(c) *Theoretical form of the density matrix.*

Figure 3: Density matrix recostruction for singlet entangled state.

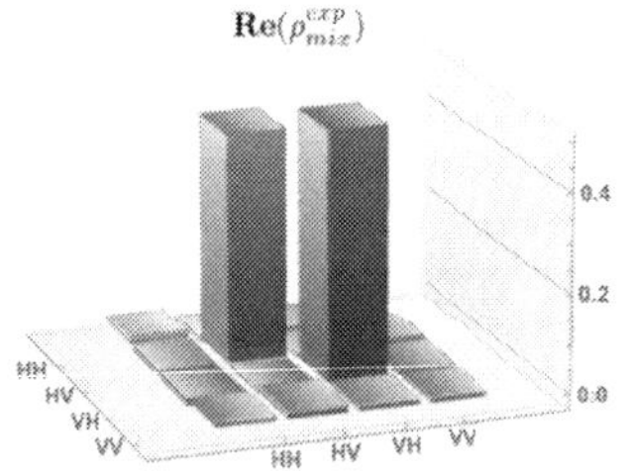

(a) *Real part of* ρ^{exp}_{mix}.

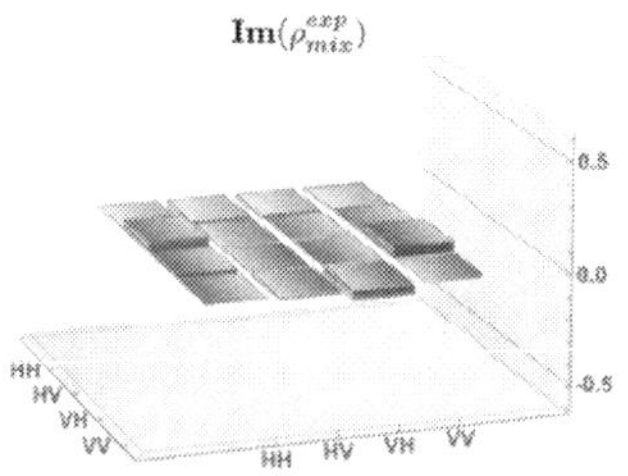

(b) *Imaginary part of* ρ^{exp}_{mix}.

$$\rho^{th}_{mix} = \begin{pmatrix} 0 & 0 & 0 & 0 \\ 0 & \frac{1}{2} & 0 & 0 \\ 0 & 0 & \frac{1}{2} & 0 \\ 0 & 0 & 0 & 0 \end{pmatrix}$$

(c) *Theoretical form of the density matrix.*

Figure 4: Density matrix recostruction for decoherent mixture.

The Uhlmann's Fidelity is defined as:

$$F = Tr\left(\sqrt{\sqrt{\rho^{exp}}\rho^{th}\sqrt{\rho^{exp}}}\right) \tag{2}$$

and the values[6] for $\rho_{|\psi_-\rangle}$ and ρ_{mix} are respectively $F = 0.975$ and $F = 0.985$.

After these tomographies, because the values of the two fidelities are close enough to 1, we are sure that we have *a priori* knowledge of our state, so we perform measurements in $\{|+\rangle, |-\rangle\}$ basis in order to estimate negativity and concurrence both for maximally-entangled state and decoherent state. For each parameter we use two estimator algorithms: a non-optimal one and an optimal one. Optimal means that the theoretical variance of the estimated parameter saturates the Quantum Cramér-Rao bound:

$$Variance \geq \frac{1}{QuantumFisherInformation}$$

Here we introduce the parameters that we want to estimate, both with the corresponding optimal and non-optimal estimators.

Negativity is defined by:

$$\mathcal{N}(\rho) = ||\rho^{T_A}||_1 - 1 \tag{3}$$

[6] The more experimental density matrix tends to the theoretical one, the more Uhlmann's Fidelity tends to 1

where: ρ^{T_A} is the partial transpose of ρ with respect to the subsystem A and $||X||_1 = Tr\sqrt{X^\dagger X}$ is the trace norm of the operator X. When ρ describes a completely separable state like the one in Fig. 4.c, $\mathcal{N}(\rho)$ is equal to 0, while for a maximally entangled state the negativity is 1.

We define the non-optimal estimator $\epsilon\mathcal{N}_1$:

$$\epsilon\mathcal{N}_1 = -4\left(P(|++\rangle) - \frac{1}{4}\right) \tag{4}$$

when $P(\bullet)$ is the probability of the event $\bullet$, and the optimal estimator $\epsilon\mathcal{N}_2$:

$$\epsilon\mathcal{N}_2 = (P(|+-\rangle) + P(|-+\rangle) - P(|++\rangle) - P(|--\rangle)) \tag{5}$$

Concurrence is defined by:

$$\mathcal{C}(\rho) = max(0, \lambda_1 - \lambda_2 - \lambda_3 - \lambda_4) \tag{6}$$

where λ_i are eigenvalues of the matrix $R = \sqrt{\sqrt{\rho}(\sigma_y \otimes \sigma_y)\rho^*(\sigma_y \otimes \sigma_y)\sqrt{\rho}}$, σ_y is the y Pauli's matrix.

Because Concurrence and Negativity have the same theoretical value, we can use the same estimators previously introducted in Eqns (4) and (5).

The results for estimation of $\epsilon\mathcal{N}_1$ are still in evaluation, while the ones for $\epsilon\mathcal{N}_2$ are shown in Fig. 5. The parameter p is defined by Eq. (1): at $p = 0$ we have the mixing state, while

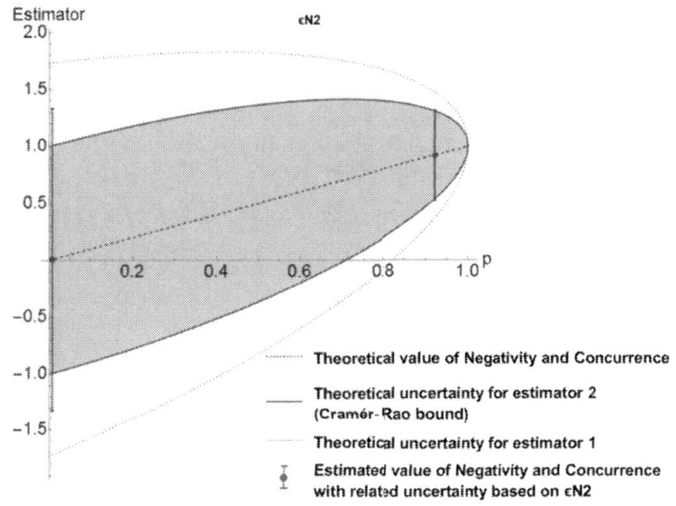

Figure 5: Negativity and Concurrence: $p = 0$ corresponds to the mixing state, while $p = 1$ corresponds to the maximally entangled state.

at $p = 1$ we have the maximally entangled singlet state. This last state is not properly at $p = 1$ in our graphics, because of small decoherence due to experimental imperfections, but, as previously explained, we can ignore this discrepancy because the Uhlmann's fidelity of ρ^{exp} is close enough to 1.

We are now working in order to evaluate intermediate points with $0 < p < 1$, when we have a partially-decoherent two-photon state with their uncertainties.

4. Conclusions

We performed an experiment comparing two different parameters (Negativity and Concurrence) able to quantify the amount of entanglement in specific two-photon states. Both Negativity and Concurrence have the same estimators and we computed two of these: an optimal one and a non-optimal one. In all cases, with our preliminary data, the optimal estimator values and their experimental uncertainties are in good agreement with the theoretical predictions. The optimal estimator shows an uncertainty that is compatible with the minimum uncertainties allowed by the Cramér-Rao bound in the case of the maximally-entangled state, while the uncertainty appears to be slightly bigger for the separable one. The uncertainties on the non-optimal estimator are currently bring evaluated. Furthermore, we are evaluating estimators and uncertainties for statistical mixtures of the singlet state and the decoherent two-photon state, in order to have a good sampling of states with density matrix of the form in Eq. (1). We are working to apply the same technique to other parameters like Log-Negativity [4] and Quantum Geometric Discord [11], and also to extend it to the case of non-maximally-entangled states, i.e. states described by:

$$\rho = p\begin{pmatrix}0 & 0 & 0 & 0\\ 0 & \cos^2\theta & -\cos\theta\sin\theta & 0\\ 0 & -\cos\theta\sin\theta & \sin^2\theta & 0\\ 0 & 0 & 0 & 0\end{pmatrix} + (1-p)\begin{pmatrix}0 & 0 & 0 & 0\\ 0 & \cos^2\theta & 0 & 0\\ 0 & 0 & \sin^2\theta & 0\\ 0 & 0 & 0 & 0\end{pmatrix} \tag{7}$$

Actually we are able to prepare experimentally this non-maximally-entangled state:

$$\rho_{th} = \begin{pmatrix}0 & 0 & 0 & 0\\ 0 & \frac{2}{5} & -\frac{\sqrt{3}}{5\sqrt{2}} & 0\\ 0 & -\frac{\sqrt{3}}{5\sqrt{2}} & \frac{3}{5} & 0\\ 0 & 0 & 0 & 0\end{pmatrix} \tag{8}$$

with Uhlmann's fidelity $F = 0.935$.

Even though the work is still in progress, we believe that our technique, being able to discriminate between optimal and non-optimal entanglemet parameters estimators, will be of widespread use in the quantum communication and computation frameworks, as well as for quantum metrology and, in general, all entanglement-based quantum technologies.

References

[1] Nielsen M A and Chuang I L *Quantum Computation and Quantum Information* (Cambridge University Press, Cambridge, England, 2000)
[2] Sciarrino F, Vallone G, Milani G, Avella A, Galinis J, Machulka R, Perego A M, Spasibko K Y, Allevi A, M Bondani and P Mataloni 2011 *Eur. Phys. J. Special Topics* **199** 111-125
[3] Fox M *Quantum Optics, An Introduction* (Oxford University Press, 2006)
[4] Vidal G and Werner R F 2002 *Phys. Rev. Lett. A* **65** 032314
[5] Wei T *et al* 2003 *Phys. Rev. A* **67** 022110
[6] Brida G, Degiovanni I P, Florio A, Genovese M, Giorda P, Meda A , Paris M G A, Shurupov A P 2011 *Phys. Rev. A* **83** 052301
[7] Migdall A, Polyakov S V, Fan J and Bienfang J C *Single-Photon Generation and Detection, Physics and Applications* (Academic Press, 2013)
[8] Mandel L and Wolf E *Optical Coherence and Quantum Optics* (Cambridge University Press, 1995)
[9] Paris M G A 2009 *Int. J. Quant. Inf.* **7** 125
[10] Dmitriev V G, Gurzadyan G Gand Nikogosyan D N *Handbook of Nonlinear Optical Crystals, third revised edition* (Springer, 1999)
[11] Darkić B, Vedral V and Brukner C 2010 *Phys. Rev. Lett.* **105** 190502

IOP Conf. Series: Journal of Physics: Conf. Series **841** (2017) 012034 doi:10.1088/1742-6596/841/1/012034

Constraining auto-interaction terms in α-attractor supergravity models of inflation

A Di Marco[1], **P Cabella**[1] **and N Vittorio**[1]

[1] Physics Department, Tor Vergata University - Cosmology Group - Via della Ricerca Scientifica, 1, 00133 Italy

E-mail: alessandro.di.marco@roma2.infn.it

Abstract. The inflationary mechanism has become the paradigm of modern cosmology over the last thirty years. However, there are several aspects of inflationary physics that are still to be addressed, like the shape of the inflationary potential. Regarding this, the so-called α-attractor models show interesting properties. In this work, the reconstruction of the effective potential around the global minimum of these particular potentials is provided, assuming a detection of permille-order for the tensor-to-scalar-ratio by forthcoming cosmic microwave background or gravitational waves experiments.

1. Introduction

In the standard cosmological model, the universe emerges from a state known as "initial singularity". Then, it goes through a quantum gravity phase in which the four fundamental interactions result unified. Finally, the gravity separates from the other forces and the universe approaches the Grand Unification phase on energy scales at least of the order of $E_{GUT} \sim 10^{16}$ GeV. In principle, after the simmetry breaking of the Grand Unification, the cosmological action (in natural units $\hbar = c = 1$) is characterized by several fields coupled to gravity (i.e. scalars, vectors, spinors,...) and between each other:

$$S[\phi, \chi, \psi, g_{\mu\nu}...] = \int d^4x\sqrt{-g}\left\{\frac{M_p^2}{2}R - \frac{1}{2}\partial_\mu\phi\partial^\mu\phi - \frac{1}{2}\partial_\mu\chi\partial^\mu\chi - h_{\phi,\chi}\phi^2\chi^2 - h_{\phi,\psi}\phi\psi\bar{\psi} + ...\right\} \quad (1)$$

where ϕ, χ are some of the scalar fields, ψ is a spinor field and $h_{x,y}$ the related coupling constants. However, it is common belief that one of the scalar fields could provide the largest contribution to the stress energy tensor $T_{\mu\nu}$ of the early universe. In this way, the cosmological action is simply:

$$S[\phi, g_{\mu\nu}] \simeq \int d^4x\sqrt{-g}\left\{\frac{M_p^2}{2}R - \frac{1}{2}g_{\mu\nu}\partial^\mu\phi\partial^\nu\phi - V(\phi)\right\} \quad (2)$$

where ϕ is the dominant field, $V(\phi)$ is the related effective auto-interaction potential, M_p is the reduced Planck mass, R is the Ricci scalar, $g_{\mu\nu}$ the metric tensor and g its determinant. If the scalar function $V(\phi)$ has a sufficiently flat region, an accelerated expansion of the early universe takes place. This expansion is often called inflation [1–4] with $a(t) \sim e^N$, where $a(t)$ is the cosmic scale factor and N is the number of e-foldings. In the simplest scenario, the scalar field that drives inflation is neutral, homogeneous, minimally coupled to gravity and canonically normalized. It is sometimes called inflaton field while $V(\phi)$ is the inflationary potential. In particular, $V(\phi)$ is characterized by an almost flat region and by a global minimum. Fig. 1

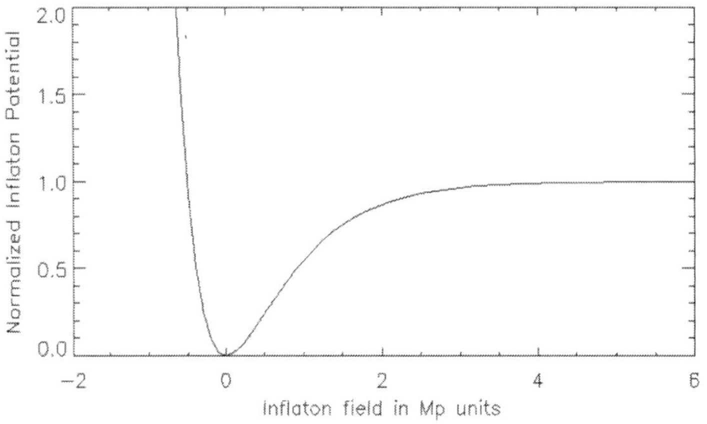

Figure 1. In the simplest scenario a scalar field is the energy source of the inflationary mechanism. Here we provide an illustration of the typical potential.

provides an example.

In the general picture, the dynamics of inflation is quite simple. The inflaton field ϕ explores the flat region that is equivalent to a "false vacuum" and $\partial_\mu\phi\partial^\mu\phi \ll V(\phi)$ in Eq. (2). In this phase, the inflationary evolution takes place. Subsequently, the field relaxes towards the real minimum, inflation stops and the reheating phase can start (see [5]): the scalar field oscillates, decays and its energy density is converted into radiation. The dynamics can be codified by appropriate parameters, i.e. the slow-roll parameters for which several definitions have been proposed. One of this is based on the use of the potential function itself. The first two slow-roll parameters of this formalism are:

$$\epsilon_V(\phi) = \frac{M_p^2}{2}\left(\frac{V'}{V}\right)^2, \quad \eta_V(\phi) = M_p^2\left(\frac{V''}{V}\right) \tag{3}$$

where $'$ indicates the derivative with respect to the field ϕ. The inflationary mechanism is crucial to resolve the puzzles of the standard Big Bang model i.e. the flatness problem, the horizon problem as well as the monopoles problem [1–5]. Moreover, inflation provides an explanation for the origin of those inhomogeneities that lead to the formation of the structures and to the intrinsic temperature fluctuations $\Delta T/T$ in the Cosmic Macrowave Background [6]. In the very early universe, the inflaton field can be considered as a quantum field with a given mean value and related quantum fluctuations. These fluctuations are naturally transmitted to the $T_{\mu\nu}$ and so one has fluctuations on the metric tensor $g_{\mu\nu}$, by the Einstein field equation. The inflationary expansion provides a magnification of these modes over the Hubble horizon $R_H = 1/H$ (where H is the Hubble parameter giving the universe expansion rate) of the epoch: the modes freeze out and become classical metric perturbations δ_i. In particular, inflation excites two types of relevant perturbations δ_i, i.e. scalar modes (δ_s) and tensor modes (δ_t, i.e. gravitational waves). It is possible to define quantities describing such perturbations: the power spectrum and the spectral index. The first, $P(k) = \frac{k^3}{2\pi^2}|\delta_i(k)|^2$, is related to the amplitude of the perturbations while the second, $n(k) = \frac{dP(k)}{d\ln k}$, describes the variation of the given δ_i with respect to the scale $k \sim 1/\lambda$. The results for the scalar sector are [6]:

$$P_s(k) = \frac{1}{8\pi^2 M_p^2} \frac{H^2}{\epsilon}\Big|_{k_*}, \quad n_s = 1 - 6\epsilon_V + 2\eta_V \tag{4}$$

while for the tensor sector are:

$$P_t(k) = \frac{2}{\pi^2} \frac{H^2}{M_p^2}\Big|_{k_*}, \quad n_t = -2\epsilon_V. \tag{5}$$

In addition, one can define the tensor-to-scalar ratio amplitude r:

$$r = \frac{P_t}{P_s} = 16\epsilon_V. \tag{6}$$

Note that all of these quantities are calculated at first order in the slow-roll parameters and at horizon crossing of the pivot scale k_*. In particular, n_s and r are the main inflationary observables that can be calculated for every given $V(\phi)$. In general, the values of n_s and r are given in terms of the remaining number of e-foldings, N_*, that is the number of e-foldings before the end of inflation with respect to the horizon crossing moment of the pivot scale k_*. This quantity is defined as:

$$N_* = \frac{1}{M_p} \int_{\Delta\phi} d\phi \frac{1}{\sqrt{2\epsilon_V(\phi)}}, \quad \Delta\phi = \phi_* - \phi_{end}. \tag{7}$$

So one has relations of the form: $n_s = n_s(N_*)$ and $r = r(N_*)$. Currently: $n_s = 0.968$, with $\sigma_{n_s} = 0.006$ and $r < 0.07$ at 95% CL [7].

2. The shape of the inflationary potential and the reconstruction method

The question related to the true shape of the eventual inflationary potential is the most important problem of inflationary cosmology. The standard way to address this problem consists in writing a lagrangian density or a model $V(\phi)$ and extracting the predictions (n_s, r). Then, one can compare the results with the cosmological observations. An alternative procedure is based on the use of the cosmological data to constrain the local shape of the inflationary potential, around the value of the inflaton field (ϕ_*) at the horizon crossing moment of the observable scales k_i [8]. This expansion is realized in terms of the inflationary observables by the Hamilton-Jacobi formalism for the inflationary dynamics. The results at first order in the quantities n_s and r, are [8]:

$$V(\phi) = \Lambda^4 \left[1 + d_1 \left(\frac{\Delta\phi}{M_p}\right) + \frac{1}{2} d_2 \left(\frac{\Delta\phi}{M_p}\right)^2\right] \tag{8}$$

with:

$$\Lambda^4 = \frac{3}{2}\pi^2 M_p^4 P_s r, \quad d_1 = \frac{1}{2}\sqrt{\frac{r}{2}}, \quad d_2 = \frac{1}{3}\left[9\left(\frac{r}{16}\right) - \frac{3}{2}(1 - n_s)\right] \tag{9}$$

where Λ is the energy scale of the inflationary process. The reconstruction technique can be also used in a different way. If in the next future a detection of inflationary gravitational waves and so of r is available, it will be possible to match the Taylor expansion of the generic function $V(\phi)$ given by Eq. (8) with the Taylor expansion of a given class of inflationary potentials. Then, one can derive constraints on the parameters that characterize the model and even on the vacuum expectation value assumed by the inflaton field, during the stretching of the modes, ϕ_* (see [9]).

3. The α-attractor models: constraining the effective potential around the true vacuum state

The superconformal α-attractor models of inflation represent a very interesting class of inflationary potentials. This class can be built in the supergravity context, and the scalar potential of the exponential version of this class is given by the following expression [10]:

$$V(\phi) = \Lambda^4 \left(1 - e^{-b\phi/M_p}\right)^2 \tag{10}$$

where b is the model parameter related to the α parameter by the relation $b = \sqrt{2/(3\alpha)}$. Furthermore, one can define the Kälher curvature of the inflaton scalar manifold as:

$$R_K = -\frac{2}{3\alpha}. \tag{11}$$

This class of potentials provides both small values of r (for small α and large curvature) and large values (up to the scale of $V \sim m^2\phi^2$). The upcoming next generation cosmological experiments could provide a detection of r of the order of 10^{-3}. Because of this, one could ask what would be the related α-attractor model and what kind of constraints one might have on its parameters. A possible solution is to use the so called matching-procedure: we can compare the generic Taylor expansion given by the Eq. (8) with the Taylor expansion of the class $V_\alpha(\phi)$. This procedure allows to put bounds on the b-parameter and on the vacuum expectation value of the inflaton field at horizon crossing ϕ_*, by the following expressions [9]:

$$b = -\frac{d_2}{d_1}, \quad \frac{\phi_*}{M_p}(b, d_1) = -\frac{1}{b}\ln\left(\frac{d_1}{2b}\right). \tag{12}$$

These quantities are useful to put bounds on several features of the model, like the parameter α, the supergravity quantity R_K and, consequently, the number of e-foldings N_*. Moreover, one can constrain the particular shape of the related effective potential $V(\phi)$ around the global minimum of $V_\alpha(\phi)$, providing constraints on the coupling constants. The minimum of the α-attractor models is localized in $\phi = 0$, so that:

$$V(\phi) \simeq \frac{1}{2}m_\phi^2\phi^2 - \frac{1}{3!}g\phi^3 + \frac{1}{4!}\lambda\phi^4 + \mathcal{O}(\phi^5)... \quad . \tag{13}$$

Here:

$$\frac{m_\phi^2}{M_p^2} = 3\pi^2 P_s r b^2, \quad \frac{g}{M_p} = 3\pi^2 P_s r b^3, \quad \lambda = 3\pi^2 P_s r b^4 \tag{14}$$

where m_ϕ is the mass of the inflaton field while g and λ are the coupling constants of the third and fourth order terms of the expansion, respectively. From a quantum field theory point of view, the three different terms correspond to the standard auto-interaction term, the three-legs interaction and fourth-legs interaction. In the present work, we simulate values of n_s and r (10^4 samples) randomly extracted from a multivariate gaussian distribution $G = G(\mu_{n_s}, \sigma_{n_s}; \mu_r, \sigma_r)$ of the form:

$$\mathcal{G}(n_s, r) = \frac{1}{\sqrt{4\pi^2(1 - \sigma_{n_s}\sigma_r\rho^2)}}\exp\left(-\frac{Q^2}{2}\right), \quad \text{with} \quad Q^2 = \left[\frac{(n_s - \mu_{n_s})}{\sigma_{n_s}} + \frac{(r - \mu_r)}{\sigma_r}\right]^2 \tag{15}$$

where: $\mu_{n_s} = 0.968$, $\sigma_{n_s} = 0.006$, $\mu_r = 0.002, 0.003$, $\sigma_r = 0.0001$. The correlation coefficient is fixed to $\rho = 0.1$ to take into account a possible and small dependency between n_s and r. Fig.

IOP Conf. Series: Journal of Physics: Conf. Series **841** (2017) 012034 doi:10.1088/1742-6596/841/1/012034

2 shows the distribution function for the value of the inflaton mass in the case of a detection of $r = 0.003$. Tab. 1 and Tab. 2 report the simulation results of the three coupling constants (in M_p-units) for the two different assumed values of r. The results show that the mass term is dominant with respect to the other two couplings and m_ϕ decreases as r increases. In addition, one can outline how the uncertainty on the constants is very small and this implies, for instance, a very precise knowledge of the mass of the inflaton field. Note that, the signal-to-noise ratio at 2-σ level is of the order of ~ 2.5 which is well calibrated by our Monte Carlo simulation. Furthermore, Fig. 3 shows the dispersion relation between the curvature R_K and the inflaton mass, again in the case of $r = 0.003$. The center of the distribution is localized around the corresponding mean values of the two quantities.

Table 1. Simulation results for the coupling constants with $r = 0.002$.

Couplings	Mean value	1-σ
m_ϕ	1.13×10^{-5}	2.18×10^{-6}
g	1.40×10^{-10}	7.82×10^{-11}
λ	1.53×10^{-10}	1.15×10^{-10}

Table 2. Simulation results for the coupling constants with $r = 0.003$.

Couplings	Mean value	1-σ
m_ϕ	1.11×10^{-5}	2.18×10^{-6}
g	1.10×10^{-10}	6.24×10^{-11}
λ	9.75×10^{-11}	7.40×10^{-11}

4. Conclusions

Knowing the slope of the inflationary potential is fundamental to understand the evolution of the universe after the accelerated expansion. After the end of inflation, there must be a period of reheating where the inflaton field takes mass, oscillates and decays to relativistic particles, disappearing from the "cosmological particle spectrum". Subsequently, the well known radiation-dominated epoch of the early universe takes place. The physics of the reheating phase is extremely complicated and depends on many factors. One of these is the shape of the inflaton potential near its minimum because this determines the properties of the "inflaton fluid". The simplest proposal for the reheat phase is called "elementary theory of reheating" [5]. In this scenario one has coherent oscillations of the inflaton field around the minimum of a simple quadratic potential $V(\phi) \sim m^2\phi^2$. These oscillations of the scalar field produce a cold gas of ϕ-particles that decay to relativistic particles (i.e. entropy). Then, they strongly interact and reach the thermal equilibrium. The duration of the reheating phase is given by the inflaton decay rate. The effective equation of state of the "reheating fluid" is simply $w_{reh} = 0$. However, the reheating era could be very complicated, as cited above, with $w_{reh} \neq 0$. In this work, we have shown that in the α-attractor models, the mass term m_ϕ^2 provides the largest contribution in the coherent oscillation phase. Nevertheless, this does not imply that w_{reh} is trivial as suggested in [11] and also for other models [12]. A possible extension of this work is based on the use of the next order expression of the coefficients d_1 and d_2 in terms of the inflationary variables. The natural consequence is the introduction of the running of the scalar spectral index defined as $dn_s/d\ln k$. This opportunity can lead to an improvement of the simulation results presented in this paper.

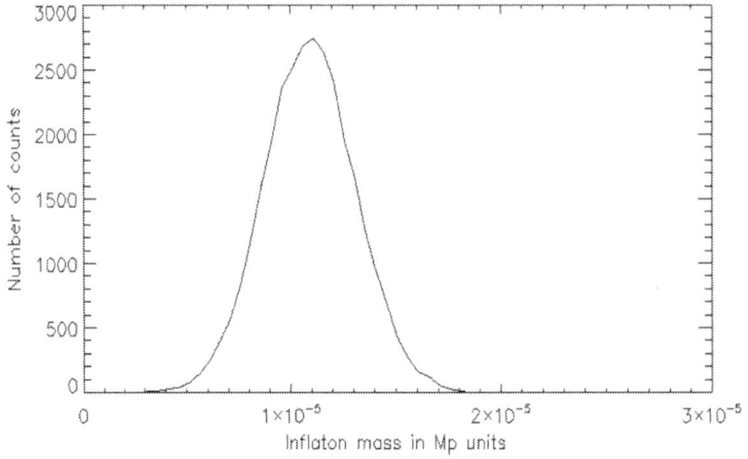

Figure 2. Distribution of the mass of the scalar field in the case $r = 0.003$.

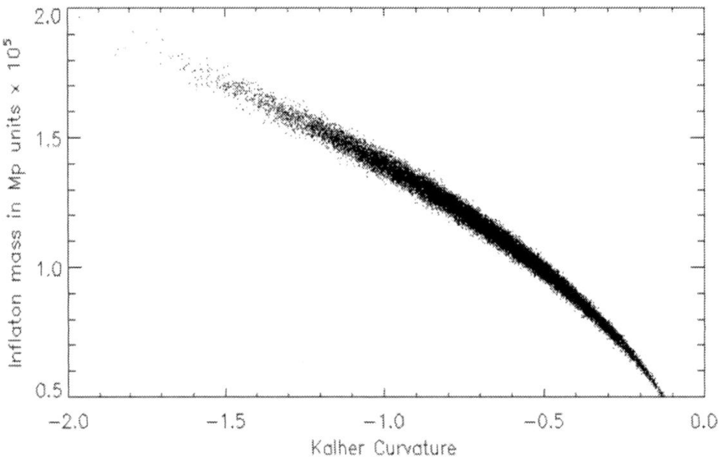

Figure 3. Dispersion relation between the mass of the scalar field and the curvature R_K.

References

[1] Guth A 1981 *Phys. Rev.* D **23** 2
[2] Linde A D 1981 *Phys. Lett.* B **108** 6 389-393
[3] Linde A D 1983 *Phys. Lett.* B **129** 3 4
[4] Albrecht A, Steinhardt P J 1982 *Phys. Rev. Lett.* **48**, 1220
[5] Albrecht A, Steinhardt P J, Turner M S, Wilczek F 1982 *Phys. Rev. Lett.* **48** 20
[6] Stewart E D Lyth D H 1993, *Phys. Lett.* B **302**, 171
[7] Planck 2015 results. XIII. Cosmological parameters 2016 A&A **594** A13
[8] Lidsey J E, Liddle A R, Kolb E W, Copeland E J, Barreiro T, Abney M 1997 *Rev. Mod. Phys.* **69** 373
[9] Di Marco A, Cabella P, Vittorio N 2017 *Phys. Rev.* D **95** 023516
[10] Kallosh R, Linde A D, Roest D 2013 *JHEP* **1311**
[11] Ueno Y, Yamamoto K 2016 *Phys. Rev.* D **93** 083524
[12] Dai L, Kamionkowski M, Wang J 2014, *Phys. Rev. Lett.* **113** 041302

IOP Conf. Series: Journal of Physics: Conf. Series **841** (2017) 012035 doi:10.1088/1742-6596/841/1/012035

Testing strong gravity with gravitational waves and Love numbers

E Franzin,[1] V Cardoso,[2,3] P Pani,[4,2] and G Raposo[2]

[1] Dipartimento di Fisica, Università di Cagliari & INFN, Sezione di Cagliari, Cittadella Universitaria, 09042 Monserrato, Italy
[2] CENTRA, Departamento de Física, Instituto Superior Técnico, Universidade de Lisboa, Av. Rovisco Pais 1, 1049 Lisboa, Portugal
[3] Perimeter Institute for Theoretical Physics, 31 Caroline Street North, Waterloo, Ontario N2L 2Y5, Canada
[4] Dipartimento di Fisica, "Sapienza" Università di Roma & INFN, Sezione di Roma1, p.zza A. Moro 5, 00185, Roma, Italy

E-mail: edgardo.franzin@ca.infn.it

Abstract. The LIGO observation of GW150914 has inaugurated the gravitational-wave astronomy era and the possibility of testing gravity in extreme regimes. While distorted black holes are the most convincing sources of gravitational waves, similar signals might be produced also by other compact objects. In particular, we discuss what the gravitational-wave ringdown could tell us about the nature of the emitting object, and how measurements of the tidal Love numbers could help us in understanding the internal structure of compact dark objects.

1. Introduction

The observations of the GW150914 and GW151226 events by the LIGO/Virgo collaborations have been interpreted as the gravitational waves emitted by a binary stellar-mass black holes merger [1, 2]. These events are characterised by three phases: the inspiral, when the distance between the two companions is large; the merger, when the two objects coalesce; and the ringdown, when the product of the merger relaxes to a stationary equilibrium configuration, with the emission of gravitational waves. The implications on theoretical physics of such mergers have been broadly discussed in [3]. Here, we discuss what gravitational-wave observations can tell us about the nature of the binary companions and the product of the coalescence.

The existence of black holes is supported by various indirect observations in the electromagnetic band, but the detection of GW150914 and GW151226 can be acknowledged as the first direct observation. In addition, the existence of event horizons — the defining property of black holes — might be probed only with gravitational-wave observations, as electromagnetic observations cannot probe the existence of event horizons, but only the existence of a light ring, *i.e.* a boundary within which photons can be trapped in circular orbits [4]. But, is there any evidence for event horizons in the detected signals?

There exist compact objects whose exterior geometry is the same as black holes but without an event horizon: they are called black hole mimickers and they include boson stars [5], wormholes [6], gravastars [7] and superspinars [8]. These objects and black holes are dark and then indistinguishable to electromagnetic instruments. On the other hand, ultra-compact objects

IOP Conf. Series: Journal of Physics: Conf. Series **841** (2017) 012035 doi:10.1088/1742-6596/841/1/012035

are generically unstable [9], and with the exception of boson stars, there exists no formation mechanism (yet) of black hole mimickers; wormholes and gravastars are rather artificial solutions to Einstein's equations.

In Section 2, we investigate the gravitational radiation emitted by a point particle in radial motion towards a traversable wormhole and we compare it to that of a Schwarzschild black hole. We show that, if the wormhole is compact enough, the distinction in the initial ringdown signal is irrelevant [10].

During the inspiral phase, at large orbital separations (low frequencies) the tidal interaction is negligible. As the orbital separation decreases (higher frequencies), the tidal interaction becomes important. In linear perturbation theory, the relation between the tidal field and the induced moment is constant, and such a constant is known as the tidal Love number.[1] Love numbers encode the information about the deformability of an object in a tidal environment and depend significantly on the object internal structure. Love numbers for black holes in general relativity are identically zero, while this is not the case for black holes in modified theories of gravity and for other exotic compact objects such as boson stars, wormholes and gravastars.

In Section 3, we compute Love numbers for minimal boson stars and gravastars and we discuss their possible detection [12].

We adopt natural units, $c = G = 1$.

2. Ringdown

In general relativity, Quasi-Normal Modes (QNMs) appear in the study of linear perturbations of black hole spacetimes [13, 14]. In most cases, the perturbation equations reduce to a linear second-order differential equation supplemented by boundary conditions at the horizon and at infinity. QNMs are the complex eigenvalues of this equation, with the real part representing the actual frequency of the oscillation and the imaginary part representing the damping.

QNMs were thought to dominate the ringdown waveform, and then the ringdown to be a piece of evidence for black holes. In fact, the black hole QNMs are deeply related to the boundary conditions required at the event horizon, while, for a horizonless object, the boundary conditions change completely, and so the QNM spectrum.

We now show that the ringdown phase should not depend on the presence of a horizon as long as the final object is compact enough to possess a light ring. For a black hole, the ringdown phase is described by the QNMs *by accident*, as the ringdown waves are carried inside the black hole because of the ingoing condition at the horizon. For another horizonless object, its relaxation consists of the light ring ringdown modes (insensitive to the boundary conditions), followed by the proper modes of vibration (read QNMs) of the object itself.

2.1. Traversable Wormholes

We study the gravitational radiation emitted by a point particle that plunges radially in a traversable wormhole and emerges in another universe.

The mimicker is built with two copies of Schwarzschild with the same mass M,

$$\mathrm{d}s^2 = -\left(1 - \frac{2M}{r}\right)\mathrm{d}t^2 + \left(1 - \frac{2M}{r}\right)^{-1}\mathrm{d}r^2 + r^2\,\mathrm{d}\Omega^2, \tag{1}$$

glued together at the throat $r = r_0 > 2M$. Such a surgery at the throat requires a thin shell of matter with surface density σ and surface pressure p. In general relativity, the matter content satisfies some energy conditions: for a perfect fluid, the null and strong energy conditions put constraints on combinations of σ and p, while the weak energy condition adds that $\sigma \geqslant 0$. In

[1] When the orbital motion approaches a resonance, the relation between the induced quadrupole moment and the tidal field is no longer constant, and the tides become dynamical [11].

order to have a traversable wormhole in general relativity, the weak energy condition must be violated, meaning that σ is negative. The null and strong energy conditions are satisfied when the throat is within the light ring [15].

In the point-particle limit, the Einstein equations reduce to a pair of Zerilli equations [16],

$$\frac{\mathrm{d}^2\psi_\ell(\omega,r)}{\mathrm{d}r_*^2} + \left[\omega^2 - V_\ell(r)\right]\psi_\ell(\omega,r) = S_\ell\,, \tag{2}$$

where r_* is the tortoise coordinate and $\ell \geqslant 2$ is the index of the spherical-harmonic expansion. Notice that the source term is different in the two universes.

The QNMs of the wormhole are defined by the eigenvalue problem associated with Eq. (2) when $S_\ell = 0$, supplemented by regularity boundary conditions. At infinity of both universes we require $\psi_\ell \sim \mathrm{e}^{\pm i\omega r_*}$, while at the throat the continuity of $\mathrm{d}\psi_\ell/\mathrm{d}r_*$ imposes either $\mathrm{d}\psi_\ell(r_0)/\mathrm{d}r_* = 0$ or $\psi_\ell(r_0) = 0$. Correspondingly, we find two families of QNMs for different values of the throat location r_0. Even in the black hole limit $r_0 \to 2M$, the wormhole QNM spectrum is dramatically different from that of the Schwarzschild case. The QNMs of the wormhole approach the real axis and become long-lived.

This behaviour is related to the Z_2-symmetry of the effective potential V_ℓ: the potential has a second barrier for negative r_*, and as long as the throat is within the light ring, wormholes can support long-lived modes trapped between the two potential wells near the light rings.

2.2. Gravitational Waves Extraction

The energy flux emitted in gravitational waves is

$$\frac{\mathrm{d}E}{\mathrm{d}\omega} = \frac{1}{32\pi}\sum_{\ell\geqslant 2}\frac{(\ell+2)!}{(\ell-2)!}\,\omega^2\left|\psi_\ell(\omega, r\to\infty)\right|^2. \tag{3}$$

In the top panels of Fig. 1 we plot quadrupolar gravitational wave energy spectra for representative values of the throat position and the energy of the particle. We observe that they coincide only at low frequencies, but are generically very different.

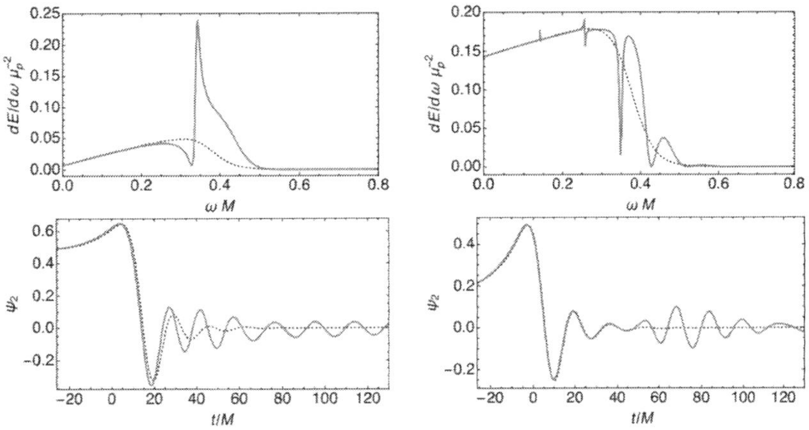

Figure 1. Top panels: $\ell = 2$ gravitational wave energy spectra for a particle crossing a traversable wormhole (red solid line) compared to a particle plunging into a Schwarzschild black hole (black dotted line) with the same energy E. Bottom panels: The corresponding gravitational-wave waveforms. Left and right panels refer to $r_0 = 2.1M$, $E = 1.1$ and $r_0 = 2.001M$, $E = 1.01$, respectively.

Given the substantial differences in both the QNMs and energy spectra, we would expect a completely different ringdown signal. Nevertheless, in the bottom panels of Fig. 1 we observe that these differences do not leave any trace in the initial ringdown waveform. The actual QNMs of the wormhole get excited only at late times and contain low energy. As the wormhole approaches the black hole limit, $r_0 \to 2M$, the initial ringdown is precisely the same as in the Schwarzschild case.

3. Love Numbers

Tidal Love numbers describe the deformability of a self-gravitating object immersed in a tidal environment. A static, spherically symmetric compact object embedded in an external tidal field, characterised by a polar and axial tidal quadrupolar moment $\mathcal{E}$ and $\mathcal{B}$, develops in response a mass and current quadrupolar moment M_2 and S_2.[2] To the linear order, the induced quadrupole moments are proportional to the tidal moments, and the constants of proportionality are the polar and axial Love numbers,

$$k_2^E \equiv -\frac{1}{M^5}\sqrt{\frac{4\pi}{5}}\frac{M_2}{\mathcal{E}_{20}}, \quad k_2^B \equiv -\frac{1}{M^5}\sqrt{\frac{4\pi}{5}}\frac{S_2}{\mathcal{B}_{20}}, \tag{4}$$

where the numerical factor is conventional, M is the mass of the object and $\mathcal{E}_{20}$ and $\mathcal{B}_{20}$ are the $\ell = 2$, $m = 0$ components of the external tidal fields written in a basis of even- and odd-parity spherical harmonics $Y^{\ell m}$. The factor $1/M^5$, that makes the Love numbers dimensionless, is non-standard. It is much more common to divide by powers of the object radius, but since the definition of radius for boson stars can be equivocal, we adopt this non-standard choice.

To calculate the mass and current moments as functions of the external tidal field, we consider static $\ell = 2$ perturbations about a static spherically symmetric background spacetime, and adopt the Geroch-Hansen formalism [18, 19]. Such perturbations can be decomposed into even- and odd-parity sectors, and each sector must be a solution to the Einstein field equations linearised about the background metric. Finally, the quadrupolar moments, and consequently the Love numbers, can be extracted by matching the exterior solution to the asymptotic behaviour of the spacetime metric,

$$g_{tt} = -1 + \frac{2M}{r} + \sqrt{\frac{4\pi}{5}}\frac{2M_2}{r^3}Y^{20} + \mathcal{O}\left(1/r^4\right) - r^2\mathcal{E}_2 Y^{20} + \mathcal{O}\left(r\right), \tag{5}$$

$$g_{t\varphi} = \frac{2J}{r}\sin^2\theta + \sqrt{\frac{4\pi}{5}}\frac{\sin\theta}{r^2}S_2\,\partial_\theta Y^{20} + \mathcal{O}\left(1/r^3\right) + \frac{r^3\sin\theta}{3}\mathcal{B}_2\,\partial_\theta Y^{20} + \mathcal{O}\left(r^2\right). \tag{6}$$

3.1. Love Numbers for Boson Stars and Gravastars

General relativistic boson stars have often been invoked as black hole mimickers, to explain dark matter halos and as generic dark matter candidates. They are self-gravitating solutions made entirely of fundamental, massive bosonic fields, described by the following action,

$$\mathcal{A} = \int \mathrm{d}^4x\,\sqrt{-g}\left(\frac{\mathcal{R}}{2\kappa} - |\partial\phi|^2 - V\left(|\phi|^2\right)\right). \tag{7}$$

Boson stars are usually classified according to their scalar potential, and their dynamics is described by the Einstein-Klein-Gordon equations. Assuming the spacetime to be spherically symmetric, the equilibrium configurations are obtained by numerically integrating the field equations supplied with regular boundary conditions at the origin and asymptotic flatness at infinity.

[2] It is relatively easy to generalise tidal fields and moments to higher-order multipoles, see [17].

Contrary to fluid stars, boson stars do not have a hard surface, as the scalar is spread out all over the radial direction. However, the configuration is highly localised in a radius $\sim 1/\mu$ and it is customary to define the effective radius R as the radius within which the 99% of the total mass M is contained. The compactness C of these objects, the ratio between mass and radius, can span from the same of planets to very close to that of black holes, *i.e.* 1/2.

Here we consider the simplest model: the free-field case, $V(|\phi|^2) = \mu^2|\phi|^2$, also known as minimal boson star, where μ is the mass of the bosonic field.

For the computation of the boson star Love numbers, we have considered only perturbations about stable equilibrium configurations. In the Regge-Wheeler gauge [20], the polar perturbations equations reduce to a system of two coupled equations, and to a single differential equation in the axial sector. We solve these equations supplemented by boundary conditions, we extract the mass and current moments by comparison with Eqs. (5) and (6), and then we compute the polar and axial Love numbers *à la* Hinderer [21],

$$k_2^E = \frac{8}{5}\left((1-2\mathcal{C})^2[2\mathcal{C}(y-1)-y+2]\right)\Big(3(1-2\mathcal{C})^2[2\mathcal{C}(y-1)-y+2]\ln(1-2\mathcal{C}) + 2\mathcal{C}\left[4\mathcal{C}^4(y+1)+2\mathcal{C}^3(3y-3)+2\mathcal{C}^2(13-11y)+3\mathcal{C}(5y-8)-3y+6\right]\Big)^{-1}, \tag{8}$$

$$k_2^B = \frac{8}{5}\frac{2\mathcal{C}(y-2)-y+3}{2\mathcal{C}\left[2\mathcal{C}^3(y+1)+2\mathcal{C}^2y+3\mathcal{C}(y-1)-3y+9\right]+3[2\mathcal{C}(y-2)-y+3]\ln(1-2\mathcal{C})}, \tag{9}$$

where $\mathcal{C} = M/r$ and $y \equiv rH'/H$ (being H the metric polar/axial perturbation function) evaluated at $r \gg R$. In the left panel of Fig. 2 we plot the polar and axial Love numbers for minimal boson stars and we observe that in the Newtonian regime, *i.e.* $M \to 0$, the polar Love number scales as $1/C^5$ while the axial scales as $1/C^4$.

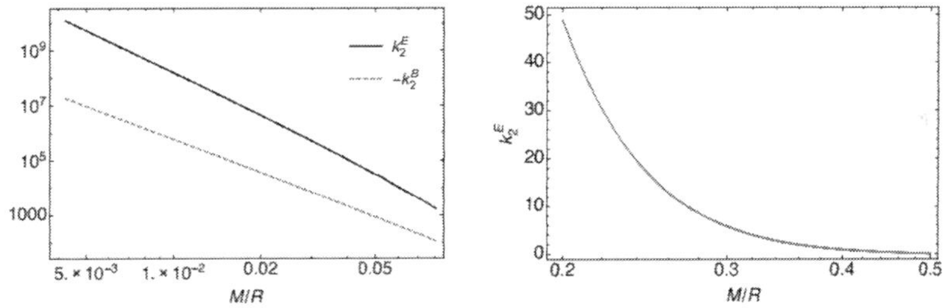

Figure 2. Quadrupolar $\ell = 2$ Love numbers for minimal boson stars (left) and thin-shell gravastars (right) as a function of their compactness.

A gravastar is another example of horizonless compact objects. Its interior is a de Sitter condensate, which is separated from the exterior Schwarzschild geometry by an intermediate region with finite thickness. In the thin-shell limit, the model is simple enough that the Love numbers can be computed analytically [22, 23].

In the right panel of Fig. 2, we see that polar Love numbers go to zero in the black hole limit and scale as $1/C^4$ for small compactness values.

3.2. Detectability

During the late inspiral phase, the tides between the binary affect the companions' gravitational field and orbital motion, and produce a measurable effect in the shape and the phasing of gravitational waves, accessible to measurements by Earth-based gravitational-wave detectors.

For neutron stars, tidal effects produce small corrections to the waveform phase which depends only on the star Love number [24].

Our results clearly show that Love numbers for boson stars and gravastar are in general not zero. When the detection of binary mergers will be ordinary, the gravitational-wave observations may be used to put upper bounds on the tidal Love numbers and constrain or rule out exotic alternatives.

4. Conclusions

To summarise, gravitational-wave astronomy has just begun and it provides a completely new instrument to explore the Universe.

In this work, we have seen that the initial ringdown signal chiefly depends on the properties of the light ring of the final object rather than its QNMs. The actual QNMs of the object are excited only at late times and typically do not contain a significant amount of energy. The events GW150914 and GW151226 strongly support the existence of light rings.

Future gravitational-wave detections should focus on extracting the late-time ringdown signal, where the actual QNMs of the final object are eventually excited and improve the signal-to-noise ratio. Such detections might eventually rule out exotic alternatives to black holes and test quantum effects at the horizon scale [25]. At the moment, the post-merger signal leaves room for alternative theories of gravity and exotic compact objects, such as gravastars and empty shells [26]. Furthermore, precise measurements of gravitational-wave signals in the earliest stages of the inspiral phase of the orbital evolution, and the subsequent extraction and comparison of the tidal Love numbers with the theoretical predictions can reveal important information about the internal structure of compact objects. This piece of information is — in principle — easy to extract as it is released cleanly well before the merger of the binary. It will shed light on the nature and the internal structure of black holes (allowing also for discrimination between general relativity and modified theories of gravity), and especially on the existence of exotic compact objects.

References

[1] Abbott B P *et al.* (LIGO/Virgo) 2016 *Phys. Rev. Lett.* **116** 061102
[2] Abbott B P *et al.* (LIGO/Virgo) 2016 *Phys. Rev. Lett.* **116** 241103
[3] Yunes N, Yagi K and Pretorius F 2016 *Phys. Rev.* **D94** 084002
[4] Abramowicz M A, Kluźniak W and Lasota J P 2002 *Astron. Astrophys.* **396** L31
[5] Liebling S L and Palenzuela C 2012 *Living Rev. Relat.* **15** 6
[6] Visser M 1996 *Lorentzian wormholes: From Einstein to Hawking* (Woodbury, USA: AIP)
[7] Mazur P O and Mottola E 2001 [gr-qc/0109035]
[8] Gimon E G and Hořava P 2009 *Phys. Lett.* **B672** 299
[9] Cardoso V, Crispino L C B, Macedo C F B, Okawa H and Pani P 2014 *Phys. Rev.* **D90** 044069
[10] Cardoso V, Franzin E and Pani P 2016 *Phys. Rev. Lett.* **116** 171101
[11] Steinhoff J, Hinderer T, Buonanno A and Taracchini A 2016 *Phys. Rev.* **D94** 104028
[12] Cardoso V, Franzin E, Pani P and Raposo G 2016 *In Press*
[13] Kokkotas K D and Schmidt B G 1999 *Living Rev. Relat.* **2** 2
[14] Berti E, Cardoso V and Starinets A O 2009 *Class. Quantum Grav.* **26** 163001
[15] Morris M S, Thorne K S and Yurtsever U 1988 *Phys. Rev. Lett.* **61** 1446
[16] Zerilli F J 1970 *Phys. Rev.* **D2** 2141
[17] Binnington T and Poisson E 2009 *Phys. Rev.* **D80** 084018
[18] Geroch R P 1970 *J. Math. Phys.* **11** 2580
[19] Hansen R O 1974 *J. Math. Phys.* **15** 46
[20] Regge T and Wheeler J A 1957 *Phys. Rev.* **108** 1063
[21] Hinderer T 2008 *Astrophys. J.* **677** 1216
[22] Pani P 2015 *Phys. Rev.* **D92** 124030
[23] Uchikata N, Yoshida S and Pani P 2016 *Phys. Rev.* **D94** 064015
[24] Flanagan É É and Hinderer T 2008 *Phys. Rev.* **D77** 021502
[25] Giddings S B 2016 *Class. Quantum Grav.* **33** 235010
[26] Cardoso V, Hopper S, Macedo C F B, Palenzuela C and Pani P 2016 *Phys. Rev.* **D94** 084031

IOP Conf. Series: Journal of Physics: Conf. Series **841** (2017) 012036 doi:10.1088/1742-6596/841/1/012036

Available evidence of a missing theory? An empirical, thermodynamical approach to quantum gravity through an alternative to dark matter

D Mamone[1]
[1]HTW, Wilhelminenhof 75a, Berlin, DE
E-mail: davide.mamone@gmail.com

Abstract. We suggest the possibility that the discrepancy between general relativity and astronomical observations in the rotational speed of flat galactic disks may be seen as a potential first evidence of a quantum gravity macroscopic effect instead of the existence of still elusive dark matter. We try to give an entropic interpretation of gravity which leads to Modified Newtonian Dynamics and in the same spirit we suggest further explorations in the strong field regime.

1. Introduction: evidence of a problem

Dozens of years of astronomical observations are telling us that something is missing in our narrative of galactic and cosmological dynamics. The missing element may be some kind of elusive dark matter and energy and/or a failure in our understanding of gravity in the weak field regime. The problem is today well known as *the missing mass*: the 95% of the mass-energy of the Universe would be invisible, still directly undetected and of unknown nature; the 70% would be dark energy (vacuum energy, cosmological constant) believed to be responsible for the universe acceleration; the remaining 25% of the supposed universal mass-energy would be the so called dark matter: an exotic kind of matter which interacts only gravitationally and is supposed to be responsible for the peculiar behaviour of the galactic rotational speed. In particular, dark matter would conspire to create a critical gravitational regime, actually a minimal critical acceleration of the order of $a_0 = 1.2 \times 10^{-10}\, m/s^2$, under which the galactic disks follow an apparent gravitational acceleration quite different from the expected one so that, the rotational speed does not decrease with the distance from the galactic center but remains constant as the gravitational force decreased as $\frac{1}{r}$ instead of $\frac{1}{r^2}$ (see Tab. 1 below).

Table 1. Rotational speed of the galactic disk.

Expected	$\frac{v^2}{r} = \frac{GM}{r^2} \;\rightarrow\; v^2 = \frac{GM}{r}$
Observed	$v^2 \propto GM \;\rightarrow\; \frac{v^2}{r} \propto \frac{1}{r}$?

IOP Conf. Series: Journal of Physics: Conf. Series **841** (2017) 012036 doi:10.1088/1742-6596/841/1/012036

Since the discrepancies were noticed someone suggested a breakdown of the gravity theory in the small field regime and/or at large distance without invoking any kind of dark matter. Following this idea Milgrom [1] proposed already in the 80s an empirical model called Modified Newtonian Dynamics (MOND) which changes Newton's law of gravity (or inertia) below the critical acceleration a_0. As an alternative to dark matter MOND resisted already tens of years of tests and succeeds in explaining the properties of an impressively large number of objects [2, 3]. Despite some attempts to get a general formulation of MOND compatible with relativity or to concile the model with some kind of dark matter, the theoretical community did not pay very much attention to MOND, neither to the possibility that such a huge discrepancy between general relativity and astronomical observations may be seen as a potential first evidence of a quantum gravity macroscopic effect. Contrary to the common idea that quantum gravity phenomenology must be restricted to extremely strong gravitational fields (like black holes) or to extreme high energies (Planck-scale) practically impossible to achieve in colliders, tiny quantum effects may play a major role in large scale phenomena and become visible due to an amplification and influence of the microscopic fabric of the vacuum space itself; one example of this idea is the search of discrepancies in the arrival times of gamma ray bursts in the case the vacuum behaved as a kind of material medium [4]. Milgrom developed his model initially from a totally empirical point of view in order to explain the strange behaviour of galactic disks in the low gravity regime; still, he tried to formulate a hypothesis that could justify the proposed modified dynamics. He suggested [5] that the vacuum space could have an influence on what we call inertia: such effect would become visible and even relevant for a gravitational field smaller than the critical acceleration $a_0 = 1.2 \times 10^{-10}\, m/s^2$ whose value is of the order of some constants of cosmological significance [2] like:

$$a_0 = c\sqrt{\frac{\Lambda}{3}} \tag{1}$$

with Λ the value of the cosmological constant. Coincidentally or not, remarks Milgrom [2], a body accelerating at a_0 from rest will approach the speed of light in the life time of the universe. (To give an idea of the order of magnitude, in order to experience a gravitational acceleration around a_0 we should go far away from the solar system at a distance 100 times its radius). But why should this critical value a_0 be so important and what kind of transition, if any, could lead to this regime?

2. An entropic force from virtual particles

According to the well known Unruh effect [6], an accelerating observer (or an observer in a curved spacetime) would perceive the vacuum full of particles with temperature

$$T = \frac{\hbar}{2\pi kc} a \tag{2}$$

where k is the Boltzmann constant, $\hbar$ the reduced Planck constant, a the acceleration and c the speed of light. Such warm particles may be considered purely virtual and the effect interpreted as a non-equilibrium description of an ill-defined quantum vacuum, due to the non conservation of the particle number in curved spacetime: still, this may reflect the difficulties to reconcile quantum field theories on flat spacetime and gravity, which is the core of the problem. From a kinematical point of view there is an interesting general interpretation of this effect: an accelerating observer creates an event horizon with loss of information, then entropy, then heat. The

IOP Conf. Series: Journal of Physics: Conf. Series **841** (2017) 012036 doi:10.1088/1742-6596/841/1/012036

corresponding formula for an accelerated observer in a spacetime with cosmological constant Λ is also known in the literature and called the Gibbons-Hawking temperature [7]:

$$T(a) = \frac{\hbar}{2\pi kc}(a^2 + a_0^2)^{\frac{1}{2}} \tag{3}$$

where we used the notation of Milgrom with a_0 related to Λ.
If we associate a vacuum temperature T_0 to the vacuum seen by a cosmological observer moving at the acceleration a_0, we could imagine to have an interaction between warm vacuum particles at two different temperatures and consider some kind of osmotic equilibrium between the two fluids with an exchange of an elementary entropy $\Delta S = 2\pi k$ at the reduced Compton length Δx of a particle of mass m. This would lead to the following entropic force (in the spirit of the Verlinde work [8]):

$$F = (T(a) - T_0)\frac{\Delta S}{\Delta x} = \frac{\hbar}{2\pi kc}\left((a^2 + a_0^2)^{\frac{1}{2}} - a_0\right)\frac{2\pi k}{\frac{\hbar}{mc}} = \mu(z)ma \tag{4}$$

where the function

$$\mu(z) = \frac{1}{a}\left[(a^2 + a_0^2)^{\frac{1}{2}} - a_0\right] = (1 + z^2)^{\frac{1}{2}} - z \tag{5}$$

depends on the ratio $z = \frac{a_0}{a}$ and interpolates between the Newtonian and the MOND regime in the following way:

$\mu(z) \sim 1$	for $z \ll 1 \quad a \gg a_0$	(Newtonian Regime)
$\mu(z) \sim \frac{1}{z} = \frac{a}{a_0}$	for $z \gg 1 \quad a \ll a_0$	(MOND Regime)

So, on the galactic disk, far away from the center, where the gravitational field becomes smaller than the cosmic acceleration a_0 the virtual particles of the vacuum would enter into the game modifying the global inertia of bodies, like the dynamics in some fluids, leading to an effective non-newtonian gravitational force:

$$m\frac{a^2}{a_0} = m\frac{GM}{r^2} \qquad \rightarrow \qquad a(r) = \frac{\sqrt{GMa_0}}{r}. \tag{6}$$

It may be interesting to notice that in the development and relativistic extensions of MOND [9–11] the potential Φ_M (where $a = \nabla\Phi_M$) follows a Newton-Poisson modified equation

$$\nabla \cdot \left[\mu\left(\frac{||\nabla\Phi_M||}{a_0}\right)\nabla\Phi_M\right] = 4\pi\rho \tag{7}$$

like a dielectric in a non-linear medium. It was suggested [5, 12, 13] that behind the dielectric analogy of MOND there is a mechanism of gravitational polarization which involves dark matter or the vacuum itself.

IOP Conf. Series: Journal of Physics: Conf. Series **841** (2017) 012036 doi:10.1088/1742-6596/841/1/012036

3. Conclusions and further directions

First of all we have to clarify that the entropic interpretation we presented should not be taken as a derivation of MOND from general principles, neither as a demonstration that a modified dynamics should arise from a quantum gravity theory. Still, we would like to stress that despite all the extreme simplifications and the empirical nature of the model, the idea is that the dark matter paradigm may keep us away from possible alternatives. Such alternative models might suggest something about the underlying, still unknown, missing theory which describes the microstructure of spacetime: historically, also quantum mechanics itself started from empirical models trying to describe the problematic experimental evidence of black body radiation. To understand which conditions should be changed or relaxed and which are the building blocks of the new theory is not usually an easy task. We suspect that the dark matter hypothesis works as modern epicycles of the old Tolemaic model, obstructing the proposal of new frameworks. If this is the case, the low field regime of gravity may deserve more investigations in order to clarify whether it comes from a large scale magnification of a quantum effect as we here suggest. General relativity is a theory about spacetime: it tells us how spacetime interacts with matter in an appropriate regime but it does not tell us which are the spacetime microstates, neither how they interact. Still, many clues point to their existence. Moreover, the idea that gravity might have a thermodynamic origin [8, 14, 15] and that macroscopic space may emerge at a course-grained limit should be taken properly into account, in our opinion. Taking seriously the thermodynamical hypothesis, we could look for possible entropic effects in the strong field regime as well. The Tolman-Ehrenfest effect [16] is a temperature gradient correlated to any gravitational field in such a way that a sort of chemical equilibrium is reached:

$$T_2\sqrt{g_{tt}(2)} = T_1\sqrt{g_{tt}(1)} \tag{8}$$

where g_{tt} is the metric tensor. Taking the second point far away from the source of the gravitational field, we find that:

$$T_2 = T_1\sqrt{1 - \tfrac{2GM}{c^2}\tfrac{1}{r}} \simeq T_1\left(1 - \tfrac{GM}{c^2}\tfrac{1}{r}\right).$$

We could speculate to write down an entropic force also in this case, reversing the logic of the Tolman-Ehrenfest derivation: without much astonishment and cautionally admitting the possible tautology, this would lead to a correction compatible with the one coming from General Relativity. Anyway, following this direction it could be interesting to consider also corrections to the entropy variations of the order of the Planck area $l_P^2 = \frac{G\hbar}{c^3}$

$$\Delta S = 2\pi k\left(1 - \frac{l_P^2}{r^2}\right). \tag{9}$$

This would lead to first order quantum corrections in agreement with standard corrections for the gravitational potential obtained with quantum-field-theoretical techniques discussed in the literature [17, 18] (with a_1 and a_2 appropriate constants):

$$V(r) = \frac{GMm}{r}\left(1 + a_1\frac{G(M+m)}{c^2}\frac{1}{r} + a_2\left(\frac{G\hbar}{c^3}\right)^2\frac{1}{r^2}\right). \tag{10}$$

These possible developments are at the present time highly speculative, without full justifications and just cited as examples of work in progress.

Following the condensed matter polarization analogy, another further direction might be to treat the spacetime as a non-linear medium, possibly a superfluid; an old but intriguing line of research [19–21].

References

[1] Milgrom M 1983, *Astrophys. J.* **270** 365
[2] Milgrom M 2001, MOND-a Pedagogical Review *Acta Phys. Polon.* B **32** 3613
[3] Famaey B and McGaugh S S 2012, *Living Reviews in Relativity* **15(10)** 1
[4] Liberati S and Maccione L 2014, *Phys. Rev. Lett.* **112** 151301
[5] Milgrom M 1999, The modified dynamics as a vacuum effect, *Phys. Lett.* A **253** 273
[6] Unruh W G 1976, *Phys. Rev.* D **14** 870
[7] Gibbons G W and Hawking S W 1977, *Phys. Rev.* D **15** 2738
[8] Verlinde E P 2011, *J. High Energy Phys.* JHEP **0411** 029
[9] Bekenstein J and Milgrom M 1984, *Astrophys. J.* **286** 7
[10] Sanders R 1997, *Astrophys. J.* **480** 492
[11] Bekenstein J 2004, *Phys. Rev.* D **70 (8)** 83509
[12] Blanchet L 2007, *Class. Quant. Grav.* **24** 3541
[13] Milgrom M 2010, *Monthly notices of the Royal Astronomical Society* **403(2)** 886
[14] Jacobson T 1995, *Phys. Rev. Lett.* **75** 1260
[15] Padmanabhan T 2009, *Rep. Prog. Phys.* **73 (4)** 6901
[16] Rovelli C and Smerlak M 2011, *Class. and Quan. Grav.* **28(7)** 075007
[17] Hamber H W and Liu S 1995, Phys. Lett. B **357** 51
[18] Donoghue J F 1994, *Phys. Rev. Lett.* **72** 2996
[19] Sakharov 1968, *Sov. Phys. Dokl.* **12** 1040 (Reprinted in 2000, *Gen. Rel. Grav.* **32** 365)
[20] Sinha K P, Sivaram C and Sudarshan E C G 1976, *Found. Phys.* **6** 65
[21] Volovik G E 2006, *Int. J. Mod. Phys.* D **15** 1987

Tree-level scattering amplitudes from the amplituhedron

L Ferro,[1] T Łukowski,[2] A Orta[1,⋆] and M Parisi[3]

[1] Arnold Sommerfeld Center for Theoretical Physics, Ludwig-Maximilians-Universität, Theresienstraße 37, 80333 München, Germany

[2] Mathematical Institute, University of Oxford, Andrew Wiles Building, Woodstock Road, Oxford, OX2 6GG, United Kingdom

[3] Center for Research in String Theory, School of Physics and Astronomy, Queen Mary University of London, Mile End Road, London E1 4NS, United Kingdom

E-mail: andrea.orta@lmu.de

Abstract. A central problem in quantum field theory is the computation of scattering amplitudes. However, traditional methods are impractical to calculate high order phenomenologically relevant observables. Building on a few decades of astonishing progress in developing non-standard computational techniques, it has been recently conjectured that amplitudes in planar $\mathcal{N} = 4$ super Yang-Mills are given by the volume of the (dual) amplituhedron. After providing an introduction to the subject at tree-level, we discuss a special class of differential equations obeyed by the corresponding volume forms. In particular, we show how they fix completely the amplituhedron volume for next-to-maximally helicity violating scattering amplitudes.

1. Introduction

Scattering amplitudes are among the most fundamental quantities in quantum field theory. The current precision frontier has risen and demands highly non trivial theoretical input. Already several decades ago it became clear that amplitudes are generally simpler than one would imagine looking at their traditional expressions in terms of Feynman diagrams. One of the earliest examples of this phenomenon is the celebrated Parke-Taylor formula for the tree-level colour-ordered scattering amplitude of n gluons in a Maximally-Helicity-Violating (MHV) configuration: parametrizing their null momenta via spinor-helicity variables $\lambda^\alpha, \tilde{\lambda}^{\dot{\alpha}}$, one finds [?]

$$A_n^{\rm MHV}(1^+,\ldots,i^-,\ldots,j^-,\ldots,n^+) = \frac{\langle ij\rangle^4}{\langle 12\rangle\langle 23\rangle\cdots\langle n1\rangle}\delta^4(p)\,, \tag{1}$$

where $\langle ab\rangle \equiv \epsilon_{\alpha\beta}\lambda_a^\alpha\lambda_b^\beta$. Much progress has been achieved in the understanding of this simplicity, especially in the context of $\mathcal{N} = 4$ super Yang-Mills (SYM), a four-dimensional superconformal theory which can be thought of as a (maximally) supersymmetric generalization of QCD: indeed, tree-level gluon amplitudes in the two theories coincide and it has been conjectured that $\mathcal{N} = 4$ SYM always provides a specific part of the QCD result, called maximally transcendental.

⋆ Speaker at the *7th Young Researcher Meeting*, 24th–26th October 2016, Torino, Italy.

IOP Conf. Series: Journal of Physics: Conf. Series **841** (2017) 012037 doi:10.1088/1742-6596/841/1/012037

The spectrum of $\mathcal{N}=4$ SYM comprises sixteen massless states organized in a CPT-self-dual supermultiplet: two gluons, eight gluinos and six scalars, all living in the adjoint representation of the colour group SU(N). In the following we will assume to have performed a colour decomposition, so that we can focus on the kinematics alone. Supplementing spinor-helicity variables with auxiliary Grassmann-odd variables η^{A} allows us to collect all states in a superfield

$$\Phi = G^+ + \eta^{\mathsf{A}}\Psi_{\mathsf{A}} + \frac{1}{2}\eta^{\mathsf{A}}\eta^{\mathsf{B}}S_{\mathsf{AB}} + \frac{1}{3!}\eta^{\mathsf{A}}\eta^{\mathsf{B}}\eta^{\mathsf{C}}\epsilon_{\mathsf{ABCD}}\,\bar{\Psi}^{\mathsf{D}} + \frac{1}{4!}\eta^{\mathsf{A}}\eta^{\mathsf{B}}\eta^{\mathsf{C}}\eta^{\mathsf{D}}\epsilon_{\mathsf{ABCD}}\,G^- \,. \tag{2}$$

Then the amplitude for the scattering of arbitrary states can be extracted from the super-amplitude $\mathcal{A}$ computed from superfields, accounting for all possible helicity configurations and admitting a decomposition in helicity sectors:

$$\mathcal{A}_n(\{\lambda_i,\tilde{\lambda}_i,\eta_i\}) = \mathcal{A}^{\mathrm{MHV}}_{n,\mathrm{tree}}\,\mathcal{P}_n(\{\lambda_i,\tilde{\lambda}_i,\eta_i\})\,, \quad \mathcal{A}^{\mathrm{MHV}}_{n,\mathrm{tree}} = \frac{\delta^4(p)\delta^8(q)}{\langle 12\rangle\langle 23\rangle\cdots\langle n1\rangle}\,, \tag{3}$$

where $p^{\alpha\dot{\alpha}} = \sum_i \lambda_i^{\alpha}\tilde{\lambda}_i^{\dot{\alpha}}$, $q^{\alpha\mathsf{A}} = \sum_i \lambda_i^{\alpha}\eta_i^{\mathsf{A}}$ are the conserved total momentum and supermomentum and

$$\mathcal{P}_n = \mathcal{P}_n^{\mathrm{MHV}} + \mathcal{P}_n^{\mathrm{NMHV}} + \mathcal{P}_n^{\mathrm{N^2MHV}} + \cdots + \mathcal{P}_n^{\mathrm{N^{n-4}MHV}}\,. \tag{4}$$

Each (next-to-)^{k}MHV contribution $\mathcal{P}_n^{\mathrm{N^kMHV}} \equiv \mathcal{P}_{n,k}$ is a monomial in the η^{A} of order $\mathcal{O}(\eta^{4k})$.

It turns out that spinor-helicity variables are yet not the best ones to make all the remarkable features of $\mathcal{N}=4$ SYM manifest. This goal is achieved by *twistor variables*: momentum supertwistors $\mathcal{Z}^{\mathcal{A}} = (\lambda^{\alpha},\tilde{\mu}^{\dot{\alpha}},\chi^{\mathsf{A}})$ [?] – where $\tilde{\mu},\chi$ are related to $\tilde{\lambda},\eta$ respectively – are particularly suited to expose a hidden, *dual* superconformal symmetry of the model in the planar limit, *i.e.* when the number of colours N has been taken to be large. The interplay of the latter with the ordinary one gives rise to the Yangian $Y(\mathfrak{psu}(2,2|4))$, an infinite-dimensional symmetry leaving tree-level scattering amplitudes invariant [?]. Tree-level scattering amplitudes admit a formulation as integrals over Grassmannian manifolds involving twistor variables [?, ?]. For instance, in terms of momentum supertwistors,

$$\mathcal{P}_{n,k} = \frac{1}{\mathrm{Vol}(\mathrm{GL}(k))}\int_{\gamma}\frac{\mathrm{d}^{k\times n}\,c_{\alpha i}}{(1\,2\,\ldots\,k)(2\,3\,\ldots\,k+1)\ldots(n\,1\,\ldots\,k-1)}\prod_{\alpha=1}^{k}\delta^{4|4}\left(\sum_{i=1}^{n}c_{\alpha i}\mathcal{Z}_i\right)\,. \tag{5}$$

Here the integration variables are the entries of a matrix C parametrizing the points in $G(k,n)$ – the set of k-planes through the origin in n dimensions – and every $(i\,i+1\,\ldots\,i+k-1)$ factor represents a *consecutive*, *ordered* $k\times k$ minor of C involving columns $i,i+1,\ldots,i+k-1$ (mod n). Choosing an appropriate integration contour, encircling a subset of the singularities, any $\mathcal{P}^{\mathrm{tree}}_{n,k}$ can be obtained as the corresponding sum of residues. For additional details, we refer the interested reader to the review [?].

The connection to Grassmannians has inspired new geometric and even combinatoric methods for studying amplitudes: on the one hand, Hodges observed in [?] that next-to-MHV (NMHV) amplitudes are volumes of polytopes in momentum twistor space; later on, the authors of [?] also showed that the residues of $\mathcal{P}_{n,k}$ are in one-to-one correspondence with on-shell diagrams, objects appearing in the so-called *positroid stratification* of the *positive Grassmannian* $G_+(k,n)$: this space is just the restriction of $G(k,n)$ to the matrices whose $k\times k$ ordered minors are all positive. Both ideas combined led to the amplituhedron proposal [?], which aims at providing a fully geometric picture of the physics of scattering, at least within planar $\mathcal{N}=4$ SYM.

In this paper we will give a short introduction to the tree-level amplituhedron, *i.e.* the object computing tree-level scattering amplitudes, focusing on how they are derived from its geometry. The symmetries of the problem will be discussed, with an emphasis on a set of PDEs called *Capelli differential equations*, which will play a special role. Finally we will present a master formula capturing all NMHV scattering amplitudes as integrals over a *dual* Grassmannian and provide a few examples. We will conclude with some outlook.

2. The tree-level amplituhedron

To define the amplituhedron we need to introduce a bosonized version of the external data encoded in momentum supertwistors $\mathcal{Z}_i^A$. We define new variables Z_i^A, whose components include the regular momentum twistor variables $z_i^a \equiv (\lambda_i^\alpha, \tilde{\mu}_i^{\dot{\alpha}})$, supplemented by a bosonized version of the fermionic components χ_i^{A}:

$$Z_i^A = \begin{pmatrix} z_i^a \\ \phi_1^{\mathsf{A}}\,\chi_{i\mathsf{A}} \\ \vdots \\ \phi_k^{\mathsf{A}}\,\chi_{i\mathsf{A}} \end{pmatrix}, \qquad \begin{matrix} i = 1, \ldots, n \\ A = 1, \ldots, m+k\,. \\ a, \mathsf{A} = 1, \ldots, m \end{matrix} \tag{6}$$

Here the ϕ_α^{A} are auxiliary Grassmann-odd parameters and the Z_i^A will be called *bosonized momentum twistors*. Moreover, n is the number of scattering particles, k the next-to-MHV degree and m an even number which takes the value four in the physical case.

Now, let us demand that the external data be *positive*, *i.e.* $Z \in M_+(m+k, n)$, where Z is the matrix whose columns are the individual Z_i and $M_+(m+k,n)$ is the set of $(m+k) \times n$ positive real matrices, *i.e.* matrices whose ordered maximal minors are positive:

$$\epsilon_{A_1 \ldots A_{m+k}}\, Z_{i_1}^{A_1} \cdots Z_{i_{m+k}}^{A_{m+k}} \equiv \langle Z_{i_1} \ldots Z_{i_{m+k}} \rangle > 0\,, \quad \text{with} \quad 1 \leq i_1 < \cdots < i_{m+k} \leq n\,. \tag{7}$$

The tree amplituhedron is now the space [?]

$$\mathfrak{A}^{\text{tree}}_{n,k;m}[Z] := \left\{ Y \in G(k, m+k) \;\; : \;\; Y_\alpha^A = \sum_i c_{\alpha i} Z_i^A \quad , \quad C = (c_{\alpha i}) \in G_+(k,n) \right\}, \tag{8}$$

namely a subspace of the Grassmannian $G(k, m+k)$ determined by positive linear combinations of positive external data. The k vectors Y_α are non-physical and will be eventually eliminated.

One can canonically define a $(k \cdot m)$-dimensional differential form $\mathbf{\Omega}_{n,k}^{(m)}$ on $\mathfrak{A}^{\text{tree}}_{n,k;m}$, demanding that it has logarithmic singularities on all boundaries of the space: this object is called *volume form*. Indeed, top-dimensional forms on any Grassmannian space must take the form

$$\mathbf{\Omega}_{n,k}^{(m)}(Y, Z) = \prod_{\alpha=1}^{k} \langle Y_1 \cdots Y_k\, d^m Y_\alpha \rangle \; \Omega_{n,k}^{(m)}(Y, Z)\,. \tag{9}$$

Here the *volume function* $\Omega_{n,k}^{(m)}$ simply depends on the auxiliary k-plane Y and the external data: it is conjectured to compute the volume of yet another space, dual to $\mathfrak{A}^{\text{tree}}_{n,k;m}$. There exist several methods to calculate the volume form. The standard one [?] requires us to triangulate the amplituhedron, *i.e.* to determine a set $\mathcal{T} = \{\Gamma_a\}$ of $(k \cdot m)$-dimensional cells of $G_+(k,n)$ whose images in the amplituhedron do not overlap and cover it completely. One then simply adds up the volume forms of all cells, which in turn admit a very simple expression in terms of local coordinates. The downside of this procedure is that $\mathcal{T}$ is not known in general, as the geometry of the (physical) amplituhedron for $k > 1$ is not well understood. In [?] another method was suggested, based on demanding regularity of the form everywhere outside of $\mathfrak{A}^{\text{tree}}_{n,k;m}$. Lastly, what we will be using is the following integral representation of the volume function [?]:

$$\Omega_{n,k}^{(m)}(Y, Z) = \int_\gamma \frac{\mathrm{d}^{k \times n} c_{\alpha i}}{(1\,2 \ldots k)(2\,3 \ldots k+1) \cdots (n\,1 \ldots k-1)} \prod_{\alpha=1}^{k} \delta^{m+k}(Y_\alpha - \sum_i c_{\alpha i} Z_i)\,. \tag{10}$$

The integral is taken over a suitable contour, in full analogy with the Grassmannian integral (??). Each residue corresponds to the volume function on a cell of the tree amplituhedron and, in order to get the proper expression for $\Omega_{n,k}^{(m)}$, we need to take an appropriate sum of them.

The tree-level "amplitude" for generic m is then calculated by integrating the canonical form in the following way:

$$\mathcal{P}^{\text{tree}}_{n,k;m}(\mathcal{Z}) = \int \mathrm{d}^{m\times k}\phi \; \Omega^{(m)}_{n,k}(Y^*, Z)\,, \tag{11}$$

where we used a δ-function to localize $\Omega^{(m)}_{n,k}(Y,Z)$ on the reference point $Y^* = \left(0_{m\times k} \,\middle|\, \mathbb{1}_k\right)^T$.

Integral (??) will be the starting point for our later derivation. In particular, we begin by considering its properties and symmetries in the case of generic m, n and k. This allows us to write down a set of differential equations satisfied by the volume function [?], which enforce constraints on $\Omega^{(m)}_{n,k}$ and can be fully solved for $k=1$. In all cases the solution admits an integral representation over the Grassmannian $G(k, m+k)$. We term this space the *dual Grassmannian* and stress that it does not depend on the value of n, in contrast to (??).

3. Capelli differential equations and NMHV volume forms

The tree-level Grassmannian integrals (??) defined in momentum twistor space possess a lot of interesting properties: in particular, they are Yangian invariant. As was shown in [?, ?] this symmetry uniquely fixes their form, up to the contour of integration. We presently lack a realization of the Yangian in the context of the amplituhedron, it is however an interesting question to ask whether it is also possible to determine the volume form directly from symmetries. The answer we provide is positive, at least in the NMHV case. For N^kMHV amplitudes with $k \geq 2$, however, known symmetries of the amplituhedron are not sufficient to completely fix the expression for the volume.

After introducing collective variables W_a^A – equal to Y_a^A for $a = 1, \ldots, k$ and to Z_{a-k}^A for $a = k+1, \ldots, k+n$ – we can show that the integral (??) enjoys the following properties:

- GL$(m+k)$ right covariance:

$$\Omega^{(m)}_{n,k}(Y\cdot g, Z\cdot g) = \frac{1}{(\det g)^k}\,\Omega^{(m)}_{n,k}(Y,Z)\,, \tag{12}$$

  for $g \in \mathrm{GL}(m+k)$, where by the right multiplication we mean $(W\cdot g)_a^A = \sum_B W_a^B g_B^A$.
- Scaling, *i.e.* $\mathrm{GL}(k)_+ \otimes \mathrm{GL}(1)_+ \otimes \cdots \otimes \mathrm{GL}(1)_+$ left covariance:

$$\Omega^{(m)}_{n,k}(h\cdot Y, \lambda\cdot Z) = \frac{1}{(\det h)^{m+k}}\,\Omega^{(m)}_{n,k}(Y,Z)\,, \tag{13}$$

  for $h \in \mathrm{GL}(k)_+$ and $\lambda = (\lambda_1, \ldots, \lambda_n) \in \mathrm{GL}(1)_+ \otimes \cdots \otimes \mathrm{GL}(1)_+$, where all transformations belong to the identity component of linear groups, $\mathrm{GL}(l)_+ = \{\ell \in \mathrm{GL}(l) : \det \ell > 0\}$.
- *Capelli differential equations* on the Grassmannian $G(m+k, k+n)$: for every $(k+1)\times(k+1)$ minor of the matrix composed of derivatives $\frac{\partial}{\partial W_a^A}$, one can check that

$$\det\left(\frac{\partial}{\partial W^{A_\nu}_{a_\mu}}\right)_{\substack{1\leq\nu\leq k+1\\ 1\leq\mu\leq k+1}} \Omega^{(m)}_{n,k}(Y,Z) = 0\;, \quad \begin{array}{c} 1 \leq A_1 \leq \cdots \leq A_{k+1} \leq m+k \\ 1 \leq a_1 \leq \cdots \leq a_{k+1} \leq k+n \end{array}. \tag{14}$$

Interestingly, the Capelli equations (??) together with the invariance and scaling properties (??), (??) were already studied intensively by the school of Gelfand [?] and also by Aomoto [?, ?]. For the $k=1$ case, relevant for NMHV amplitudes, the solution of the above problem can be inferred from the results presented in [?] and reads

$$\Omega^{(m)}_{n,1} = \int_0^{+\infty} \left(\prod_{A=2}^{m+1} \mathrm{d}s_A\right) \frac{m!}{(s\cdot Y)^{m+1}} \prod_{i=m+2}^{n} \theta(s\cdot Z_i)\,, \tag{15}$$

where GL$(1+m)$ covariance of (??) enabled us to fix $m+1$ variables $\{Z_1, \ldots, Z_{1+m}\} = \mathbb{1}_{1+m}$. Here $s \cdot W_a := W_a^1 + s_2 W_a^2 + \ldots + s_{1+m} W_a^{1+m}$ and the volume function is thus expressed as an integral over a region in the dual Grassmannian $G(1, 1+m) = \mathbb{RP}^m$.
A few comments are in order. The integrand depends on the number of particles only through the θ-functions, shaping a domain of integration $\mathcal{D}_n^{(m)}$ where no singularities are present. Indeed,

$$s \cdot Y = s \cdot (c_i Z_i) = c_i \, (s \cdot Z_i) > 0 \,, \tag{16}$$

since $s \cdot Z_i > 0$ and Y is inside the amplituhedron, *i.e.* $c_i > 0$. Positivity of the external data implies that $\mathcal{D}_n^{(m)}$, bounded by the $(m-1)$-dimensional subspaces $\ell_{Z_i} : s \cdot Z_i = 0$, is convex. Moreover, the choice of frame allowed by the GL$(1+m)$ covariance implies that eventually we will have to uplift our results to be functions of $(1+m)$-brackets. Finally, an important feature of our master formula is that it can be evaluated without referring to any triangulation of the integration domain.

4. Examples

We now briefly present a few examples showing that we can reproduce the known results [?]

$$\Omega_{n,1}^{(2)} = \sum_{i=2}^{n-1} [1\, i\, i+1] \qquad , \qquad [i\, j\, k] \equiv \frac{\langle i\, j\, k\rangle^2}{\langle Y\, i\, j\rangle \langle Y\, j\, k\rangle \langle Y\, k\, i\rangle} \,, \tag{17}$$

$$\Omega_{n,1}^{(4)} = \sum_{i<j} [1\, i\, i+1\, j\, j+1] \,, \quad [i\, j\, k\, l\, m] \equiv \frac{\langle i\, j\, k\, l\, m\rangle^4}{\langle Y\, i\, j\, k\, l\rangle \langle Y\, j\, k\, l\, m\rangle \langle Y\, k\, l\, m\, i\rangle \langle Y\, l\, m\, i\, j\rangle \langle Y\, m\, i\, j\, k\rangle} \,. \tag{18}$$

Although (??) and (??) were obtained triangulating the amplituhedron, we stress that our formula does not rely on this fact at all. Scattering amplitudes in planar $\mathcal{N}=4$ SYM correspond to $m=4$, nevertheless we find it advantageous to study formula (??) first in the two-dimensional toy model with $m=2$.

At three points there are no θ-functions and the volume function is to be evaluated integrating over the full positive quadrant of the (s_2, s_3) plane: we find $\Omega_{3,1}^{(2)} = (Y^1 Y^2 Y^3)^{-1}$, which correctly lifts to the invariant $[1\,2\,3]$. At four points, $\theta(s \cdot Z_4)$ implies that $\mathcal{D}_4^{(2)}$ is the restriction of $\mathcal{D}_3^{(2)}$ to the region above the line ℓ_{Z_4}. In general, moving from $n-1$ to n points means one θ-function constraint more, which in turn removes a wedge $\mathcal{W}$ from the integration domain $\mathcal{D}_{n-1}^{(2)}$ (see figure ??). Since

$$\int_{\mathcal{W}} \frac{\mathrm{d}s_2\, \mathrm{d}s_3}{(s \cdot Y)^3} = -[1\, n-1\, n] \,, \tag{19}$$

the general formula (??) holds.

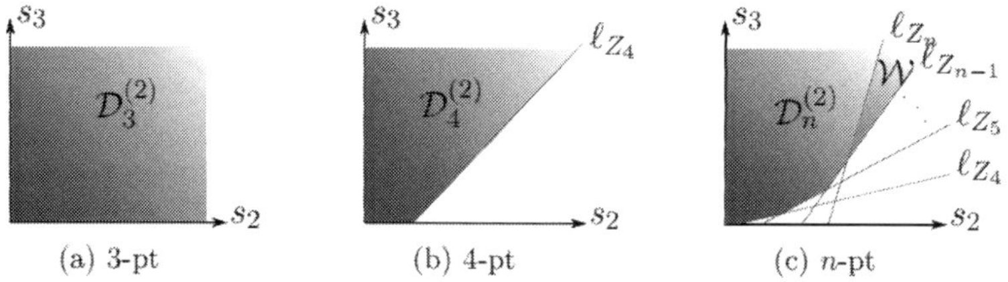

Figure 1: The $m=2$ toy model results in a nutshell.

Moving on to $m=4$, we cannot visualize the four-dimensional integration domains $\mathcal{D}_n^{(4)}$ anymore. The exceptions are the two lowest-n cases, for which it is useful to project everything

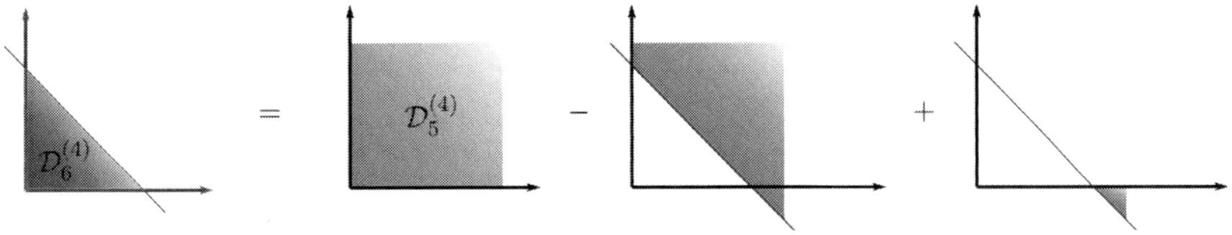

Figure 2: Domains of integration for six points and $m = 4$.

on the (s_2, s_4) plane. The simplest volume function occurs for five-particle scattering and $\Omega_{5,1}^{(4)} = [1\,2\,3\,4\,5]$, as expected. The six-point calculation is subtler, due to the presence of one θ-function, but still yields the correct result

$$\Omega_{6,1}^{(4)} = [1\,2\,3\,4\,5] + [1\,2\,3\,5\,6] + [1\,3\,4\,5\,6]\,. \tag{20}$$

Remarkably, a convenient integral representation exists for every invariant appearing in equation (??), allowing for a graphical representation of it, shown in Fig. ??. Many more details on these derivations can be found in [?].

5. Discussion and outlook

In this paper we gave a brief introduction to the amplituhedron idea, a recently proposed approach to the calculation of planar scattering amplitudes of $\mathcal{N} = 4$ SYM. We studied the symmetries of the problem at tree-level and explained how they lead to the novel formula (??), relevant for NMHV amplitudes. These can now be computed without any reference to triangulations of the amplituhedron. Rather, they admit an integral representation over a dual Grassmannian space $G(1, 1+m)$. This suggests a natural generalization to higher-k amplitudes, leading to a framework where the volume functions are integrals over the dual Grassmannian $G(k, m+k)$. For $k > 1$, though, symmetry constraints are not enough to fix the final formula completely. Further studies are needed and it would be particularly interesting to understand whether Yangian symmetry could be directly realized in the bosonized momentum twistor space.

References

[1] Parke S J and Taylor T R 1986 *Phys. Rev. Lett.* **56** 2459–60
[2] Hodges A 2013 *J. High Energy Phys.* JHEP05(2013)135
[3] Drummond J M, Henn J M and Plefka J 2009 *J. High Energy Phys.* JHEP05(2009)046
[4] Arkani-Hamed N, Cachazo F, Cheung C and Kaplan J 2010 *J. High Energy Phys.* JHEP03(2010)020
[5] Mason L and Skinner D 2009 *J. High Energy Phys.* JHEP11(2009)045
[6] Elvang H and Huang Y t 2015 *Scattering Amplitudes in Gauge Theory and Gravity* (Cambridge: Cambridge University Press)
[7] Arkani-Hamed N, Bourjaily J L, Cachazo F, Goncharov A B, Postnikov A and Trnka J 2012 *Scattering Amplitudes and the Positive Grassmannian* (Cambridge: Cambridge University Press)
[8] Arkani-Hamed N and Trnka J 2014 *J. High Energy Phys.* JHEP10(2014)030
[9] Arkani-Hamed N, Hodges A and Trnka J 2015 *J. High Energy Phys.* JHEP08(2015)030
[10] Ferro L, Lukowski T, Orta A and Parisi M 2016 *J. High Energy Phys.* JHEP03(2016)014
[11] Drummond J M and Ferro L 2010 *J. High Energy Phys.* JHEP12(2010)010
[12] Korchemsky G and Sokatchev E 2010 *Nucl.Phys.* B **839** 377–419
[13] Gelfand I 1986 *Dokl. Akad. Nauk SSSR* **288** 14–18
[14] Aomoto K 1975 *J. Fac. Sci. Univ. Tokyo, Sect. IA Math.* **22** 271–229
[15] Kita M and Aomoto K 2011 *Theory of Hypergeometric Functions* (Tokio: Springer)
[16] Gelfand I M, Graev M I and Retakh V S 1992 *Uspekhi Mat. Nauk SSSR* **47(4)** 3–82
[17] Arkani-Hamed N, Bourjaily J L, Cachazo F, Hodges A and Trnka J 2012 *J. High Energy Phys.* JHEP04(2012)081

The emergent gravity's entropy Ansatz from an augmented variational principle

M Tuveri[1], L Fatibene[2] and M Ferraris[2]

[1] Dipartimento di Fisica, Università di Cagliari & INFN, Sezione di Cagliari
Cittadella Universitaria, 09042 Monserrato, Italy
[2] Dipartimento di Matematica, University of Torino (Italy), via C. Alberto 10, 10123 Torino, Italy & INFN Sezione Torino – Iniziativa Specifica QGSKY, via P. Giuria, 10125 Torino, Italy

E-mail: matteo.tuveri@ca.infn.it

Abstract. In the last decade, physical and geometrical investigations about the relationship between horizon thermodynamics and gravitational dynamics suggest that gravity could be an emergent phenomenon. Among the others, Padmanabhan's theory of "emergent gravity" focus on the concept of spacetime as an effective macroscopic description of a more fundamental microscopic theory ("atoms of spacetime") at the Planck scales, thus aiming to reconcile the large scale description of gravity and the small one of quantum physics. However, some mathematical aspects of this approach are still not clear as, for example, the derivation of Einstein equations from a suitable entropy functional (and not from an action as in standard General Relativity). Thus in this work, a direct and non-trivial link between Padmanabhan's entropy used in emergent gravity and standard General Relativity action is established. To do that, Augmented Variational Principles and Kerr-Schild metrics will be used. We shall also discuss how this link accounts for the details of the variation of Padmanabhan's action based on gravitational entropy. It will also clarify the role of the background metric used in Padmanabhan's functional and its non-dynamical role.

1. Introduction

Quantum gravity is one of the most intriguing and puzzling field in physics. The need to find such a theory becomes obvious when one studies black holes in which the action of gravity is relevant at least down to Planck scales. Moreover, the possibility to associate quantities as temperature and entropy to black holes horizon (the three dimensional surface which surrounds black holes in spacetime) is a clear manifestation of the existence of some quantum microscopic origin of gravity and spacetime itself [1, 2, 3, 4].

Although there are well developed proposals for a theory of quantum gravity, as string theory or loop quantum gravity, a completely satisfactory and accepted theory is yet to come. All this fundamental theories are affected by the presence of intricate mathematical difficulties which are rather complicated to solve. Thus, in the last years, big efforts have been done in order to develop alternative approaches to get information about the quantum nature of gravity. One of these has been developed by Padmanabhan and it is a thermodynamical approach called "emergent gravity" [5, 6, 7].

Since in standard thermodynamics quantities as temperature and entropy arise from the existence of some internal degrees of freedom of the system, according to Padmanabhan, the

IOP Conf. Series: Journal of Physics: Conf. Series **841** (2017) 012038 doi:10.1088/1742-6596/841/1/012038

possibility to attribute some entropy to spacetime is an indication of some coarse-grained structure of spacetime at Planck scales. Given a certain region of spacetime, it is possible to define an entropy functional in terms of suitably parametrized microscopic degrees of freedom. In particular, the variation of this functional leads to Einstein equations. However, this procedure is not completely clear from a mathematical (as well as physical) point of view. In fact, we know that the standard procedure to find some field equation starting from a functional is to define a Lagrangian, or a suitable action. By minimizing the latter with respect to some dynamical field, one obtains the equations of motion of the system. This is true for every general covariant theory as in the case of General Relativity, where the dynamical field is the metric. However, an essential part of the definition of covariant theories is that any field appearing in the functional has to be treated as a dynamical field and eventually determined by field equations. That is not the case for Padmanabhan's functional, where the two fields appearing in the functional, the metric and some lightlike vector parametrizing the microscopical degrees of freedom cited above, are treated, respectively, as a fixed and a dynamical field (see [5]).

The aim of this paper is both to clarify this procedure by giving an explicit mathematical definition of the entropy ansatz by finding an action with the same structure of Padmanabhan's entropy ansatz and to find a way to mimic the dynamic of the metric with the same mathematical objects used by Padmanabhan. We use a mathematical tool called *Augmented Variational Principle* [8] applied for a special class of metrics called Kerr-Schild metrics [9, 10].

The paper is organized as follow: in Sec. 2 we will give a brief review on the emergent gravity approach. In Sec. 3 we will introduce the Augmented Variational Principle and in Sec. 4 we will show how one can recover Padmanabhan's entropy functional in that context. In order to do that, Kerr-Schild metrics will be used. Finally, in Sec. 5 we will summarize our results, leaving some open questions about this work to be studied in future.

2. Emergent Gravity

Emergent gravity is a classical approach to quantum gravity with the goal to understand which are the fundamental geometric quantities one should use to develop a complete theory of quantum gravity. The name may be unfortunate since it does not suggest that one derives a mesoscopic description from an underlying microscopic description, but it has become common in that specific case, so we adopt it. The main idea of this approach is that horizons are totally observer dependent, so one can perceive them and study their thermodynamical properties, even if there is no black hole (see [5]). In fact, we know that, for a Schwarzschild black hole, the surface $r = 2Gm$ defines the horizon, $\mathcal{H}$. One can compute the so–called surface gravity κ on $\mathcal{H}$ which corresponds to the gravitational acceleration at the horizon. Therefore, in view of equivalence principle, starting from a flat spacetime, an observer sitting at spatial infinity can be accelerated along one direction, e.g. the $x-$direction, with an acceleration κ and, as a consequence, he will perceive a horizon. This kind of observers are called the *Rindler* observers and they will attribute to this horizon every physical properties as the observers in presence of a real horizon of a real black hole region. This suggests to modify the Hamilton principle to obtain Einstein equations from quantities defining the horizons, namely the (lightlike) normal unit vector n^a to the horizon, and related to thermodynamics.

The starting point is to define an entropy functional, the Padmanabhan's ansatz for the total entropy of the system as a function of the lightlike vectors n^a is [5]:

$$S[n^a] = -\frac{1}{8\pi G}\int_V d^4x\sqrt{-g}\left[2P^{cd}_{ab}\nabla_c n^a \nabla_d n^b - 8\pi G\left(T_{ab} + \lambda(x)g_{ab}\right)n^a n^b\right] \tag{1}$$

where G is the Newton constant, g is the determinant of the (Rindler) metric tensor and the lightlike vectors n^a are normal to the Rindler horizon, $\mathcal{H}$, and they parametrize the spacetime internal degrees of freedom. The tensor P^{cd}_{ab} has the same symmetries of the Riemann tensor, it depends on the theory one assumes to describe gravitational dynamics and it satisfies $\nabla_c P^{cd}_{ab} = 0$. The Lagrange multiplier $\lambda(x)$ is put in the functional to take into account that the vectors n^a are lightlike. Finally, T_{ab} is the stress-energy tensor describing the matter contribution to the total action and it is conserved, viz. $\nabla_a T^{ab} = 0$.

Conversely to what happens in General Relativity, this entropy functional depends on the lightlike vectors n^a, which are the dynamical field one has to study in order to describe the dynamics of spacetime. The metric acts as a fixed background. The marvellous fact is that the minimization of the entropy with respect to the vectors n^a leads to "Einstein-like" field equations:

$$G_{ab} = 8\pi G\left(T_{ab} + \frac{\Lambda}{8\pi G} g_{ab}\right), \tag{2}$$

where Λ is a cosmological constant arising as an integration constant.

In what follows, we will clearly show how one can recover this result in General Relativity by shedding light on the role of the metric (and that of any other dynamical field) in the variational principle used by Padmanabhan.

3. Augmented Variational Principle

The Augmented Variational Principle (AVP) [8] has been introduced with the aim to generalize previous works on general covariant theories [11, 12], to solve in a natural way the problem of the so–called "Katz anomalous factor" [13] (the latter related to the definition of the mass of a black hole without adding by hand any term in the Lagrangian, as in [11]). Moreover it tries to generalize the concept of black hole entropy without using Killing vectors or Killing horizons, as for example in [11].

By using the AVP it is also possible to obtain the field equations and the relative conserved quantities for generic gauge covariant theories, i.e. the amount of conserved quantity which has to be expended to drive a physical system from one configuration of the system, $\bar{g}_{ab}$, to another one, g_{ab}.

In general we need two different metrics, g_{ab} and $\bar{g}_{ab}$, the latter considered as a reference metric. We want to study the relation between the dynamics of these two different configurations of the system. In any general covariant theory, this can be achieved by writing the Lagrangian and then by minimizing it with respect to suitable dynamical fields. Specifically, in General Relativity the Lagrangian is [8]

$$L = \sqrt{-g}R - \sqrt{-\bar{g}}\bar{R} + \nabla_c\left(\sqrt{-g}g^{ab}w^c_{ab}\right), \tag{3}$$

where $\sqrt{-g}$ ($\sqrt{-\bar{g}}$) is the square root of the determinant of the (reference) metric and the third term is a surface term, it is built up by the contraction between the metric tensor and the difference of the traceless connections of the two metrics [8]. The pure divergence is fundamental in the Lagrangian since, for example, it contributes to the definition of the Noether charges with an extra term which corrects the value of the black hole energy, correcting the anomalous factor cited above.

As usual, the dynamic of the system arises from the variation of the Lagrangian with respect to the metric tensor, thus leading to

$$\delta L = \sqrt{-g}G_{ab}\delta g^{ab} - \sqrt{\bar{g}}\bar{G}_{ab}\delta\bar{g}^{ab} + \nabla_c\left[\sqrt{-g}\delta\left(g^{ab}w^c_{ab}\right)\right], \tag{4}$$

where G_{ab} is the Einstein tensor. In order to have a well posed variational principle, the metric and the tensor w^c_{ab} have to satisfy the following conditions:

$$g_{ab} - \bar{g}_{ab}|_{\partial V} = 0, \qquad w|_{\partial V} = 0, \tag{5}$$

where V is a $d-$dimensional part of the manifold and ∂V its boundary.

4. Augmented Variational Principle in Kerr-Schild metrics

Given two metrics g_{ab} and $\bar{g}_{ab}$ and a covector ζ_a, a Kerr-Schild metric [10, 14] has the following form

$$g_{ab} = \bar{g}_{ab} + \zeta_a \zeta_b, \tag{6}$$

where $\bar{g}_{ab}$ can be set as a reference metric. The vectors ζ^a are light-like vectors defined everywhere in spacetime and they are the same for both g_{ab} and $\bar{g}_{ab}$, i.e. $\zeta_a = \bar{\zeta}_a$.

One nice feature of Kerr-Schild metrics is that they linearize Einstein's gravity [15] and, for example, they include charged and rotating black holes and stars, de Sitter and anti-de Sitter space-times [16, 17]. Recently, Kerr-Schild metrics have been used in some approach to quantum gravity [18, 19].

The Kerr-Schild metrics represent the natural candidate to satisfy the conditions imposed by AVP method, since they consist in two different metrics, g_{ab} and $\bar{g}_{ab}$, in which one can be set a reference field. In particular, from eq. (6), as the difference between the two metrics depends on the null vectors ζ^a. This allows us to relate the dynamics of $g_{ab} - \bar{g}_{ab}$ to the one of the null vectors and to establish a direct and clear link between Padmanabhan's ansatz in eq. (1) and standard General Relativity action. Therefore, by inserting the Kerr-Schild ansatz (6) into the augmented Lagrangian given in eq.(3) we have [20]

$$R = \bar{R} + \nabla_a \left(2\zeta^a \nabla_b \zeta^b\right) - 2P^{cd}_{ab} \nabla_c \zeta^a \nabla_d \zeta^b, \tag{7}$$

where $P^{cd}_{ab} = \frac{1}{2}\left(\delta^c_a \delta^d_b - \delta^d_a \delta^c_b\right)$ for General Relativity and ∇ is the covariant derivative with respect to the total metric g_{ab}. Without loss of generality, we can set $\bar{g}_{ab}$ as the Minkowski metric, then $\bar{R} = 0$ and the augmented gravitational action becomes [1]

$$A^*_{grav} = \int_V d^4x \left[\left(-2\sqrt{-g} P^{cd}_{ab} \nabla_c \zeta^a \nabla_d \zeta^b\right) + \nabla_c \left(\sqrt{-g} 2 P^{cd}_{ab} \zeta^a \nabla_d \zeta^b\right)\right] \tag{8}$$

where V is a certain volume of spacetime. For the matter action we consider the contribution of a perfect fluid described by an energy-momentum tensor T_{ab}. The integral curves of ζ^a must be null geodesics of g_{ab}, thus $\zeta^a \nabla_a \zeta_b = 0$ (which in turns satisfies the null energy condition). Then the matter action can be written as

$$A^*_{matt} = \int_V d^4x \sqrt{-g} T_{ab} \zeta^a \zeta^b, \tag{9}$$

with the matter Lagrangian density satisfying the null-energy condition and it is conserved.

One can easily realize that by using Kerr-Schild metrics, the variation of the action with respect to the metric g_{ab} now becomes a variation with respect to the null vectors ζ^a [20]. In fact, the variation of the total Kerr-Schild metric is a transformation of the type

$$\delta g^{ab} = g^{ab} - \bar{g}^{ab} = \zeta^a \zeta^b. \tag{10}$$

Moreover, the condition $\zeta^a = \bar{\zeta}^a$ over the null vectors ζ^a leads to

$$\delta \zeta^a = \zeta^a - \bar{\zeta}^a = 0, \tag{11}$$

thus we can add a Lagrange multiplier $\lambda(x)$ to the action to take into account the condition $\zeta_a \delta \zeta^a = 0$. Finally, the variation of the total action $A^*_{tot} = A^*_{grav} + A^*_{mat}$ with respect to the null vectors ζ leads to the following field equations

$$G_{ab} = 8\pi G \left(T_{ab} + \frac{\Lambda}{8\pi G} g_{ab}\right), \tag{12}$$

[1] This choice is due to simplify the variation of the total action. However, one can consider a Ricci scalar built up on a generic background metric and then minimize the action with respect to the null vectors ζ^a by keeping the background as fixed. In this way, the result in eq. (7) is still valid.

where Λ is an integration constant taking part of its value from the Ricci scalar R and part from the Lagrange multiplier $\lambda(x)$ [20]. It plays the role of a cosmological constant, whose value is completely determined by the geometrical features of the system. We notice that the equations of motion in eq. (12) do not explicitly depend on ζ^a, so the relation between the latter and the null vectors n^a used by Padmanabhan is irrelevant (even if it can be found, see [20]). The minimization of the augmented action leads to some "Einstein-like" field equations as expected from standard General Relativity. Notably, the cosmological constant Λ appears as an integration constant (as in Padmanabhan's derivation, see [5]), instead of being the result of the presence of some term added by hand in the action.

5. Conclusions and open questions

In this work we have established a nontrivial relation between Padmanabhan's entropy action and the AVP. Since the entropy functional in eq.(1) has the same expression of the augmented action shown in eq.(8), now it becomes clear how the Einstein field equations can arise from the variation of the entropy. As a consequence of our results, the AVP could give a better and clearer mathematical foundation to the physical intuition of emergent gravity. Remarkably, if one takes the on-shell solution ($R = 0$) and then evaluates the surface term in eq. (8), one finds $A^*_{grav} = \frac{1}{4}A_{\mathcal{H}}$, where $\mathcal{H}$ is the boundary of the region over we are integrating [20]. This is something already known in General Relativity, for example for the entropy of a black hole [5, 11], and it confirms the goodness of our results.

However, there is some point which needs some further comment. Our results depend on the choice of the Kerr-Schild metric and, as far as we know, we do not know if this is the most general solution of Einstein equations. However, as noted before, the validity of the Kerr-Schild ansatz is wide and it covers a lot of physical solutions, so we are encouraged to deepen this topic. Moreover, we want to stress the fact that a natural way to incorporate the presence of a cosmological constant in the Hilbert-Einstein action is then found. This is not obvious and one should investigate about the value of this constant in a cosmological context by comparing it with the one already obtained by the ΛCDM model or by Padmanabhan himself [21]. Finally, recently Padmanabhan and collaborators have obtained a result similar to the one shown in eq.(8) for the gravitational action but in a more complicated framework, by considering quantum corrections to General Relativity [22, 23, 24, 25]. Is it possible to incorporate these corrections in the AVP Lagrangian in order to extend this approach at least at a semi-classical level? All these points will be the subjects of future works in this field.

6. Acknowledgements

This article is based upon work from COST Action (CA15117 CANTATA), supported by COST (European Cooperation in Science and Technology). We acknowledge Professor M. Cadoni for useful discussions about these topics and the contribution of INFN (Iniziativa Specifica QGSKY), the local research project Metodi Geometrici in Fisica Matematica e Applicazioni (2015) of Dipartimento di Matematica of University of Torino (Italy).

References

[1] Bekenstein J D 1972 *Lett. Nuovo Cim.* **4** 737–740
[2] Bekenstein J D 1973 *Phys. Rev.* **D7** 2333–2346
[3] Bekenstein J D 1974 *Phys. Rev.* **D9** 3292–3300
[4] Hawking S W 1971 *Phys. Rev. Lett.* **26** 1344–1346
[5] Padmanabhan T 2010 *Rept. Prog. Phys.* **73** 046901 [0911.5004]
[6] Padmanabhan T 2015 *Mod. Phys. Lett.* **A30** 1540007 [1410.6285]
[7] Padmanabhan T 2006 *Int. J. Mod. Phys.* **D15** 2029–2058 [gr-qc/0609012]
[8] Fatibene L, Ferraris M and Francaviglia M 2005 *Int. J. Geom. Meth. Mod. Phys.* **2** 373–392 [math-ph/0411029]

[9] Kerr R P and Schild A 1965 *Atti del convegno sulla relatività generale; problemi dell'energia e onde gravitazionali* 173
[10] Kerr R P and Schild A 1965 *Proc. Symp. Appl. Math.* **17** 199
[11] Iyer V and Wald R M 1994 *Phys. Rev.* **D50** 846–864 [gr-qc/9403028]
[12] Fatibene L, Ferraris M, Francaviglia M and Raiteri M 1999 *Phys. Rev.* **D60** 124013 [gr-qc/9902065]
[13] Katz J 1985 *Classical and Quantum Gravity* **2** 423
[14] Stephani H, Kramer D, MacCallum M A H, Hoenselaers C and E H 2003 *Exact Solutions of Einstein's Field Equations* 2nd ed (Cambridge University Press, Cambridge UK)
[15] Xanthopoulos B C 1978 *J. Math. Phys* **19** 1607
[16] Gibbons G W, Lu H, Page D N and Pope C N 2005 *J. Geom. Phys.* **53** 49–73 [hep-th/0404008]
[17] Burinskii A 2012 [arXiv:1212.5595v1]
[18] Burinskii A 2009 [0903.2365]
[19] El-Menoufi B K 2016 *JHEP* **05** 035 [1511.08816]
[20] Tuveri M, Fatibene L and Ferraris M 2016 [1604.08067]
[21] Padmanabhan T 2016 *Int. J. Mod. Phys.* **D25** 1630020 [1603.08658]
[22] Kothawala D 2013 *Phys. Rev.* **D88** 104029 [1307.5618]
[23] Kothawala D and Padmanabhan T 2014 *Phys. Rev.* **D90** 124060 [1405.4967]
[24] Stargen D J and Kothawala D 2015 *Phys. Rev.* **D92** 024046 [1503.03793]
[25] Kothawala D and Padmanabhan T 2015 *Phys. Lett.* **B748** 67–69 [1408.3963]